Building Construction Methods and Systems

Ramazan Sarı · Ekrem Bahadır Çalışkan

Building Construction Methods and Systems

Principles, Requirements and Application Details

Springer

Ramazan Sarı
Department of Architecture
Antalya Bilim University
Döşemealtı, Antalya, Türkiye

Ekrem Bahadır Çalışkan
Department of Architecture
Ankara Yıldırım Beyazıt University
Esenboğa, Ankara, Türkiye

ISBN 978-3-031-50042-8 ISBN 978-3-031-50043-5 (eBook)
https://doi.org/10.1007/978-3-031-50043-5

This Springer imprint is published by the registered company Springer Nature Switzerland AG
The registered company address is: Gewerbestrasse 11, 6330 Cham, Switzerland

Paper in this product is recyclable.

Preface

The book presents practical information about the design and construction of building projects by addressing the principles of each method, unveiling background factors for requirements, and state-of-the-art application details. Science and technology provide thousands of construction materials, vast construction methods, and various construction equipment and tools for realizing diverse architectural and engineering design projects. From market perspectives for new participants, the current construction practices are chaotic, having wide material and method options with globally available traders. On the other hand, within this global market, there is a growing awareness and need for practical information among society and new participants in the industry about general and globally available construction methods and technologies. Rather than focusing on materials, available construction methods and technologies were described in the book content concerning their classification systems. The subjects and topics are represented in a well-structured hierarchy supported by clear and narrative figures. The book's content aims to inform general design and application principles of construction methods and technologies without diving into engineering calculations and formulas to keep the content easily understandable by all AEC practitioners and participants. Instead, state-of-the-art construction applications were explained to unveil the logic and application requirements at the background of systems and methods. The book content could also be a guidebook for undergraduate students in the AEC industry.

Antalya, Türkiye
Ankara, Türkiye

Ramazan Sarı
Ekrem Bahadır Çalışkan

Acknowledgements

This book has been finished with five years of work by the authors. The majority of the drawings and illustrations belong to the authors. Furthermore, the authors took particular support from the following people and companies and thus would like to express their gratitude.

We are particularly grateful to Prof. Dr. Salih Ofluoğlu for his guidance and mentorship in printing the manuscript.

We would like to express our deepest gratitude for their contributions to Fidanlar Construction Company, Asiltürk Project Company, AGK Company, ETÜ Construction Works and particularly to Naswood and Özsarı Çelik for sharing their experiences and practical knowledge.

We would like to state our greatest pleasure to Ali and Azize Katırcı for inspiring the authors to start writing the manuscript.

The authors are so lucky to have a great family. We would like to thank our kids, Meryem, Batuhan, and Asya for their unconditional love, care, and support, which have always been a source of strength and motivation.

The authors reserve their deepest gratitude to their beloved wives and closest friends. Özden and Melda's boundless support, sacrifices, and encouragement have propelled the authors forward even in extremely challenging times and have been the cornerstone of this arduous journey.

We would like to express our gratitude to Dear Fehmi and Sevgi Özden for their unconditional love, care, support and encouragement to complete the manuscript.

Contents

About the Authors

Ramazan Sarı received his B.Arch. in Architecture from Middle East Technical University, Faculty of Architecture. He earned his M.Sc. and Ph.D. in the Building Science Graduate Program of Middle East Technical University, Faculty of Architecture. Currently, he works as an instructor at Antalya Bilim University. Research interests include knowledge management and representation, construction methods, building information modeling (BIM), and web ontology models.

Ekrem Bahadır Çalışkan received his B.Arch. in Architecture from Middle East Technical University, Faculty of Architecture. He earned his M.Sc. and Ph.D. in the Building Science Graduate Program of Middle East Technical University, Faculty of Architecture. Currently, he works as an instructor at Ankara Yıldırım Beyazıt University. Major research interests include knowledge management, construction methods, machine computing, and typological studies.

Abbreviations

AEC	Architecture, Engineering, and Construction
AIA	American Institutes of Architects
AM	Additive manufacturing
BIM	Building Information Modeling
CAD	Computer-aided design
CAT	Chamber of Architects of Turkey
CC	Contour crafting
CMU	Concrete masonry unit
CO_2	Carbon dioxide
DCP	Digital construction platform
DED	Direct energy deposition
EAF	Electric arc furnace
EBM	Electron beam melting
FGP	Fiber glass plastic
FT	Fully tempered
GFRP	Glass fiber reinforced plastic
HBU	Hollow brick unit
HDF	High-density fiberboard
HSS	Hollow steel sections
HT	Heat strengthened
IGU	Insulating glass unit
IVE	Immersive virtual environment
IVM	Immersive virtual model
LOM	Laminated object manufacturing
LVL	Laminated veneer lumber
MDF	Medium-density fiberboard
MIT	Massachusetts Institute of Technology
OSB	Oriented strand board
PBF	Powder bed fusion
PSL	Parallel strand lumber
PVC	Polyvinyl chloride

RIBA	Royal Institutes of British Architects
SHS	Selective heat sintering
SLM	Selective laser melting
SLS	Selective laser sintering
UAM	Ultrasonic additive manufacturing
UK	United Kingdom
USA	United States of America
UV	Ultraviolet
WLF	Wood light frame
WW-II	World War II
WWM	Welded wire mesh
WWR	Welded wire reinforcement

Chapter 1
Introduction to Building Construction

Abstract The introduction chapter explains the need and necessity of the book content in the Architecture, Engineering and Construction (AEC) industry by giving reference to the project delivery approaches in the United States of America (USA), United Kingdom (UK) and Turkey. General terminologies, classification of building components and construction site preparations are introduced.

1 Introduction

Building construction is a discipline that dates back to ancient times, and its development was always in parallel with civilizations. Till the last centuries, building construction materials and methods were almost the same. The built environment was dominated by locally available stone, brick, wood, and adobe. The project participants consisted of building owners, architects, and constructors. Locally available traders delivered the construction materials due to a lack of transportation options from far distances. Building typology was limited regarding the needs and requirements of the user.

On the other hand, technological improvement, especially in the last centuries, brought new architectural necessities, building construction materials, and systems for realizing various building types. Furthermore, advances in transportation options make the many construction materials, equipment, and systems globally available anytime and anywhere. Thousands of construction materials available in the current market could be applied using several construction methods and tools.

The involvement of electricity and other technological affordances has increased the number of project participants to complete a building project. The project stakeholders were involved: an architect for architectural design, a civil engineer for structural design, a mechanical engineer for mechanical design, an electrical engineer for electricity design, a landscape architect for landscape design, various advisors concerning different expertise fields, traders for the supply of the materials,

R. Sarı and E. B. Çalışkan, *Building Construction Methods and Systems*,
https://doi.org/10.1007/978-3-031-50043-5_1

contractor, and constructors for construction and manufacturer for delivery of products. Therefore, the building project became so complicated that a single discipline could not be completed alone.

The expansion and involvement of many stakeholders establish the architecture, engineering, and construction (AEC) industry, devoting a significant ratio of the annual income of nations. Disciplinary expertise reflected through the building design and construction phases makes the building project so sophisticated that the work process must be intertwined. The resulting design and construction process presents substantial efficiency when applying proper project management practices. Project management practices are important for establishing collaboration and good communication among project participants. Since each discipline specializes within its specific building design and construction field, it is hard for project management practices to provide the same language, understanding, and viewpoint among project participants. Although a good communication and collaboration environment was presented, the project participants could not understand their design and construction-oriented needs and requirements. This situation may cause an unnecessary conflict of interest among project stakeholders, enabling cost and time overrun and decreasing motivation and satisfaction of project design and construction teams.

This situation occurs especially among the design team and construction team. Designers' viewpoint for building projects is generally assuring good design for meeting occupant needs and requirements and design-specific satisfactions. The capital of the designer is the idea of design reflected through the building project's volumes, surfaces, materials, and textures. On the other hand, construction teams try to construct the project using a simple method in less time by using less labor as much as possible. The capital of the construction team is labor, time, and cost spent for construction. These differences in viewpoints of design and construction teams prevent the establishment of good collaboration and effective communication. Designers force the construction team to implement the design idea fully. At the same time, the construction team forces the designer to construct the project using less labor, simple methods, and easy solutions.

The educational background of the design team is incomparably higher than that of the construction team in many countries. However, since the design team is so specialized in disciplinary-specific fields and areas, they overlooked the concerns of construction teams. In other words, the design and construction team has not clearly understood their concerns. This handicap could be achieved by presenting building construction systems and technologies in a simple but practical way so that the design team can learn construction methodologies to support their design projects with reasonable and practical implementation solutions without compromising design quality.

The book content supports the design and construction team by explaining state-of-the-art construction practices. The subjects in the book do not dive into engineering calculations to make the topics easily understandable by many AEC practitioners. Furthermore, this book covers the project development stages, represented in Figs. 1, 2 and 3. The figures reveal that in each type of project development, at least two project stages require information about the topics explained in this book. Thus, the reasons mentioned above and the existing project development stages motivated the author of this book to establish the content for the AEC industry requirements.

Figure 4 expresses the construction order of a building. The construction of components remaining below grade level is called "below grade construction," whereas the above part is called "above grade construction." Construction work, referencing studies below and above grade, is classified into three categories. These are superstructure due to being above grade level substructure due to remaining

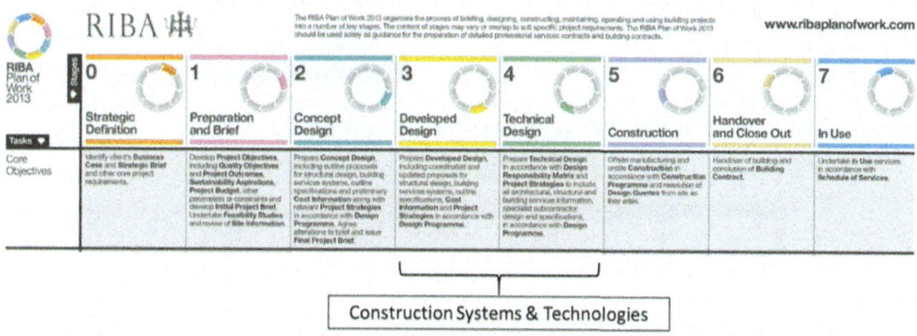

Fig. 1 The project phases that Building Construction Systems and Technologies content are required according to the RIBA project development phases. Retrieved from Davies and Davies (2020)

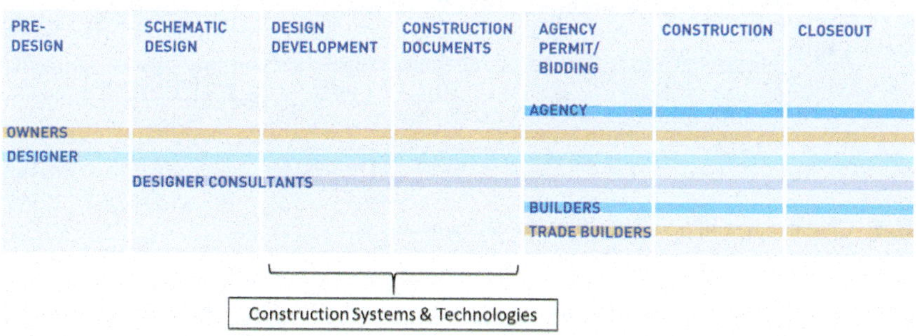

Fig. 2 The project phases that Building Construction Systems and Technologies content is required according to the AIA project development phases. Retrieved from AIA California Council (2007)

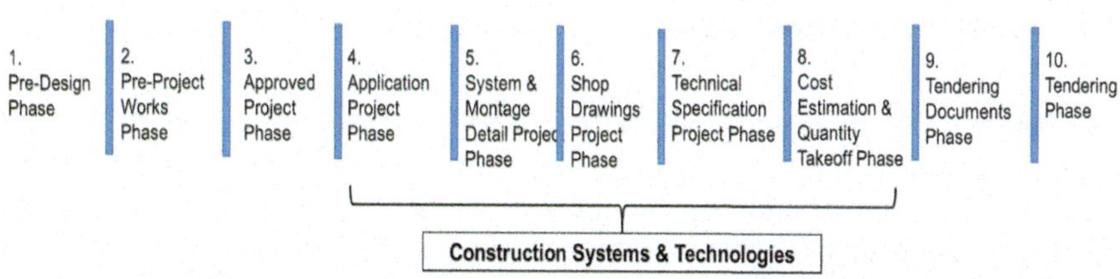

Fig. 3 The project phases that Building Construction Systems and Technologies content are required according to the CAT project development phases. *Source* CAT (2011)

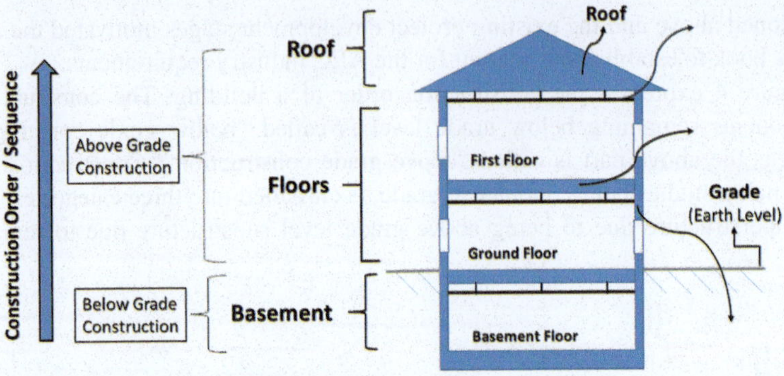

Fig. 4 Construction sequence of a building and its main components

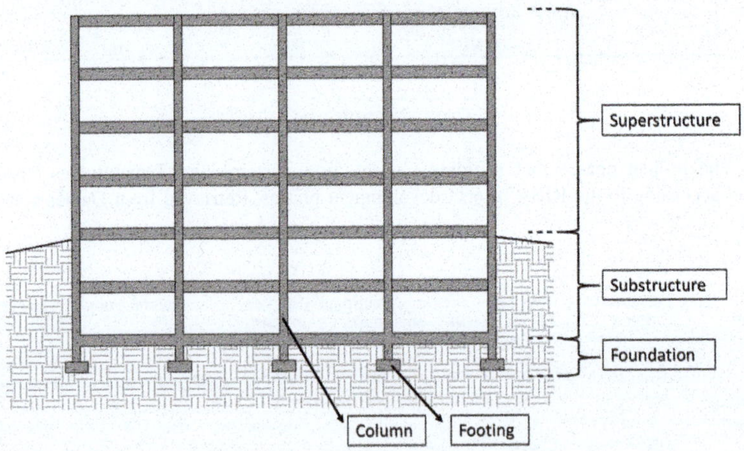

Fig. 5 Classification of building structural components

between superstructure and foundation, which were graphically demonstrated in Fig. 5.

1.1 General Terminologies

The building design provides all the necessary information to meet the client's requirements and public health, safety, and welfare. Architecture is the state of art and science of designing a building regarding the conditions mentioned earlier. Building construction is assembling materials and bringing together related systems to form a building. The environment is the surroundings made of either natural or man-made or a combination of both, while the built environment is created by people with or

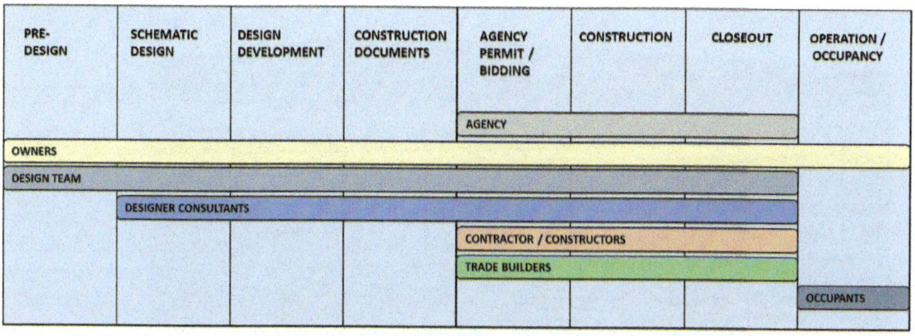

Fig. 6 Involvement of project participants in project development stages

without the aid of the natural environment. Project participants are the persons or entities having roles and responsibilities in the design and construction of a building. These are owner/client, architect, engineer, contractor, trader, and agency in general (Fig. 6).

1.2 Classification of Building Components

Building elements/components could be classified as (i) primary building components, (ii) secondary building components, and (iii) finishings. Each division is explained in the following sub-sections.

1.2.1 Primary Building Components

Primary elements are components of the superstructure above the substructure, excluding secondary building elements, finishes, services, and fittings (Fig. 7).

1.2.2 Secondary Building Elements

Secondary elements are the completion and finishing of the building structure, including around and within openings in primary elements (Fig. 8).

- Doors and windows.
- Balustrades and railings.
- Covering system above the subfloor.

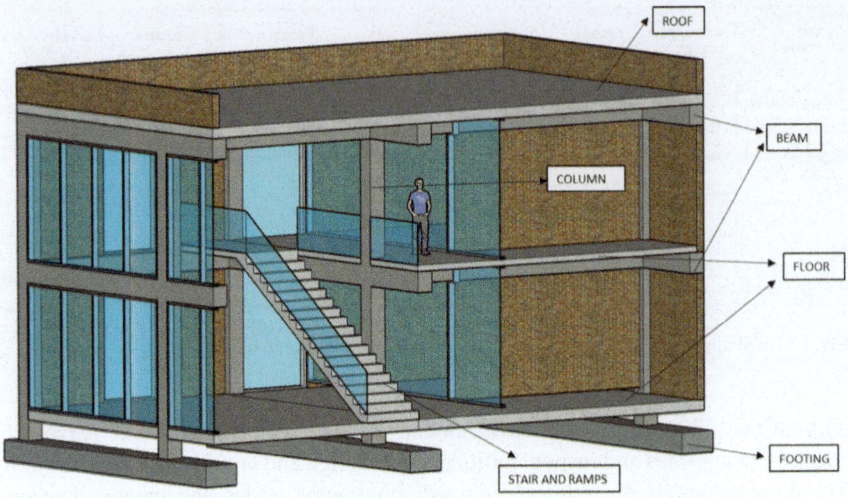

Fig. 7 Illustration of primary building components

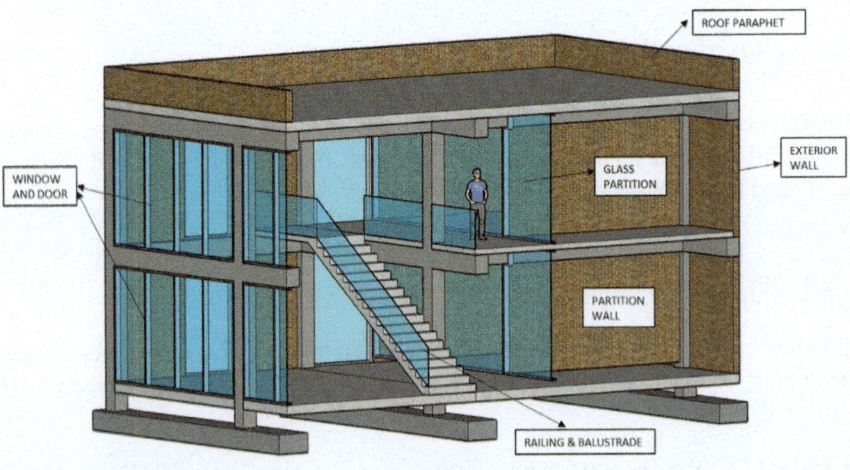

Fig. 8 Illustration of secondary building components

1.2.3 Finishings

The treatment to finalize the surface of the building members is called finishing. Finishing occurs in the roof, floor, walls, ceiling, stairs and ramps, and other primary building elements (Fig. 9).

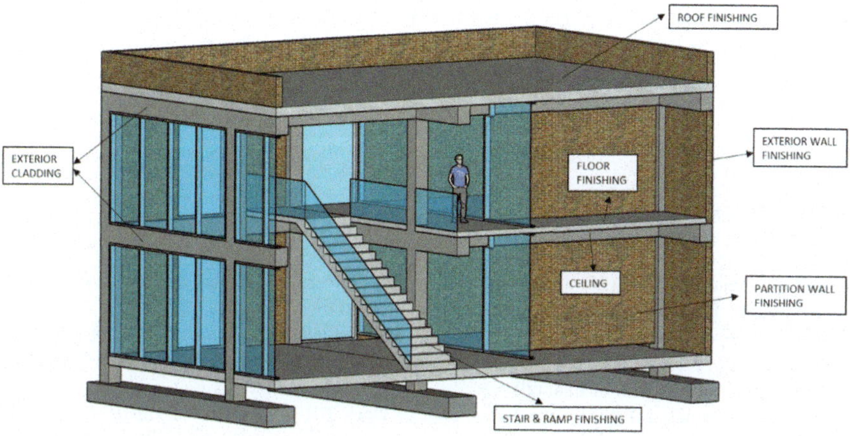

Fig. 9 Illustration of finishing building components

1.3 Starting a Building Construction

Before starting construction, there are a couple of preparation steps. The built environment has two basic considerations. These are environmental considerations and physical considerations. Environmental considerations are the primary design inputs for the design team and include the following aspects:

- Planning requirements.
- Building regulations.
- Land restrictions by vendor or lessor.
- Availability of services.
- Local amenities, including transport.
- Subsoil conditions.
- Levels and topography of land.
- Adjoining buildings or land.
- Use of building.
- Daylight and view aspect.

Physical considerations, on the other hand, directly affect the architectural design of the project and cover the following aspects:

- Natural contours of land.
- Natural vegetation and trees.
- Size of land and/or proposed building.
- The shape of the land and/or proposed building.
- Approaches and access roads and footpaths.
- Available Services.
- Natural waterways, lakes, and ponds.

- Restrictions such as rights of way, tree preservation, and ancient buildings.
- Climatic conditions created by surrounding properties, lands, or activities.
- Proposed future developments.

References

AIA California Council. (2007). Integrated project delivery: A guide. *American Institute of Architects*, 1–62. https://doi.org/10.1016/j.autcon.2010.09.002

CAT. (2011). *Tmmob mimarlar odasi mimarlik hizmetleri şartnamesi ve en az bedel tarifesi (28.12.2011)*. http://www.mimarist.org/include/uploads/2015/11/mimarlik-hizmetleri-sartnamesi-en-az-bedel-tarifesi.pdf

Davies, I., & Davies, I. (2020). The RIBA plan of work 2013. *Contract Administration*, 10–11. https://doi.org/10.4324/9780429347177-2

Chapter 2
Substructure and Foundation Systems

Abstract All structures and buildings connect with the earth to transfer loads and get reactions. The design and process of settlement in and on the ground differs from the building and conditions of the site. This chapter illustrates below-grade construction methods and systems, including site works, excavation, and foundations.

1 Below Grade Construction

Below grade construction covers the investigation of site works, soil types, excavation, and foundation types.

1.1 Site Works

Site works start with site investigation to search for checking and ensuring that the project site and building project characteristics are compliant with each other. This could be achieved by considering site elements, local conditions, and regulations affecting the project site. Besides, regarding the site information represented in Fig. 1. The following questions should be inquired for site investigation:

- Determination of features of surroundings such as location, road, facilities, footpaths and rights.
- Site dimensions and levels.
- Observation of surface characteristics, i.e., trees, steep slopes, existing buildings, rock outcrops, and wells.
- Local codes and regulations affect building form and characteristics.
- Investigation of subsoil conditions.
- Paying attention to flood potential, possibilities for drainage of water table, capping of springs, filling of ponds, diversion of streams and rivers.
- Inquiring data about underground and overhead services, proximity to the site, and whether they cross the site.

R. Sarı and E. B. Çalışkan, *Building Construction Methods and Systems*,
https://doi.org/10.1007/978-3-031-50043-5_2

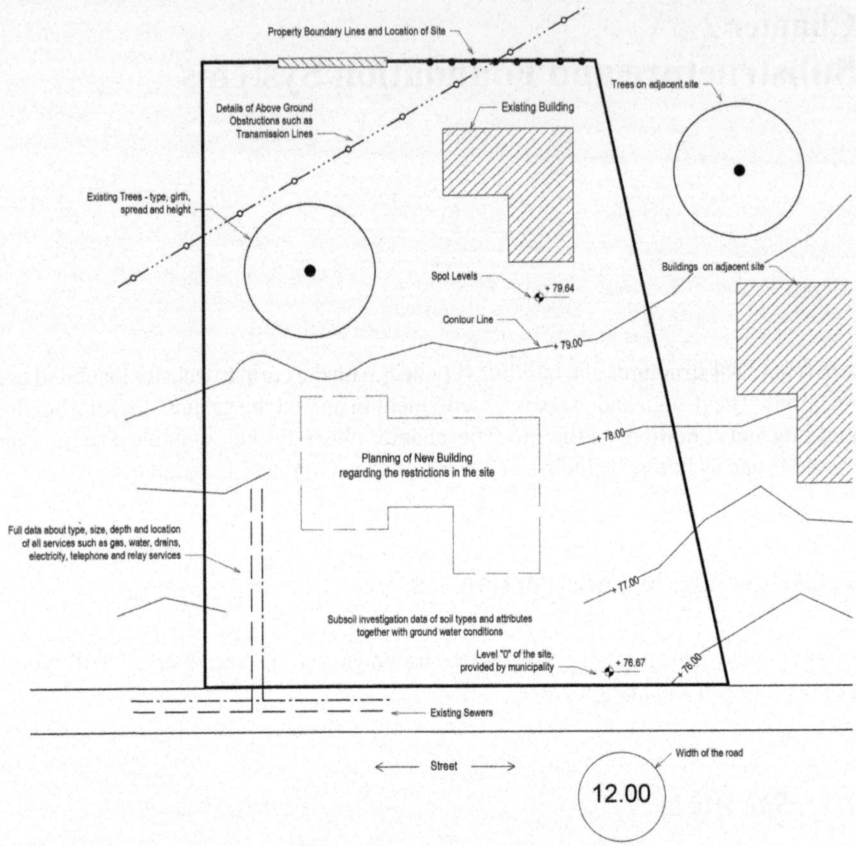

Fig. 1 Requirements in a site work. Adopted from Bridges (2010)

- Noting suspicious factors such as filled ground, cracks in the ground, subsidence due to mining, and any cracks existing in the buildings.
- Consider the neighborhood scale and character of buildings concerning the proposed new development.
- Decide on the best location for building concerning environmental conditions and sustainability of building functions.

Besides aforementioned site inspections, the following key features must also be taken into account:

- The site investigation aims to collect and record all necessary data that can be used during design and construction.
- Assessment of potential hazards to safe and healthy.
- Assessment of trees (restrictions to protect trees).
- Flood potential.
- Soil types.

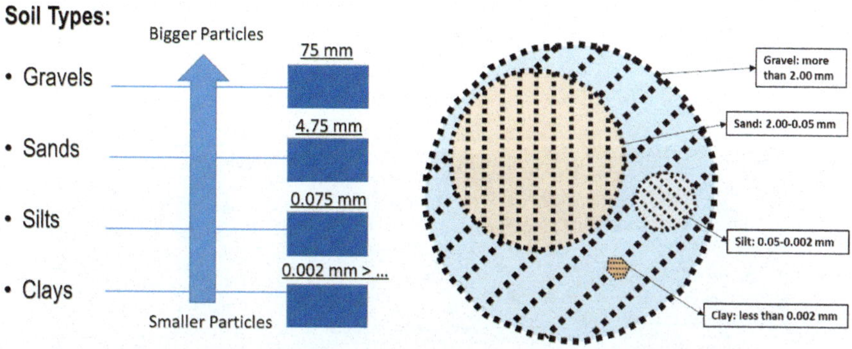

Fig. 2 Soil types and their classification considering their particle size

- Vegetation.
- History of the site (Greenfield or brownfield).

1.2 Soil Types and Attributes

Earth's crust consists of two minerals: (i) rocks and (ii) soils. In terms of structural reliability, rocks are preferable to soils. However, not all parts of the earth's crust consist of rocks. Building weight and earth load-bearing capacity must be convenient with each other. Accepting that the building weight is the same, various foundation types are applied for building project construction in different soil types. Thus, special foundation and footing systems were introduced to be used.

Depending on their particle sizes, soils are classified as gravel, sand, sill, and clay, as represented in Fig. 2. An increase in particle size increases the air and moisture in the soil. Air and moisture are two deformative elements for construction foundations. Furthermore, a decrease in the particle size increases the density of the soil, and the soil load-bearing capacity is particularly increased.

1.3 Excavation

Excavation is a process of removing unwanted soils or rocks in the construction site in order to prepare the construction site to (i) construct the foundation, (ii) basement, (iii) underground utility lines, and (iv) grade the ground surface.

Grading, on the other hand, is a process of moving earth from one location of the site to another and changing the existing land surface in compilation with the desired finished surface configuration. There are two types of grading:

- **Rough Grading** is an excavation work for foundations, basement, and utility trenches as indicated in Fig. 3.

Fig. 3 Excavation work in a construction site (1) and rough and fine grading (2)

- **Fine Grading** is a process of completing the landscape design at the end of the construction.

Excavation works executed in soil-based construction sites make it necessary to apply excavation supports. There are two types of support to be used for this purpose. These are (i) self-supporting sloped excavation and (ii) deep vertical cut supporting.

1.3.1 Self-Supporting Sloped Excavation

Self-supporting sloped excavation types have a natural slope so that unwanted soil does not flow into the construction site upon the slope's surface. Self-supporting sloped excavation does not require any cost to bear the soil. Especially it is preferable in road constructions pass through the mountainous regions. Two sloping methods are available to be used (Fig. 4):

- Open excavation with a uniform slope.
- Benched excavation.

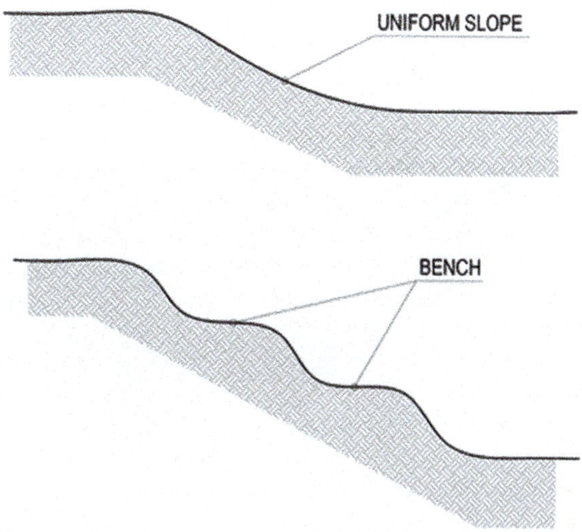

Fig. 4 Open excavation with uniform slope and benched excavation

1.3.2 Deep Vertical Cut Supporting Methods

The following methods can be used as deep vertical cut supporting methods in soils:

- **Sheet Piles**: Corrugated or special bent sheets are driven into the soil by a punching machine, as Fig. 5 represents. Then, the earth in the construction site was removed.
- **Cantilevered Soldier Piles**: The noise and vibration created by driving soldier piles into the soil are disadvantages of sheet piles. In cantilevered soldier piles, vertical studs (soldier piles) are driven into the soil with a certain distance in between them, as indicated in Fig. 6. Then, the gap between these studs is filled with wood strips. The soldier piles bear all the lateral load from the earth rather than wood strips for filling the gap.

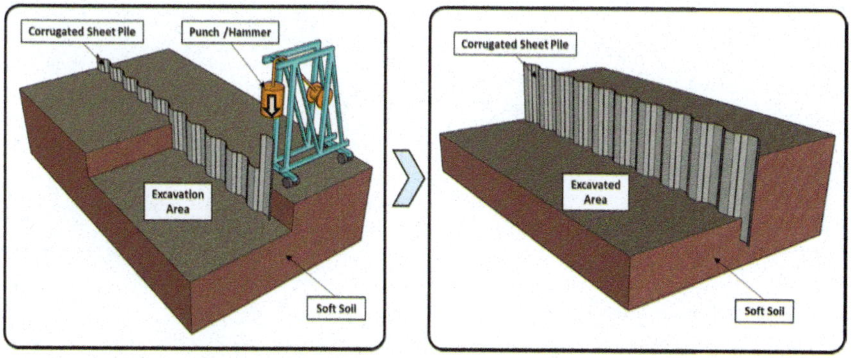

Fig. 5 Sheet piles as deep vertical cut supporting

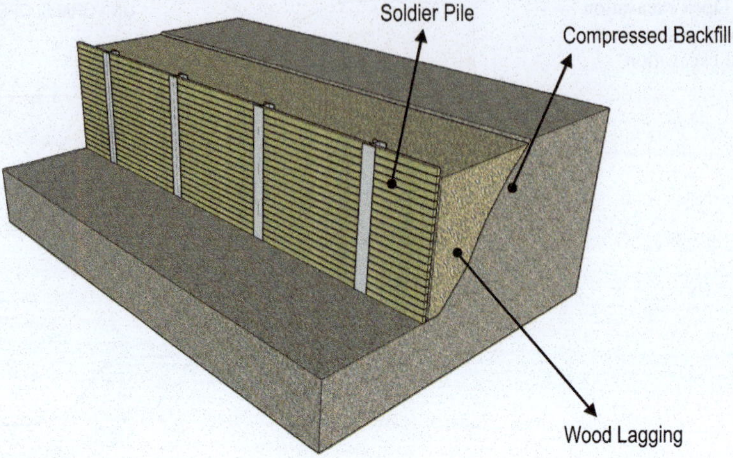

Fig. 6 Schematic representation of cantilevered soldier pile deep vertical cut supporting method

- **Anchored Soldier Piles**: For deeper excavations, the anchored soldier piles method can be used in which soldier piles are tied back (anchored) to the ground, as depicted in Fig. 7.
- **Contiguous Bored Concrete Piles**: When deeper excavation is needed, and neighbor buildings are preventing the usage of anchored soldier piles, then the contiguous bored concrete pile method, where closely spaced concrete piles, as indicated in Fig. 8, can be used.
- **Secant Piles**: The gap between the concrete piles in the contiguous bored pile excavation support system may cause low water resistance performance for foundations and the basement due to water and moisture leakage from the earth within

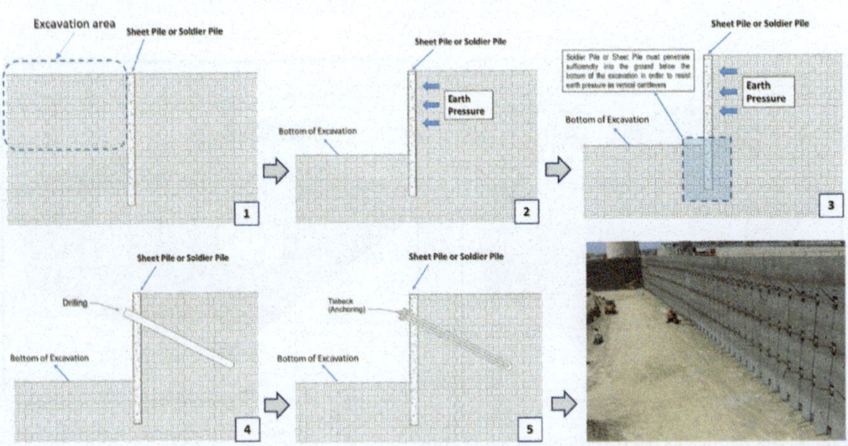

Fig. 7 Anchored soldier piles application process

Fig. 8 Contiguous bored pile construction steps and a real image

the gaps. Then, the secant pile method can be used. Concrete piles intersect with each other, as indicated in Fig. 9, to not only prevent occurrences of the gap among the piles but also provide much more rigid vertical support to the excavated area.

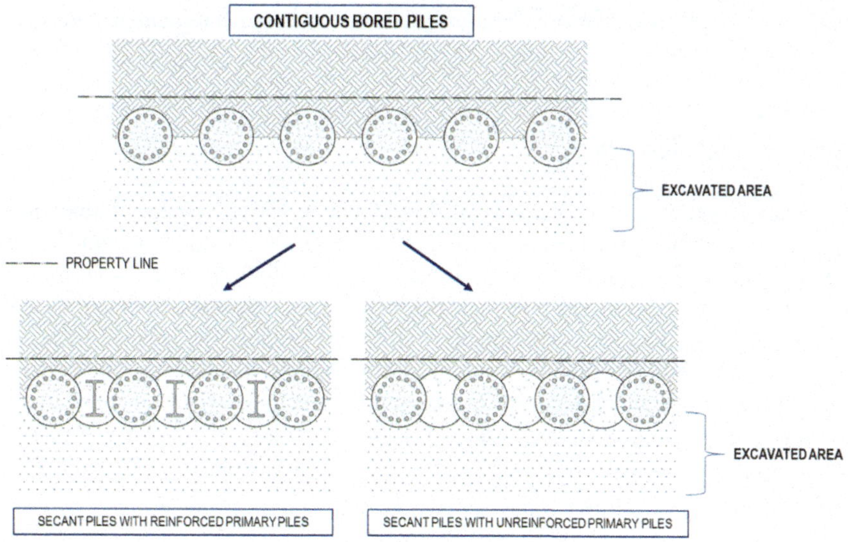

Fig. 9 Comparison of contiguous bored piles and secant pile methods

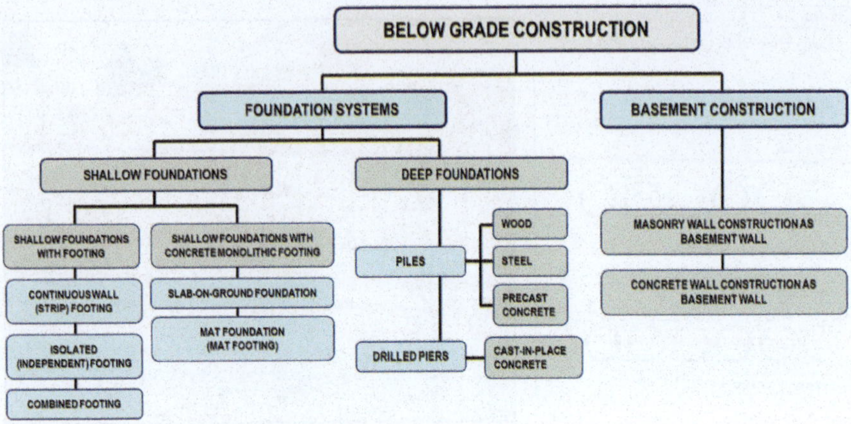

Fig. 10 The topics discussed are below-grade construction

1.4 Below Grade Construction

Below-grade construction includes foundation systems and a basement. The following phases present foundation and basement construction methods and systems in detail. Overall categorization demonstrated and discussed in this chapter are stated in Fig. 10.

1.4.1 Foundation Types

Foundation types are classified as (i) shallow foundations and (ii) deep foundations. Shallow foundations are further divided into (a) shallow foundations with footing and (b) shallow foundations with concrete monolithic footing, while deep foundations are categorized as (1) piles and (2) drilled piers. Shallow foundations with footing are classified as (i) continuous wall (strip) footing, (ii) isolated (independent) footing, and (iii) combined footing, while shallow foundations with concrete monolithic footing are classified as (a) slab-on-ground foundation and (b) mat foundation (mat footing). Piles are further classified regarding the material of piles as wood, steel, and precast concrete. Similarly, cast-in-place concrete is the only option for drilled piers regarding the applicable material.

1.4.2 Shallow Foundation

Shallow Foundations with Footing

Shallow foundations are placed and constructed at a relatively short depth from ground level and directly based on the surface of the soil stratum. The soil stratum

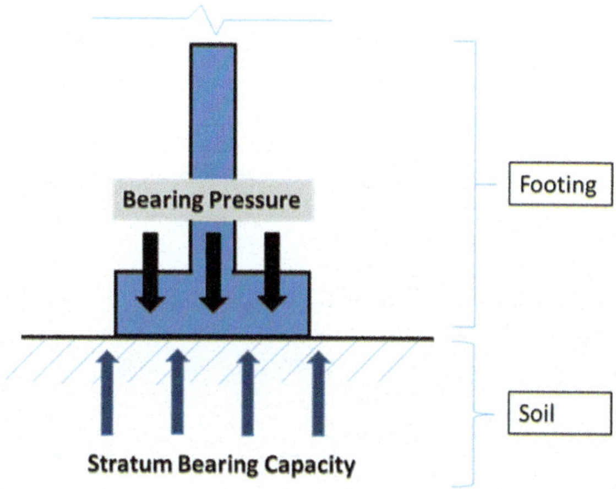

Fig. 11 The soil stratum must bear the loads in footing

is located below the excavated area where the foundation of the building and soil touch each other, and related load transfers occur, as indicated in Fig. 11. Due to its lower cost, shallow foundation, where as much as possible, is preferable to any of the deep foundation methods. The limitation to not using a shallow foundation comes when the soil's load-bearing capacity is insufficient to carry all required building load. Thus, shallow foundation is generally applicable to low-to-mid-rise buildings. The footing is directly located upon the soil in a shallow foundation, as indicated in Fig. 12; therefore, this does not mean buildings with basement floors are accepted as deep foundations. The methods considered under the shallow foundation category are (i) continuous wall (strip) footing, (ii) isolated (independent) footing, and (iii) combined footing.

- **Continuous Wall (Strip) Footing**: The construction method generally consists of a footing and load-bearing (foundation) wall, as Fig. 13 indicates. Strip footing has far better foundation settlement performance than isolated footing due to uniform load distribution over the soil stratum. Thus, load-bearing capacity and resistance to foundation settlement of strip footing are more than isolated footing.
- **Isolated (Independent) Footing**: The construction method generally consists of an independent footing and a column, as indicated in Fig. 14. Isolated footing is applicable when the bearing capacity of the soil stratum is high or the building project has a lightweight design so that foundation settlement is prevented.
- **Combined Footing**: Combined footing consists of strip and isolated footing characteristics in its construction method. Some of the applicable samples are illustrated in Fig. 15.

Shallow Foundation with Concrete Monolithic Footing

A shallow foundation with concrete monolithic footing could be categorized as a slab-on-ground (grade) and mat foundation.

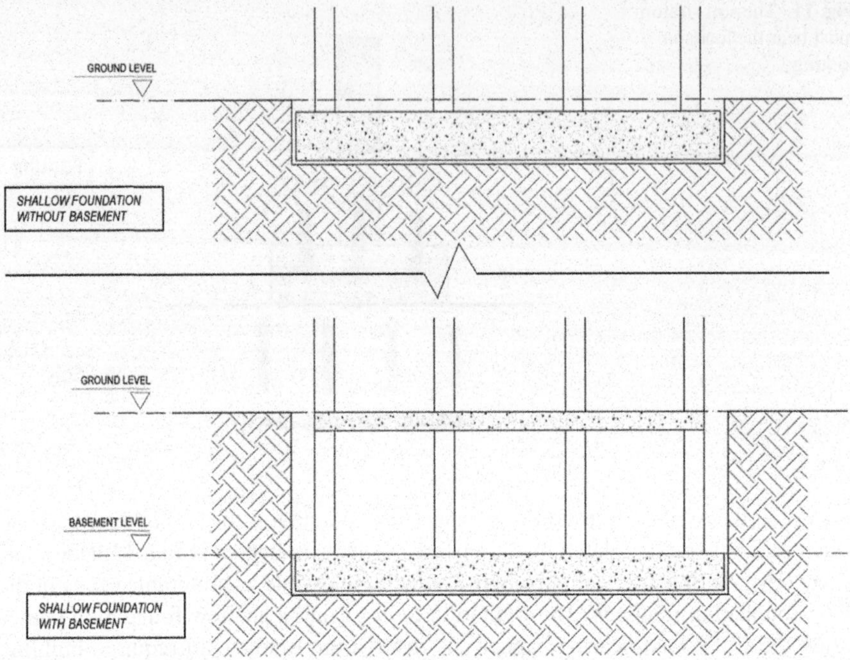

Fig. 12 Shallow foundation with and without basement floor

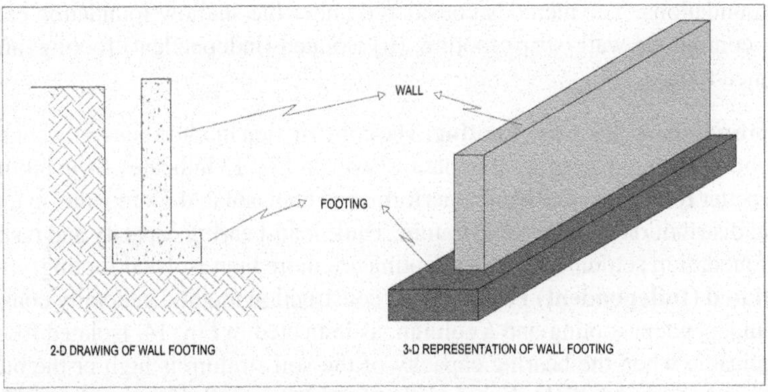

Fig. 13 Continuous wall (strip) footing

- **Slab-On-Ground (Grade) Foundation**: The foundation is directly the slab at the ground. There may be strip footing incorporated with a slab under the structural axis. It is used when building weight or soil load-bearing capacity is low (Fig. 16).
- **Mat Foundation (Mat Footing)**: Mat foundation consists of a flat slab thicker enough to bear all building weights and loads as a unique and solid building component. Mat foundation is used when soil load-bearing capacity is low. Its

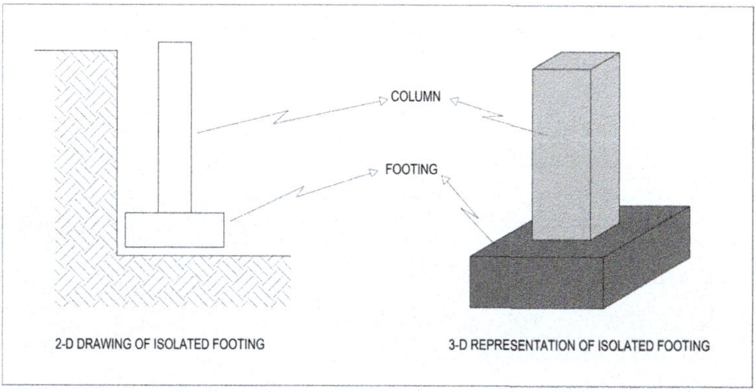

Fig. 14 Isolated (independent) footing

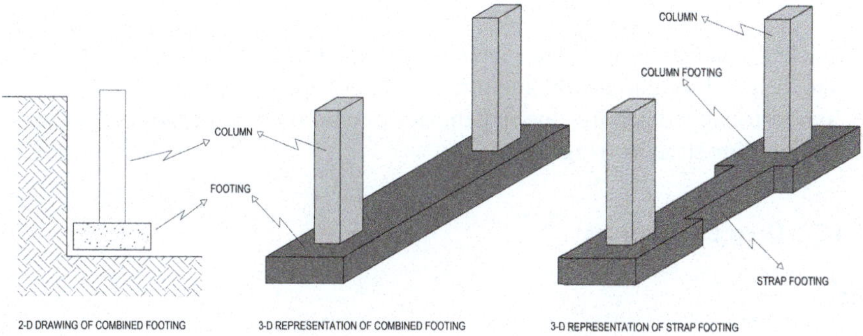

Fig. 15 2-D and 3-D illustrations of combined footing

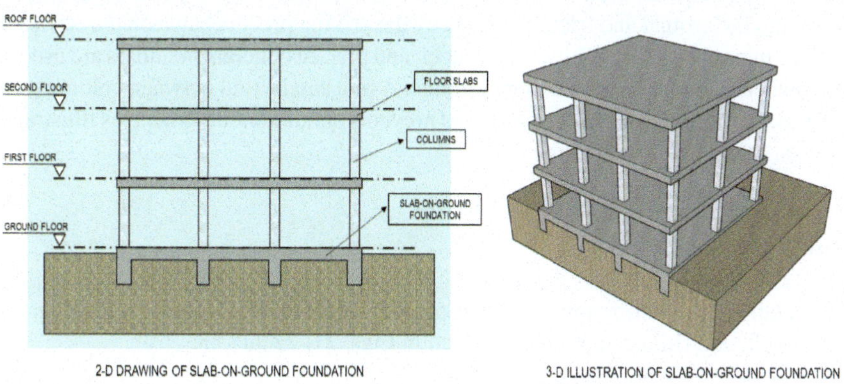

Fig. 16 2-D and 3-D illustrations of slab-on-ground (grade) foundation

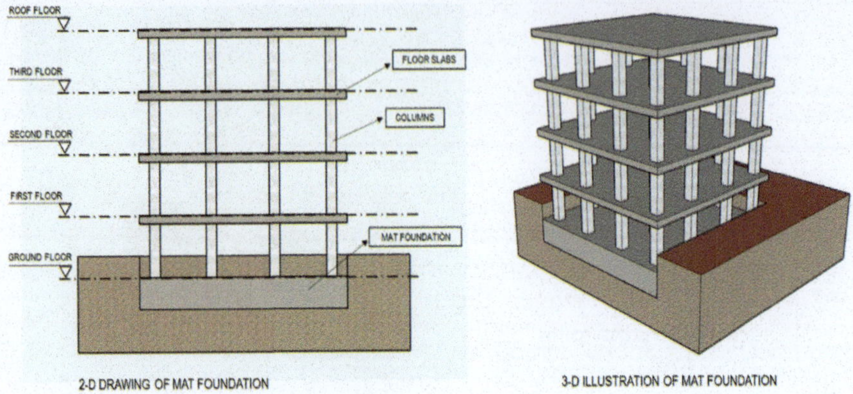

Fig. 17 2-D and 3-D illustrations of mat footing

thickness and resulting weight increase cause foundation settlement. Furthermore, the load and weight distribution upon the building structure must be distributed symmetrically; otherwise, the settlement could become a complete inclination of the building structure (Fig. 17).

1.4.3 Deep Foundations

Piles

The pile member is driven into the ground by an external force, as Fig. 18 indicates. This process requires the pile member to be rigid enough to resist punching impact to drive into the ground and friction caused by the driving process between the earth and the pile member surface. Thus, wood, steel, and precast concrete members are used as piles when piles are located under columns. A pile cap is used between columns and piles to uniformly distribute the load and prevent foundation settlement, as illustrated in Figs. 19 and 20.

Drilled Piers

The method of construction of drilled piers consists of three steps: (i) drilling, (ii) formwork, and (iii) casting, as illustrated in Figs. 21, 22 and 23.

Piles Versus Drilled Piers

Piles cause vibration and noise to drive the pile into the earth due to the punch of the piles with a hammer. Therefore, piles cannot be used in urban environments. On the other hand, piles could be used in marine and coastal sites. Piles necessitate capping

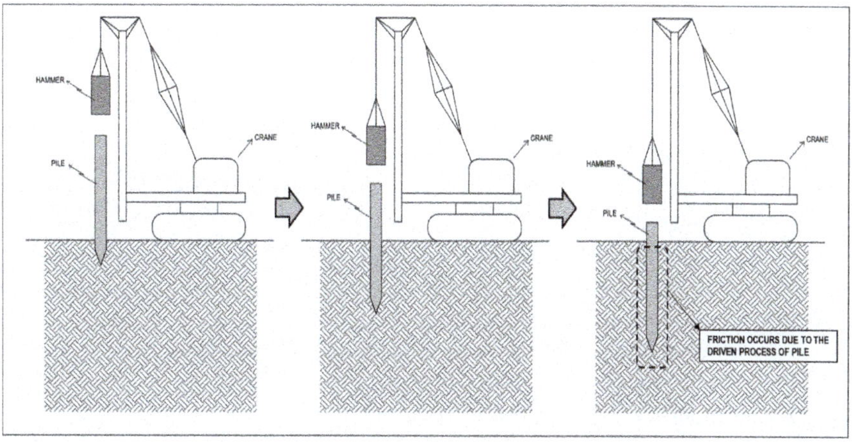

Fig. 18 Schematic illustration of pile driving process into the ground with a hammer machine

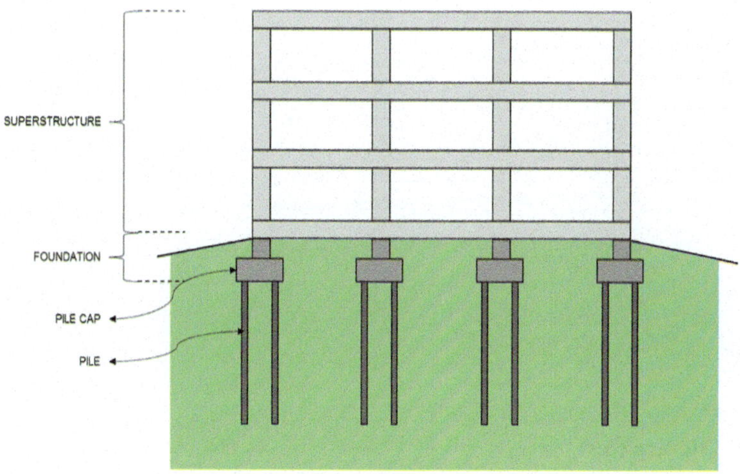

Fig. 19 Pile, pile cap, foundation, and superstructure of a building project

to eliminate foundation settlement. Drilled piers necessitate a drilling process in the ground. Therefore, it is not proper to be used in marine and coastal sites. After the drilling process, reinforcement and concrete pouring follow the process. Thus, no noise or vibration occurs on the construction site. It could be used in urban areas and does not require capping. The cost and technology required for drilling pier system is particularly cheaper than the piling process (Table 1).

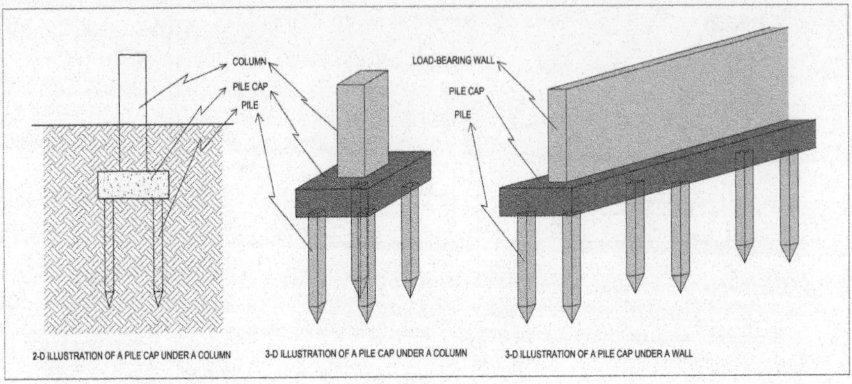

Fig. 20 Pile cap application under a column and a wall

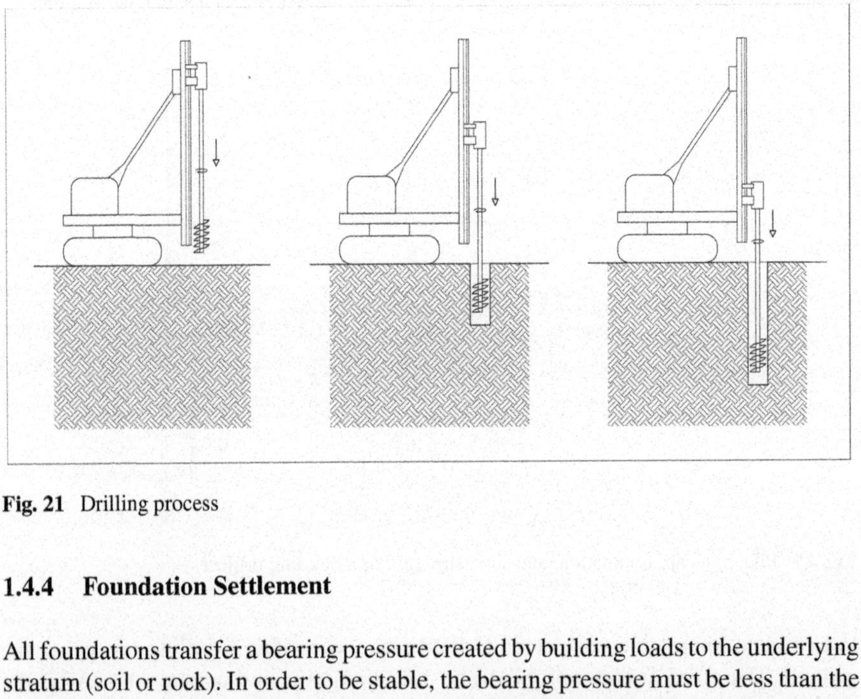

Fig. 21 Drilling process

1.4.4 Foundation Settlement

All foundations transfer a bearing pressure created by building loads to the underlying stratum (soil or rock). In order to be stable, the bearing pressure must be less than the stratum's allowable bearing capacity, as represented in Fig. 24. Otherwise, there may be distortion on the foundation, which will affect the entire building structure. This situation is called a foundation settlement. Although foundation settlement is more tolerable than steel structures for reinforced concrete and masonry structures, the foundation must resist the settlement to design the structure. Furthermore, the bearing capacity of rocks is generally more than soils; therefore, rocky areas are preferable to soil areas for construction. There are four types of foundation settlement regarding

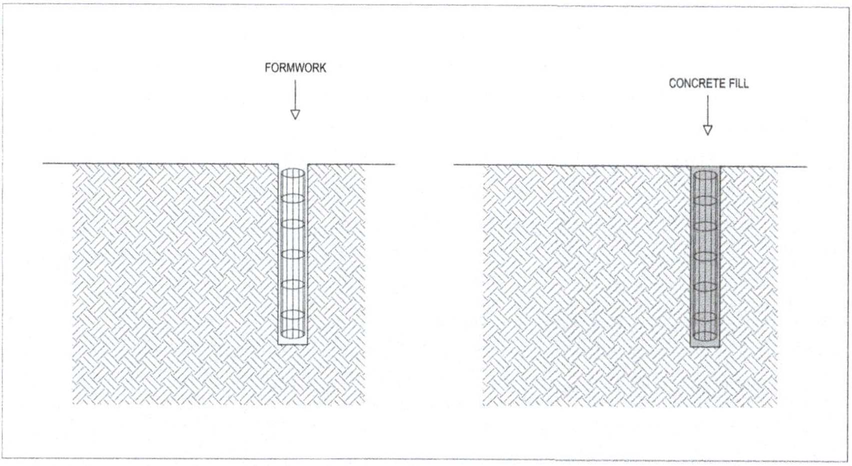

Fig. 22 Formwork placement and concrete fill into the drilled hole

Fig. 23 Drilled piers application in a construction site

Table 1 Comparison of pile and drilled piers

Pile	Drilled Piers
Suitable for marine and coastal sites	Due to water, no drilling process is available
Piles are driven into the surface (earth or water)	Piers necessitate the drilling process
Piles cause noise and vibration	Drilling piers have less environmental impact than piles
Vibration and noise are handicaps for piles to be used in the urban environment	Piers are more suitable for urban areas than piles due to the vibration and noise caused by piles
Pile necessitates capping to prevent foundation settlement	Piers does not necessitate capping
The cost and required technology for the pile are more than piers	The cost and required technology for piers are less than piles

their characteristics and occurrences: (i) immediate settlement, (ii) consolidation settlement, (iii) uniform settlement, and (iv) differential settlement.

- **Immediate Settlement**: In coarse-grained soils, settlement is immediate; in other words, settlement occurs when the load is applied as soon as possible, as illustrated in Fig. 25. Immediate settlement does not represent any risk or hazard for the building structure and is thus acceptable for construction.
- **Consolidation Settlement**: In fine-grained (particularly clayey) soils, part of the settlement is immediate, and the remainder (called consolidation settlement) occurs over several months or years, as stated in Fig. 26.
- **Uniform Settlement**: The entire structure is settled by the same amount as in Fig. 27.
- **Differential Settlement**: In differential settlement, as indicated in Fig. 28, the settlement of footings is irregular, and this situation creates structural problems due to the deformation in the structural frame when adequate tolerances are not provided during the design of the structural system.

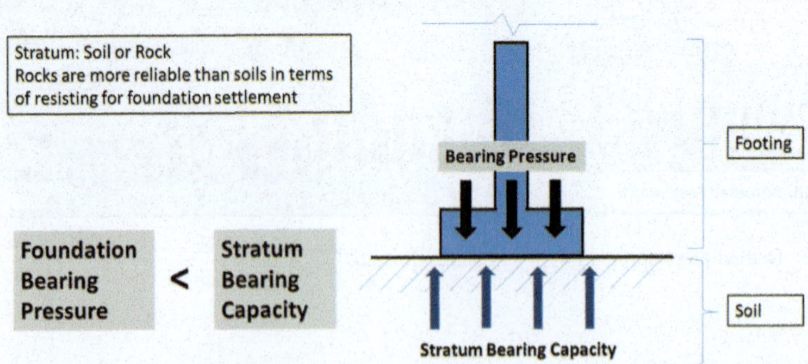

Fig. 24 The relation of foundation settlement and stratum load-bearing capacity

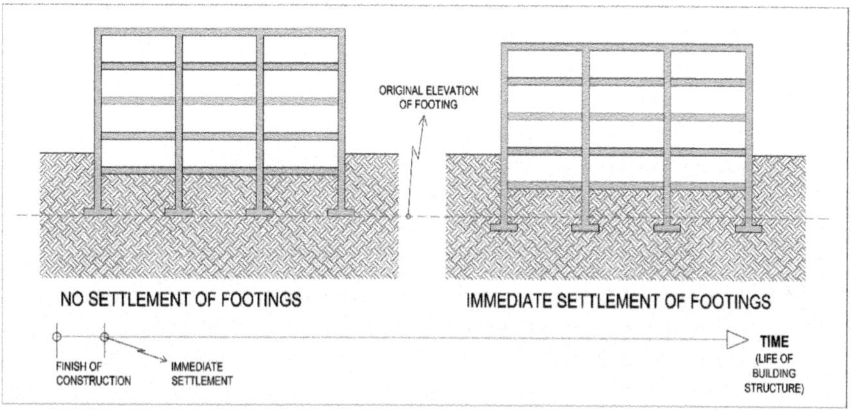

Fig. 25 Immediate settlement

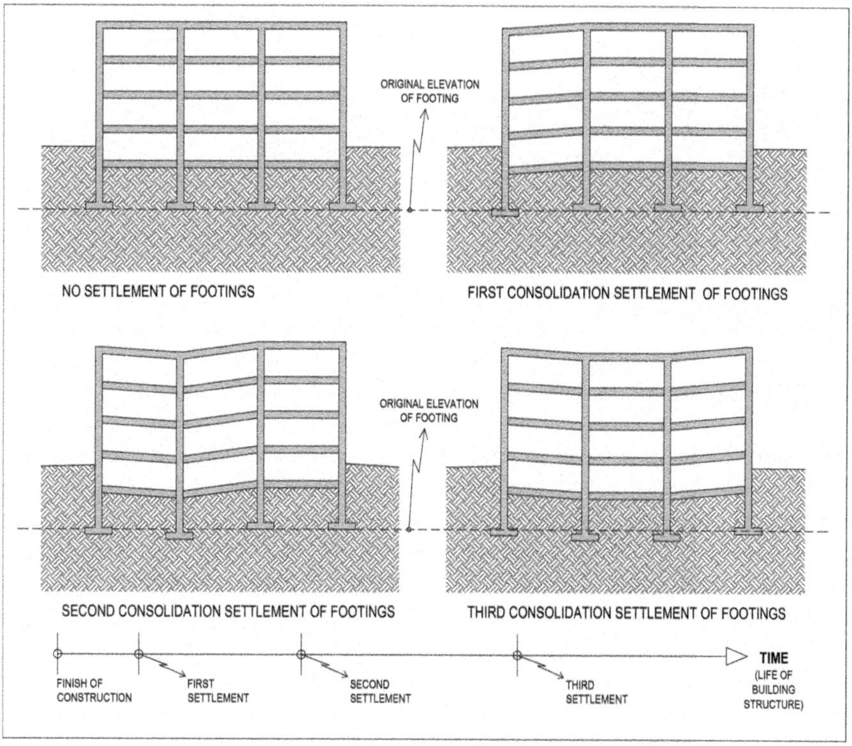

Fig. 26 Process of consolidation settlement

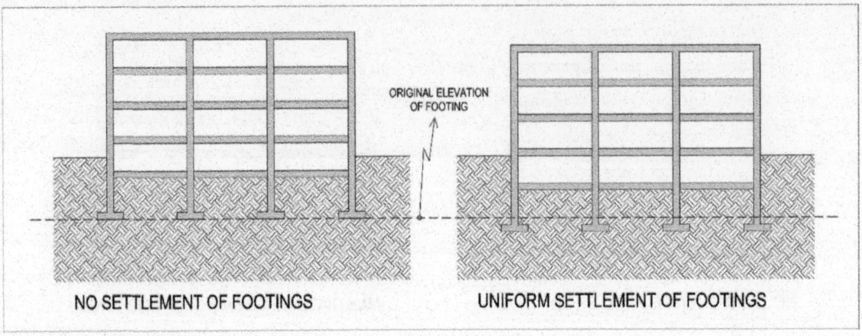

Fig. 27 Uniform settlement

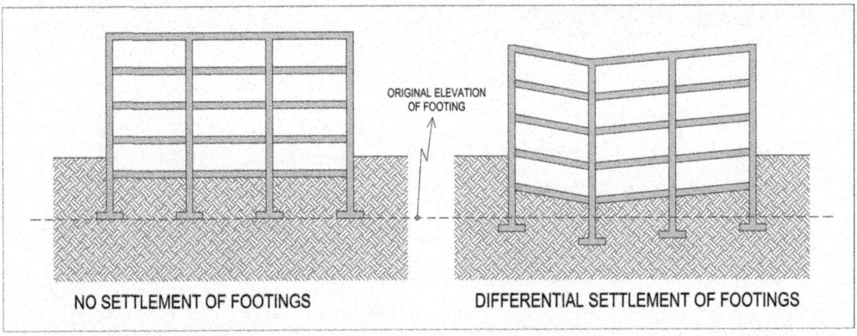

Fig. 28 Differential settlement

1.4.5 Foundation Drainage

Foundation drainage (French Drain) relies on the mechanical properties of the cohesion forces of the water. It simply consists of a drainpipe surrounded by a drainage medium. The drainage medium content is the clean crushed stone or grave, as illustrated in Fig. 29. The foundation drainage allows the control of the groundwater and moisture content around the building foundation. Therefore, the drainage line must be enveloped around the building foundation.

1.4.6 Foundation Design Inputs

A foundation consists of at least the following members:

- **Excavation Support**: After excavation work is completed, appropriate excavation support should be provided, such as self-supporting sloped excavation or deep vertical cut supporting.

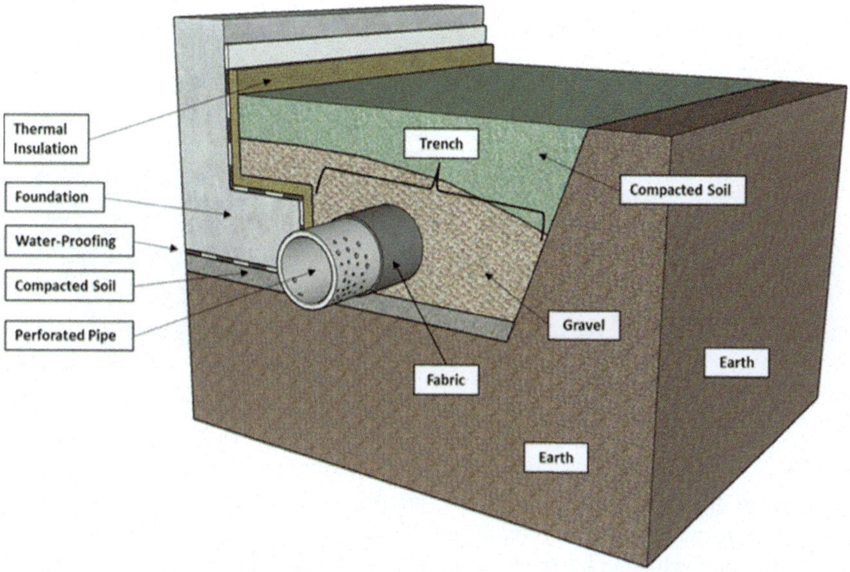

Fig. 29 Diagrammatic illustration of a drainage system

- **Foundation Drainage**: An appropriate foundation drainage system should be applied around the entire foundation to remove unneeded water and moisture inside the soil.
- **Footing Type**: An appropriate footing type must be implemented on the construction site regarding the soil type and building weight. The construction budget is another important factor in the decision of footing type.
- **Necessary Underlayment for Footing and Grading Work**: Regarding the selected footing type, appropriate grading should be done before installing the footing formwork. Fine grades, gravels, and crushed stones are some of the available materials to be used as an underlayment. Using smaller particles for grading provides better building pad performance and a smoother surface for insulation materials.
- **Water and Moisture Insulation of Foundation**: Water as liquid and moisture as gas are two harmful natural materials in our built environment. Water and moisture deteriorate the footing materials and decrease the structural performance of foundations as the water and moisture penetrate the foundation. Therefore, appropriate water-proofing members and moisture (vapor) retarder materials should be applied around the foundation. The process of wrapping up the entire foundation to protect against water and moisture is called *"bundling."*
- **Thermal Insulation Work**: Thermal insulation in the foundation of the building is necessary, especially in cold climates. This is because concrete includes water as an ingredient, and when the temperature goes below 0 degrees, the concrete's water transforms into ice. The transformation of water to ice causes a volume

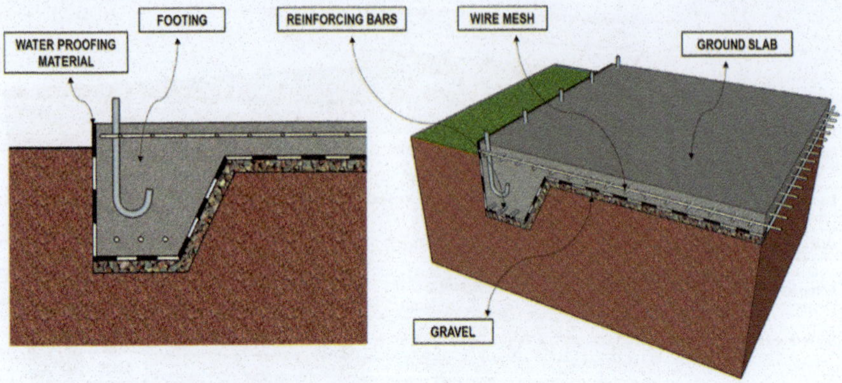

Fig. 30 2-D and 3-D illustrations of slab-on-ground foundation

increase due to this process, concrete cracks, which cause deformation in structural performance. In order to prevent the transformation of water to ice, thermal insulation is provided for foundations. Regarding the climatic conditions, thermal insulation material and thickness can be changed. However, for thermal insulation for the foundation, it should be noticed that insulation material will always be affected by building weight and soil pressure.

- **Protection of Thermal Insulation**: Soil consists of numerous materials and minerals chemically harmful to insulation material. Thus, a covering material or a protection system must be provided for thermal insulation materials applied in foundations to protect thermal insulation.

1.4.7 General Details About Foundations

Some of the simple foundation applications are discussed and explained in this section.

Slab-On-Ground (Grade) Foundation

After excavation and grading, soil is laid on the surface and compressed. Waterproofing material is applied to compressed soil. The formwork of the foundation is installed at the site. Following the application of reinforcing bars (rebar), concrete is poured. The formwork is removed after curing and drying of concrete, and proper foundation drainage is applied. For the last step, soil is infilled into the excavated area. Figure 30 represents the resulting foundation procedure as a 2-D and 3-D illustration.

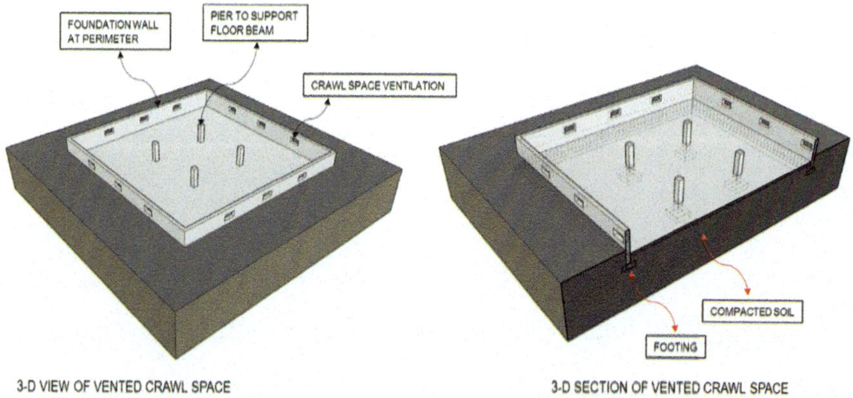

Fig. 31 3-D view, 3-D section, and a real image for exterior foundation walls, interior piers, and an elevated ground floor

Exterior Foundation Walls, Interior Piers (Short Columns), and an Elevated Ground Floor

Following the excavation and grading, soil is laid and compressed. Then, waterproofing material is applied to isolate the footing from water and moisture. After completion of formwork and rebar, concrete is poured. The exterior border of the foundation consists of a concrete strip wall. If interior columns are out of the project's exterior wall, these columns are supported by isolated or strip footing. The ground level is elevated from the foundation, and some air inlets are presented at the exterior wall. The resulting foundation design allows an increase in the thermal performance of the building. Therefore, this type of foundation is widely used in tropical climatic regions (Fig. 31).

Wall and Column Footing and Ground Floor Slab

The overall foundation design is similar to the exterior foundation walls—interior piers; on the other hand, the ground floor is directly based on the soil (Fig. 32).

Grade Beams, Drilled Piers, and Ground Floor Slab

When soil load-bearing capacity is lower than building loads, the foundation is supported by drilled piers. These piers carry foundation–grade beams to present a safe foundation for the ground floor. Regarding the relation between soil stratum and building loads, the piers are applied underneath the grade beam with certain intervals, as illustrated in Figs. 33 and 34.

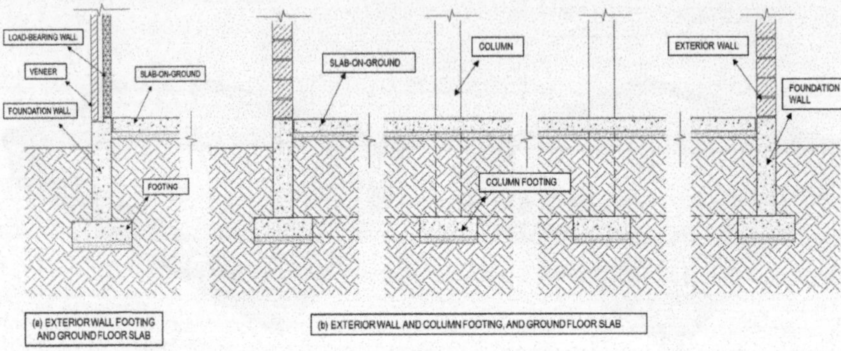

Fig. 32 Wall and column footing and ground floor slab

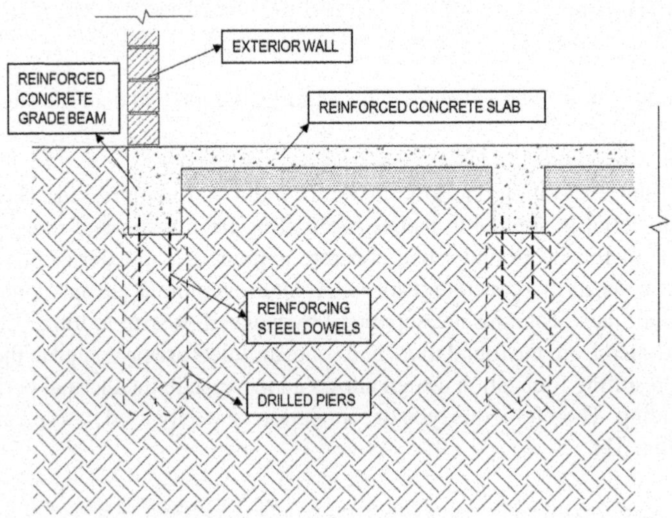

Fig. 33 Grade beams—drilled piers and ground floor slab

Frost-Protected Shallow Foundation

The frost line indicates the temperature level in the earth that the earth above at the frost line can be frost in cold weather. The frost line is measured by local authorities and published in codes and regulations. For example, the frost line in Moscow could be 76 cm, while in Alaska, it could be 252 cm. Theoretically, shallow foundations cannot be used in cold climatic conditions due to insufficient depth to pass the frost line. However, incorporating appropriate thermal insulation at the foundation's perimeter makes using a concrete slab-on-ground foundation possible in cold climatic conditions. A sample design for the frost-protected shallow foundation is illustrated in Fig. 35. The crucial thing is that thermal insulation must be deep enough to pass the

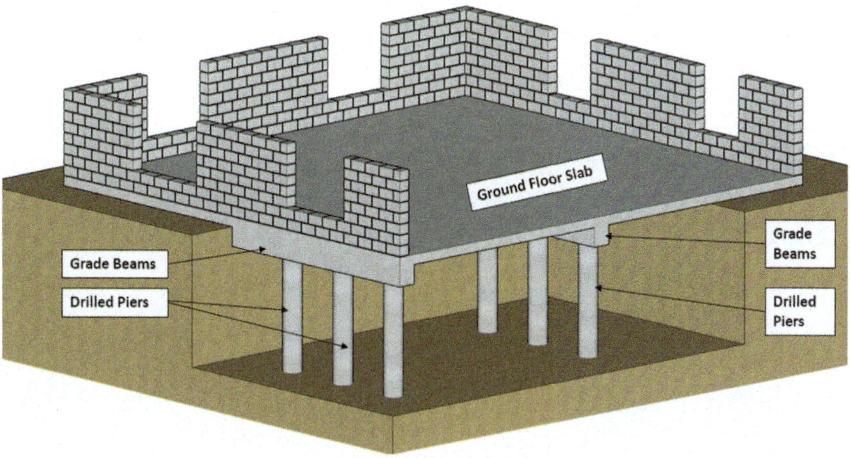

Fig. 34 3-D illustration of grade beams—drilled piers and ground floor slab

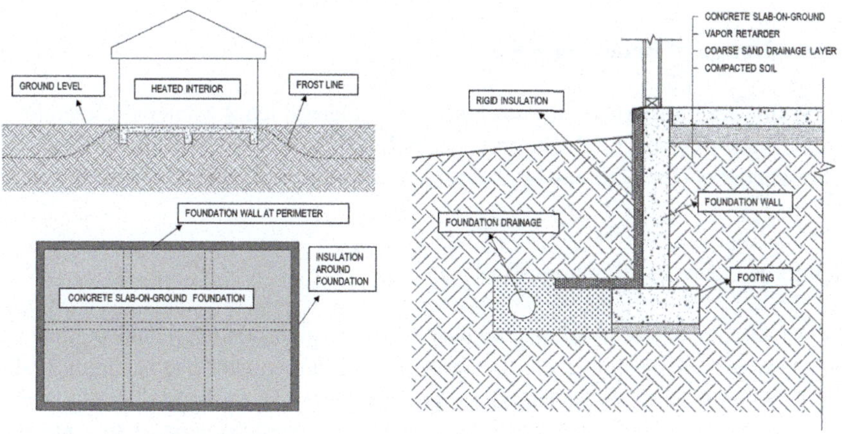

Fig. 35 An example of a frost-protected shallow foundation

frost line. Otherwise, there may be cracks in the concrete due to the transformation of water and moisture inside the concrete to ice. It is a water-specific characteristic that the transformation of water to ice causes an increase in the water's volume. This situation causes cracks and deformation of concrete.

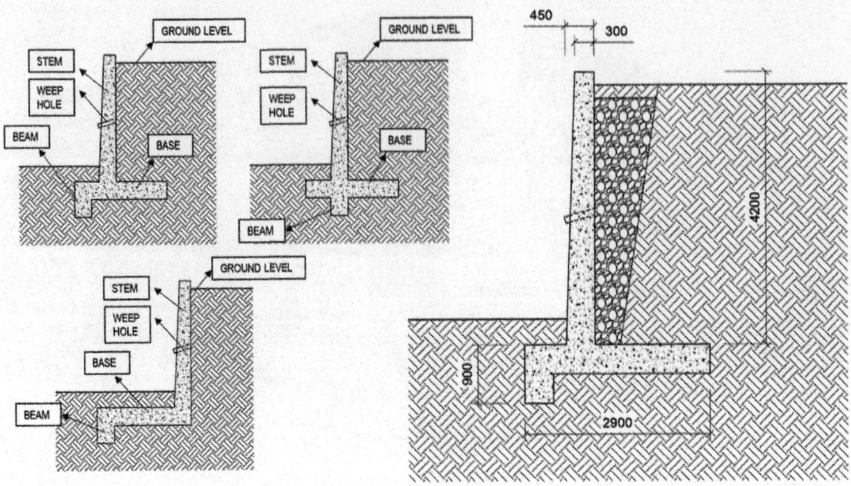

Fig. 36 Reinforced concrete retaining wall terminologies and different types

1.4.8 Design of Retaining Walls

Retaining walls are structural walls built to resist lateral loads. Generally, a retaining wall is made of stone, brick masonry, or reinforced concrete. The following sections represent details for stone and brick masonry and reinforced concrete retaining walls.

Reinforced Concrete Retaining Wall

Figure 36 represents a sample section illustrating terminology and application requirements for designing and constructing reinforced concrete retaining walls. Since the wall must resist lateral loads caused by the earth, besides the wall's strength, the foundation must also have enough strength to support the retaining wall. Therefore, a couple of reinforced retaining wall examples are presented in Fig. 36.

Masonry Retaining Wall

The behavior of earth and design principles of retaining masonry walls are the same with reinforced concrete retaining walls. On the other hand, since masonry construction is heavier and more vulnerable than reinforced concrete to lateral loads, the masonry wall must be thicker than reinforced concrete. Furthermore, the foundation of the masonry retaining wall must also be thicker and heavier than the foundation of the reinforced concrete retaining wall. Its foundation could support reinforced

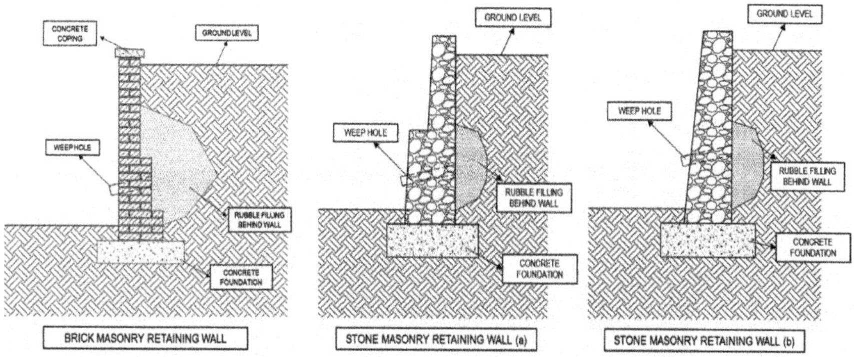

Fig. 37 Various types of brick and stone masonry retaining wall

concrete retaining walls. Therefore, increasing the thickness of masonry retaining walls increases the wall's strength (Fig. 37).

1.4.9 Basement Construction

In the aforementioned sections, foundation systems and foundation walls were introduced. Based on these principles and application details, basement construction detail is presented in the following sections. Due to being below the ground level and affected by soil lateral loads, the basement wall structure is similar to a retaining wall. Furthermore, retaining walls and basement walls are generally made of either masonry or reinforced concrete regarding the economic and structural requirements. The bearing capacity of masonry walls is low, although the dead load (weight of the structure) is high, where these characteristics are far more preferable in reinforced concrete basement walls. Therefore, when higher load-bearing capacity and low dead load are required, reinforced concrete is preferable to masonry basement walls.

Masonry Basement Wall

A sample brick masonry wall in a basement is presented in Fig. 38. Special attention must be given to wall connection and wall-to-ground floor slab connection. Water isolation solution at these two connections must prevent water and moisture leakage inside the structure. Otherwise, water and moisture decrease the duration of the building structure and deform both structure and human health.

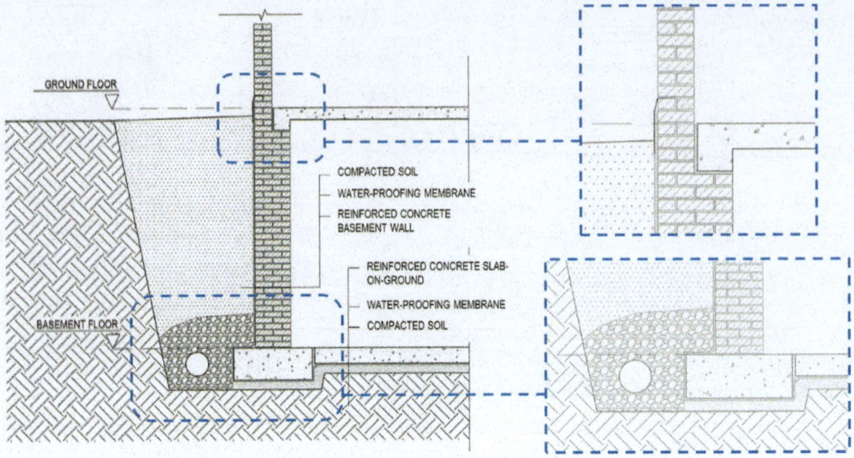

Fig. 38 Brick masonry basement wall detail

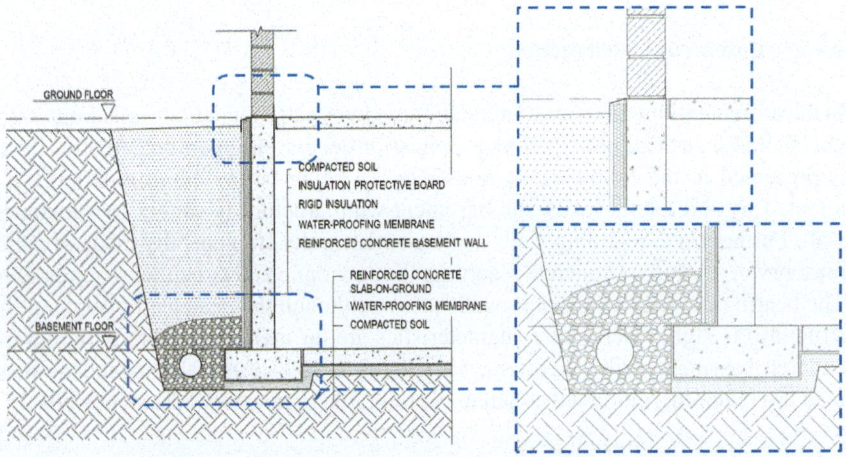

Fig. 39 Reinforced concrete basement wall detail

Reinforced Concrete Basement Wall

Similar to the masonry basement wall, two connections at the reinforced concrete basement wall necessitate special attention to prevent water leakage. These are the foundation-to-wall and wall-to-ground floor slab connections. A sample section detail study is presented in Fig. 39.

Reference

Bridges, A. H. (2010). Building construction handbook. *The Construction Net*. https://doi.org/10.4324/9780203476826_chapter_9

Chapter 3
Superstructures

Abstract Superstructures are the main body of any structure and building stand. They are classified according to structural systems and components that work due to the order of the same loads and forces of nature. However, each typology has special attributes and characteristics reflecting diverse construction and structural performance behaviors. This chapter explains the types of superstructures and their details.

1 Concrete Superstructures

Concrete has started to be widely used in the last centuries. This section introduces the history and development of concrete, its mixture, properties, application methods, and systems.

1.1 Lime and Lime Types In Construction

Before the invention of Portland cement, lime-sand mortar was the only available masonry mortar. Lime is a cement (binder) used for thousands of years in masonry mortar to bind stone and brick units in the wall. Lime is made from limestone, widely available in the earth's crust. Due to its easy production by simply heating limestone, lime has been used in masonry mortar, plaster, stucco, and soil stabilization.

1.2 Pozzolanic Reaction—Roman Concrete and Mortar

Romans found that when lime and volcanic ash were mixed, the mixer, by the involvement of sand and water, gave a mortar that quickly set was stronger and more durable than lime-sand mortar. The greater durability was due to the pozzolanic reaction of

R. Sarı and E. B. Çalışkan, *Building Construction Methods and Systems*,
https://doi.org/10.1007/978-3-031-50043-5_3

lime and volcanic ash mixed with water. The mixture has resulted in a water-resistant or hydraulic cement that does not dissolve in water. The Pantheon in Rome is one of the well-known examples of Roman concrete. The historical records stated that Roman concrete was used for a while, and then the formula got lost or forgotten. Following the disappearance of Roman concrete buildings, the buildings were started to be constructed by masonry.

1.3 Modern Concrete and Its Basic Ingredients

Modern concrete was started by the invention of Portland Cement in 1824 by Joseph Aspidin, a British stonemason. Early concrete applications did not involve reinforcement, thus seen and accepted as liquid stone where instead of cut, the stone was formed using concrete. Concrete performs well at compressive strength but is weak at withstanding stresses. The use of inexpensive steel rods as reinforcement within the concrete by François Hennebique in the late 1870s brought high tensile strength to concrete, allowing the use of concrete as horizontal structural materials. Thus, modern concrete starts with reinforcing bars inside the concrete.

The first examples of reinforced concrete were applied to industrial buildings. Due to its high compressive strength, many military buildings were made of reinforced concrete during World War—II (WW-II). On the other hand, the wider use of concrete was after WW II. Many large cities were demolished in Europe, and they were immediately renewed. Reinforced concrete was the best option for the AEC industry because of its rapid application, allowance for design freedom, and durability. Reinforced concrete still dominates the built environment. On the other hand, there is a growing argument and criticism due to its high contribution to global CO2 emissions and lack of proper recycling opportunities. Researchers and practitioners have given their special attention for the last decades to overcome the sustainability problems of concrete.

The basic ingredients of reinforced concrete mixture are cement, aggregate, and water. The following sub-sections explain the basic ingredients of the concrete mixtures.

1.3.1 Cement and Cement Types

Portland cement was patented in 1824 by Joseph Aspidin, a British stonemason. His method was to make limestone and clay burned in a kiln. The end product is hydraulic cement, set and gained much more quickly than lime. Regarding their characteristics and application area, there are five types of Portland cement:

- **Type I—General Purpose Portland Cement**: type I is used where there are no special requirements.

- **Type II—Moderate-sulfate-resistant and moderate-heat-of-hydration Portland cement**: It combines the properties of type IV and V moderately.
- **Type III—High-early-strength Portland cement**: It is generally used in precast elements and where high early strength is required, such as in cold weather. Economic considerations may require precast element formwork being used as frequently as possible.
- **Type IV—Low-heat-of-hydration Portland cement:** Type IV is used in massive civil engineering structures such as dam walls or bridge piers. Due to its limited usage, Type IV cement is produced by manufacturers only on special request.
- **Type V—Sulfate-resistant Portland cement:** Sulfur causes the decomposition of concrete structures for several years. Thus, Type V cement is required in an environment with high sulfur content.

1.3.2 Air Entrained Portland Cement

Placing concrete in forms requires a much greater amount of water in a mix (called the water of convenience) than required for the chemical reaction between Portland cement and water. An inadequate amount of water in concrete does not allow concrete to be fluid enough to flow into the form and fill it. This results in honeycombed concrete. The excess water in a concrete mix evaporates as it stiffens, leaving voids. These voids contain **entrapped air**.

Because of the entrapped air, concrete is porous and tends to absorb rainwater. During freezing weather conditions, the absorbed water turns into ice and expands. If the concrete is critically saturated and undergoes cycles of freezing and thawing, it will spill unless air voids are present to dissipate the pressure caused by the increased volume. To reduce freeze–thaw damage, tiny air particles are introduced into a concrete mix, called **air entrainment**. As the absorbed water in concrete expands on freezing, the entrained air relieves the pressure. Air-entrained concrete is commonly specified in exposed concrete elements subjected to freeze–thaw cycles.

Using air-entraining admixture is the preferred way of obtaining air entrainment, or in the market, air-entrained Portland cement contains air-entraining chemicals as its integral part.

1.3.3 Concrete Mixture

Concrete consists of **Aggregate** (coarse and fine aggregate) as a matrix or filter and **Portland cement–water paste** as an adhesive. These materials provide a hard, rock-like, durable, fire-resistant, and relatively inexpensive substance. It is used universally because every region of the earth has the raw materials necessary to produce it. The technology associated with its production and use is fairly simple. Unlike other structural materials, concrete can be formed to any desired shape. An entire structure can be formed monolithically regardless of shape or size.

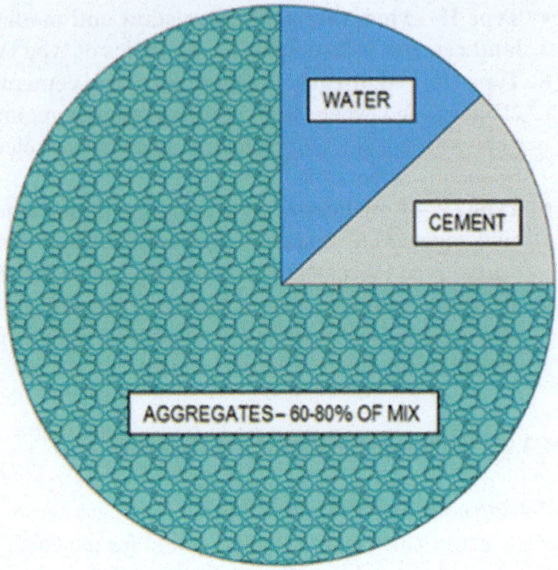

Fig. 1 Basic ingredients of a concrete mixture

60–80% of the concrete volume is aggregate; the remainder is Portland cement. The aggregate must consist of different-sized particles, referred to as size gradation. Proper gradation ensures that smaller particles fit within the voids created by larger particles so that the entire concrete mass is relatively dense. If aggregate size variation is not provided, the voids will be filled with Portland cement. Since Portland cement is more expensive than aggregate, using different-sized aggregate provides economic benefits (Fig. 1).

The concrete industry divides the aggregate into (i) fine aggregate and (ii) coarse aggregate.

- Fine aggregate: Sand, but more precisely, it is that material of which 95% passes through a No.4 sieve and cannot pass through a No.100 sieve.
- Coarse aggregate: Aggregate of which 95% is retained on a No.4 sieve. It consists of either crushed stone or gravel. Gravel is more efficient to be used, but crushed stone is more economical to be used.

1.4 Classification of Concrete

Regarding structural usage, concrete is classified as structural concrete and insulating concrete. Structural concrete is then further divided into two categories: normal-weight and lightweight. The following sections are based on structural concrete and its characteristics (Fig. 2).

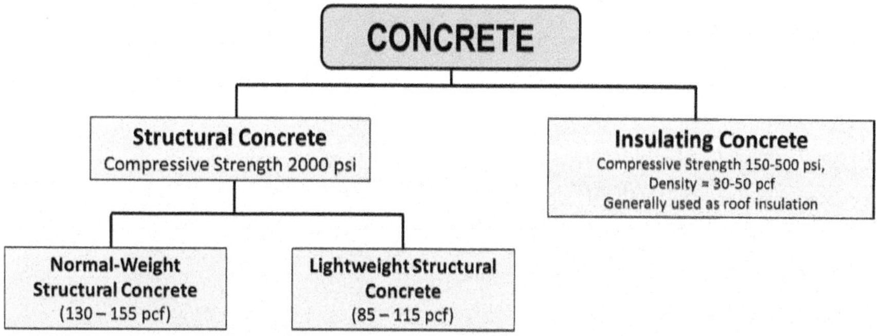

Fig. 2 Classification of concrete

Non-structural concrete is used for roof insulation; thus, it is called insulating concrete. On the other hand, structural concrete is used in structural frames such as beams, slabs, columns, and load-bearing walls.

- **Normal-weight structural concrete**: Consisting of normal-weight aggregate. Normal-weight structural concrete costs less than lightweight concrete, but the dead load is much more than lightweight concrete.
- **Lightweight structural concrete**: Consisting of lightweight aggregate. Preferable in high-rise buildings due to reduction in dead loads. However, lightweight concrete has limitations on the strength that cannot be used in columns. Therefore, in high-rise structures, lightweight concrete is used in floor beams and slabs where the floor is just carrying its loads, while normal-weight concrete is used in columns where columns have to carry all loads on upper floors (Fig. 3).

1.5 Design Considerations for and Properties of the Concrete

1.5.1 Quality of Water

The water that will be used for the concrete mixture must be clean. This is because Portland cement gains its binding property from its reaction with water. When foreign particles interrupt this reaction, the quality of concrete is decreased. Drinkable water can be used, while seawater causes corrosion of reinforcing bars due to containing the salt (Fig. 4).

1.5.2 Fresh (Plastic-State) Concrete Properties—Workability (Slump) of Concrete

The workability of concrete has the characteristic that concrete can be placed and compacted with minimum loss of consistency and homogeneity. Concrete that is not

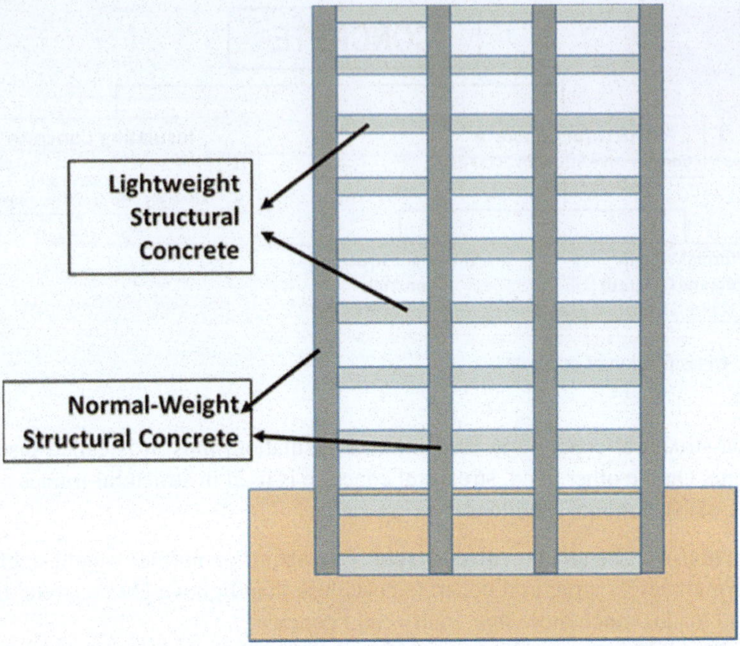

Fig. 3 Lightweight and normal-weight structural concrete usage in a building structure

workable is referred to as harsh concrete. In order to obtain a workable concrete, the concrete mixture should consider the following situations:

- The aggregate should be well-graded.
- Large aggregate reduces workability.
- The aggregate should not be too angular in shape.
- Due to its naturally round particles, Gravel gives greater workability than crushed stone.
- Sand obtained from riverbeds (not containing salt) or mined from sand pits gives greater workability than that obtained by crushed and pulverized stone.
- An adequate amount of water is necessary.
- A commonly used method for measuring the workability of concrete is slump.
- The concrete mixture is dumped into a truncated cone fully. Then, the cone is removed with a slow upward motion.
- As soon as the cone is removed, the concrete settles.
- The slump of concrete is a measure of settlement in inches or cm.

Fig. 4 A chemical reaction occurs when water and cement come together, and cement gains its binding property

1.5.3 Hardened Concrete Properties—Compressive Strength of Concrete

The two most important properties of hardened concrete are durability and compressive strength. A sample concrete mixture cast in the site is dumped into a cylinder. After 24 h of passing, the cylinders are transferred to a testing laboratory. The cylinder boxes are removed, and concrete cylinders remain in the moist chamber until they are tested for failure. After 28 days, the cylinders are removed from the moist chamber. Then, the cylinders are applied to pressure to measure their compressive strength. The pressure amount causing the crush of the cylinder is accepted as the compressive capability of the sample.

1.6 Making Concrete

Concrete is made by mixing, of course, fine aggregates and Portland cement and water. Although there are concrete plants –computer-controlled facilities to prepare the concrete mixture, it can be produced manually by involving and mixing the

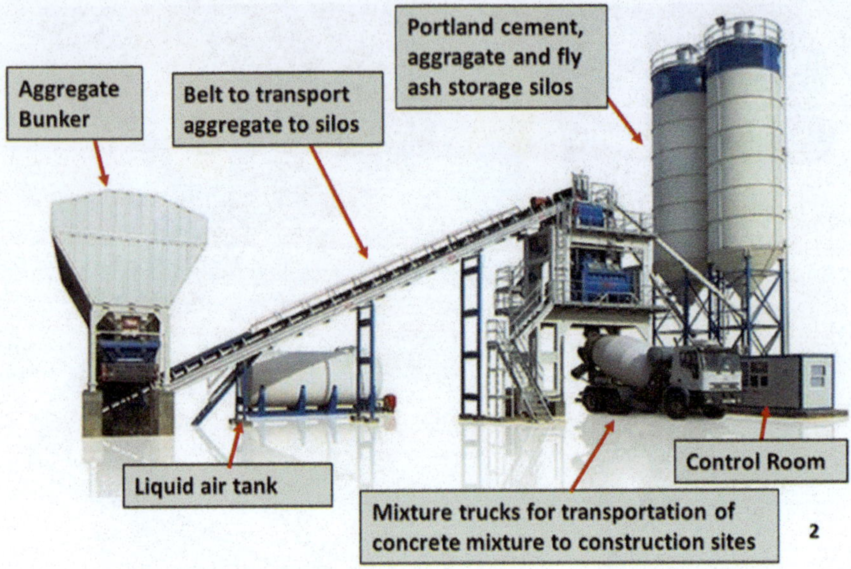

Fig. 5 A concrete plant

ingredients with human power. On the other hand, the second method risks not presenting a homogeneous mixture; thus, it is not preferable (Fig. 5).

1.7 Placing and Finishing the Concrete

After adding water to the dry concrete mixture, the concrete starts to set within a few hours. Once the concrete has been placed in the form, it must be consolidated. Consolidation is the process of compacting concrete to ensure that it has no voids and air pockets. Workers use a steel rod into the concrete—up and down and some sideways motion to consolidate the concrete. In general, a high-frequency power-driven vibrator is employed. Excessive vibration must be avoided because it leads to particle separation and concrete bleeding. For the same reason, concrete must be carefully placed in the forms, not dropped from an excessive height.

After compacting the concrete, its exposed surfaces are finished while still plastic. The exposed surfaces are those that are not covered by the formwork. The following methods are used to surface exposition:

- **Strikeoff (screeding)**: The strikeoff (screeding) process is applied to the concrete surface following the concrete pouring. This study makes the concrete surface a straight level compared to the surface after concrete pouring.
- **Floating (Darbying)**: Immediately following the strikeoff operation, the concrete surface is floated. Floating is usually done by a hand or bull float, which further

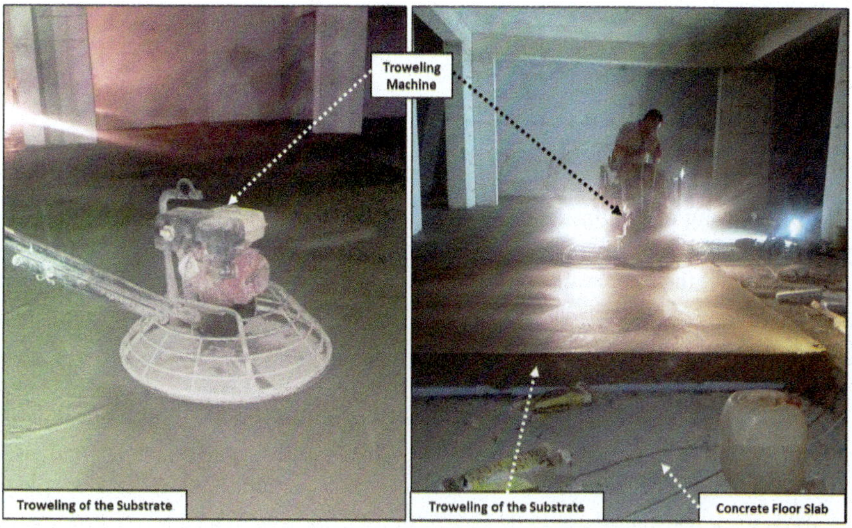

Fig. 6 Troweling application of the screed over the concrete slab

smoothes the surface. For non-air-entrained concrete, floats are usually made of wood. For air-entrained concrete they are generally made of aluminum or magnesium alloys.

- **Troweling**: Striking and floating are the only operations needed for surface exposition. Troweling is done to achieve a smoother surface required for finishings such as carpet or floor tiles. The concrete should have stiffened to do troweling (Fig. 6).

1.8 Water Content in Concrete

Concrete and other mixes made from Portland cement gain their strength due to the reaction of Portland cement with water, referred to as the hydration of Portland cement. The water-cement ratio (w–c ratio) for complete hydration should be 0.40. However, for the workability of concrete, this ratio may be increased to 0.55 or 0.70.

1.9 Duration of Hydration Reaction—Curing of Concrete

The hydration reaction begins as soon as water and Portland cement come into contact, but the rate at which this reaction proceeds is extremely slow. Concrete takes up to 6 months or longer to gain its full strength. However, approximately 80% of concrete strength develops in 28 days. Until gaining enough strength, concrete is needed to be watered to provide enough continuous hydration reaction. Providing moisture to concrete continuously for hydration is called ***curing of concrete.***

1.9.1 Methods for Field Curing of Concrete

Concrete curing occurs faster when the surface interacts with the weather than the content within the formwork. This situation causes dryness of the exposed surface while liquid of the remaining content. In order to control the dryness of the exposed surface, some applications are applied as depicted below:

- ***Keeping concrete wet with water:*** regarding the evaporation time, the exposed concrete surface remains wet by regularly feeding the surface with water.
- ***Covering concrete with a plastic sheet*** (to retain concrete moisture): The exposed surface is covered with a plastic sheet. The evaporated moisture is condensed within the plastic sheet, and the exposed surface remains wet. After a couple of days, the plastic sheet could be removed.
- ***Using curing compounds:*** This consists of spraying concrete with a liquid membrane. A liquid membrane is like a plastic sheet. However, it becomes an integral part of the concrete surface. Curing compounds should not be used if there will be further finishing treatment to the concrete surface.

1.10 Influence of Temperature on Hydration

The hydration of Portland cement is greatly affected by temperature. Below 13 C, the rate of hydration decreases significantly. The use of Type III Portland cement is helpful under these circumstances. However, low-temperature concreting should be avoided because a significant reduction in concrete strength occurs if the water in the concrete freezes within a few hours of the concrete's placement. That is why concrete is generally not placed if the surrounding air temperature is below 4.5 C or is expected to fall below 4.5 C. If concrete must be placed when the temperature is below 4.5 C, the following precautions must be taken:

- Warming the aggregates and water
- Insulating the forms
- Surrounding the concrete in a heated enclosure
- Using Type III Portland cement.

Particular precautions are also required if concreting is to be done in hot weather:

- Wetting the aggregates with water to cool them evaporative.
- Building a protective cover to shield against direct solar radiation and wind.
- Concreting after sunset
- Use chilled water, crushed ice, or liquid air to decrease the temperature of the concrete mixture.

1.11 High-strength Concrete

1.11.1 Water Reducing Concrete Admixtures

Since reducing the water content in concrete increases its strength, the concrete industry began to find ways of reducing the amount of water without decreasing the workability of concrete. Chemicals are used for plasticizers or water-reducing admixtures (WRA). WRA increases concrete slump by 10% to 15%. WRA enables stronger concrete when added to the mixture. Superplasticizers are more efficient chemical admixtures as high-strength water reducers (HRWRs). A HRWR reduces water requirements by up to 30%. Usage of HRWR enables much more concrete strength of approximately 50,000 psi, considering that until 1960, the highest concrete strength used in buildings was 5000 psi.

1.11.2 High-Strength Concrete

Until the last decades, only steel was widely used in high-rise buildings taller than 40 stories. However, regarding the improvement in the concrete industry to pump it to higher altitudes with the help of water-reducing concrete admixtures, reinforced concrete became a prominent structural system for tall buildings. Furthermore, the following facts enable concrete to be widely used in tall structures competing with steel:

- Concrete has inherent fire resistance, while steel requires additional fire protection, increasing the construction cost.
- Concrete framed structures in tall buildings do not necessitate extra structure for lateral loads such as wind and earthquakes.
- The vibration frequency of tall buildings under lateral loads (wind, earthquake) for concrete is lower than steel. High vibration frequency decreases the occupancy comfort.

Other ingredients of high-strength concrete are fly ash and silica fume:

- **Fly Ash**: It is a waste product produced by coal-fired power stations. Because fly ash particles are microscopic, they densify concrete, filling in voids between Portland cement particles. The densification of concrete increases its strength and durability. Small particles of fly ash also increase concrete's workability and reduce the permeability and creep of concrete. The main benefit of fly ash is its pozzolanic properties. During the manufacturing of Portland cement, a certain amount of lime remains called free lime. Fly ash reacts with free lime, converting it into hydraulic cement and strengthening concrete.
- **Silica Fume**: Silica fume is a waste product from the silicon industry. The particle size is extremely fine. More than 100 silica fume particles occupy the same space as one particle of Portland cement. Silica fume is an expensive concrete admixture.

It has excellent pozzolanic properties; like fly ash, it reacts with free lime to convert it into a hydraulic cement.

1.12 Steel Reinforcement

Concrete is much weaker in tension than in compression. Its tensile strength is approximately 10% of its compressive strength. Therefore, concrete is generally used with steel reinforcement, which provides tensile strength in a concrete member. Steel is the ideal material to complement concrete because the thermal expansion of both materials is the same. Steel also bonds well with concrete. A chemical bond occurs when concrete and steel come together in the mixture.

Additionally, as water evaporates from concrete, it shrinks and grips the steel bars, making a mechanical bond. The mechanical bond is enhanced by reinforcing bars—rebars with surface deformations. Excessively rusted rebar should not be used. Plain rebar is used as dowels at expansion joints to enable expansion movement and as spiral reinforcement in round columns.

1.12.1 Welded–Wire Reinforcement

Welded Wire Reinforcement (WWR) is a prefabricated reinforcing steel available in rolls or mats. They are generally used in ground-supported slabs (or pavement) and steel deck-supported slabs where the reinforcement requirement is smaller than those needed for reinforced concrete suspended slabs.

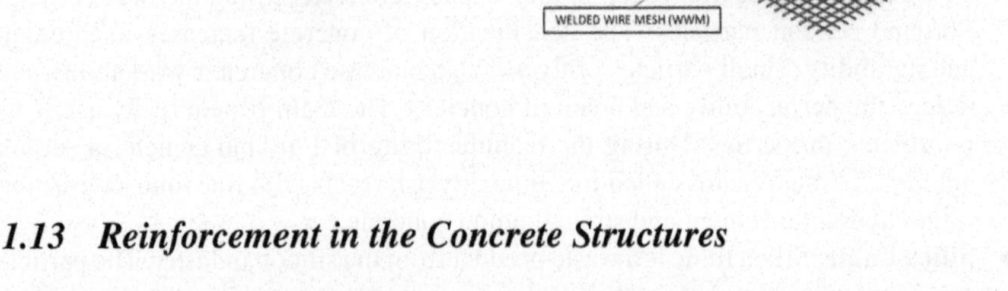

1.13 Reinforcement in the Concrete Structures

Concrete without steel reinforcement is called plain concrete. Plain concrete represents the same structural characteristics as stone. Thus, the ancient (Roman concrete) and initial concrete examples were in the architectural form of arches, vaults, or domes. Two methods of reinforcing concrete with steel are (i) steel reinforcing bars and (ii) pre-stressed concrete.

1.13.1 Steel Reinforcing Bars

In most concrete members subjected to bending (beams and slabs), reinforcing bars are positioned where tensile stresses are likely to be produced. Because steel is also much stronger than concrete in compression, reinforcing bars are also used to provide compressive strength (columns, walls, and some beams). A concrete member containing reinforcing bars is called a ***reinforced concrete member.*** If the entire structure consists of reinforced concrete members, the structure is referred to as a ***reinforced concrete structure.***

1.13.2 Pre-stressed Concrete

The compression is introduced in regions of the member where tension is expected to be produced by the loads. The magnitude of compression produced by pre-stressing is controlled to ensure that it reduces or cancels the tension created by the loads. Pre-stressing of a member is obtained using pre-stressing cables called ***strands*** made of high-strength steel wires. Pre-stressing can be accomplished in two ways:

- **Pre-tensioning**: The strands are tensioned between two fixed abutments using hydraulic jacks.
- **Post-tensioning**: The strands are tensioned after the concrete has been placed. The strands for post-tensioning are encased in sleeves placed in the form before the concrete is placed.

The combinations of strands, sleeves, end anchorages, and so on are referred to as tendons. The tendons are laid within the concrete forms in the same way as the reinforcing bars. Pre-stressing reduces the dimension of structural members, resulting in smaller dead loads. However, the pre-stressing cost is higher than rebars. Furthermore, pre-stressing requires a more skilled labor force.

1.14 Formworks of Concrete

Concrete and reinforcement must be contained in molds, referred to as ***formwork.*** The formwork for elevated slabs and beams must be supported on vertical supports called ***shores*** (Fig. 7).

On average, the concrete costs nearly 25–30% of the cost of the structure (including placing and finishing of concrete), and steel reinforcement costs approximately 20–25% (including laying reinforcement in the forms, making reinforcement

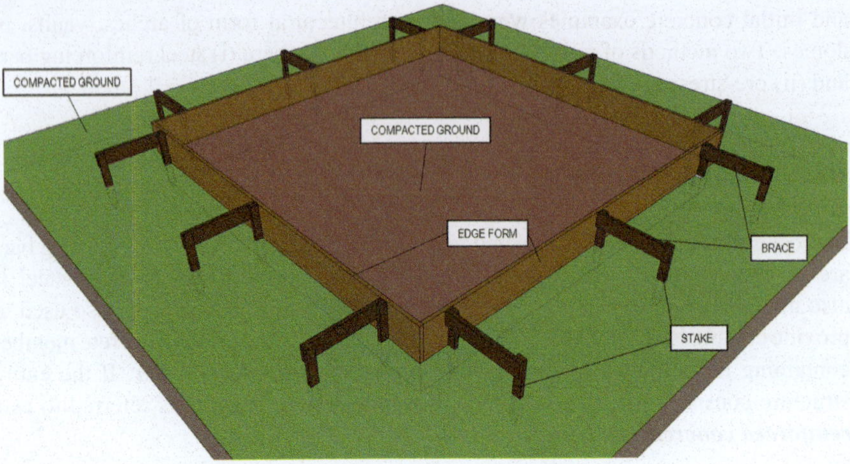

Fig. 7 Formwork for concrete slab-on-ground slab

cages, etc.). The remaining 50–55% of the cost of the structure is in formwork. In complex geometry structures, the cost of formwork may exceed 75% of the total cost. The high cost of formwork is the primary handicap for applying shell roofs (Fig. 8).

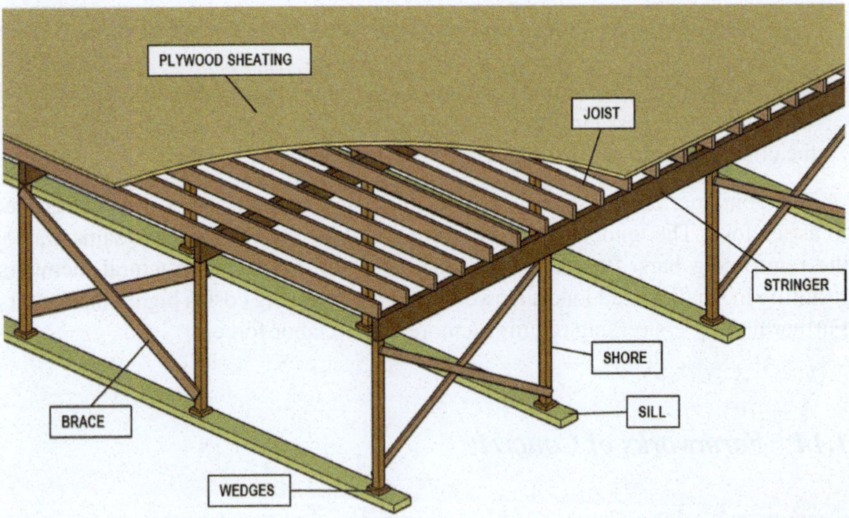

Fig. 8 Formwork and shores for an elevated concrete slab

1.14.1 Materials Used for Formwork

Criteria for selecting formwork materials are cost, strength, reusability, durability, ease of assembly, and weight. The available formwork materials in the current industry are wood and plywood, steel, aluminum, and glass-fiber-reinforced plastic (GFRP). The cheapest materials are wood, plywood, steel, aluminum, and GFRP. In the following images, some sample formworks are illustrated (Fig. 9).

1.15 Principles of Reinforcing Concrete

There are three basic members of reinforcing concrete. These are stirrups, hangers, and tension reinforcement. Commonly used stirrup forms are illustrated in Fig. 10. Tension reinforcement is necessary, while hanger bars are used in case of more compression requirements in beam design. Furthermore, due to load and force distribution along the beam, stirrups are dense at the column and beam connection, as depicted in Fig. 11. The number of tension bars could be changed according to the tensile strength requirements of the beam. Two layered tension reinforcement in a beam design is represented in Fig. 12.

Fig. 9 Wood and plywood formwork and shores

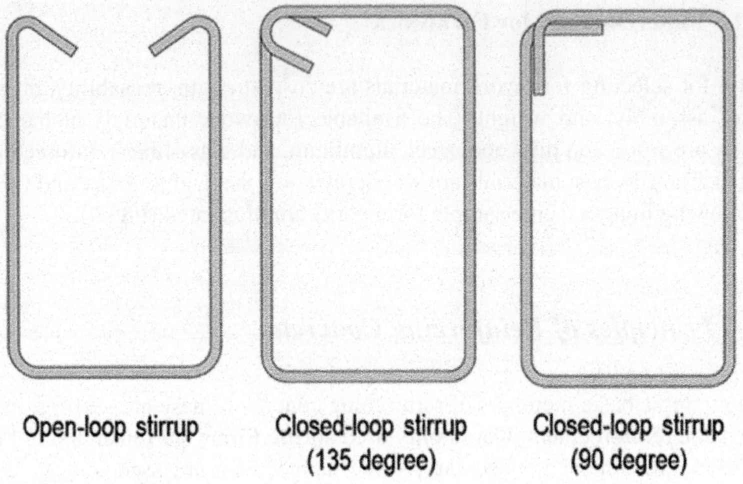

Fig. 10 Commonly used stirrups

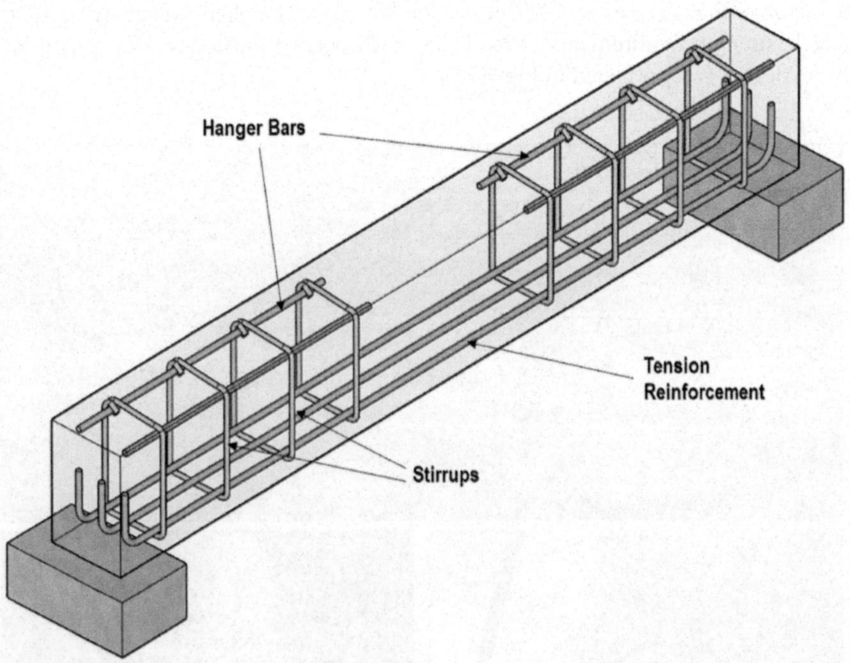

Fig. 11 Illustration of stirrups, hanger bars, and tension reinforcement in a concrete beam (Fig. 12)

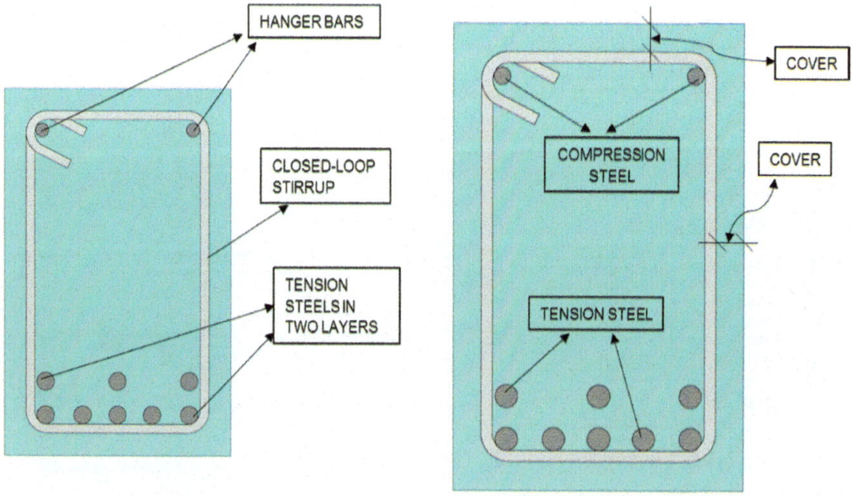

Fig. 12 Steel reinforcement in a concrete beam

1.16 Reinforcement and Formwork for Columns

Reinforcement of concrete columns is different from beam. The stirrup in the beam is called a tie in the column reinforcement. A sample column reinforcement is illustrated in Fig. 13. Unlike beams, reinforcement longitudinal bars are located equally and symmetrically within the column. This is because the compression enforces the column deformation on all sides while the deformation occurs in the gravity direction in the beam.

1.17 Reinforcement and Formwork for Concrete Walls

Concrete walls have a high structural rigidity when the proper reinforcement is applied. Since the concrete is poured from top to bottom, the increase in the formwork height requires particular formwork. Form ties are required to prevent the formwork's bending due to the liquid concrete's excessive weight during the pouring process. A sample formwork for the concrete wall is illustrated in Fig. 14.

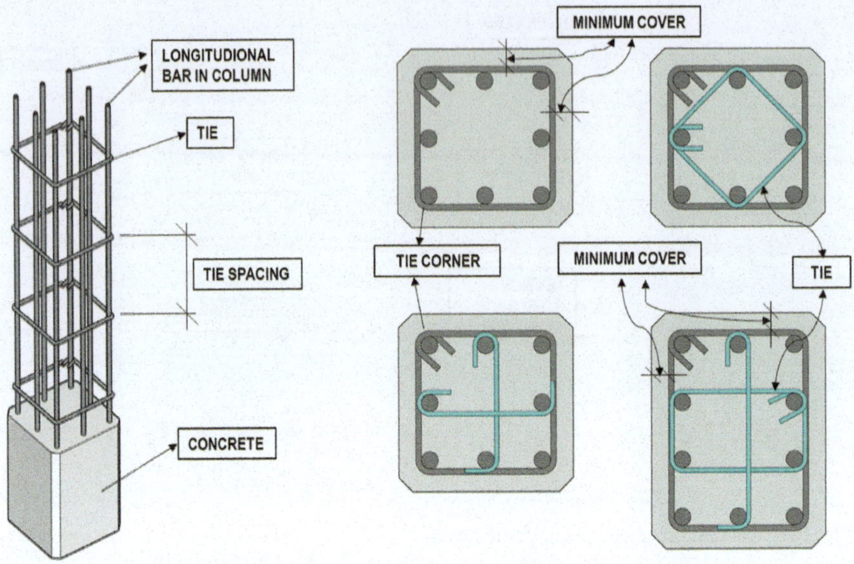

Fig. 13 Reinforcement for columns

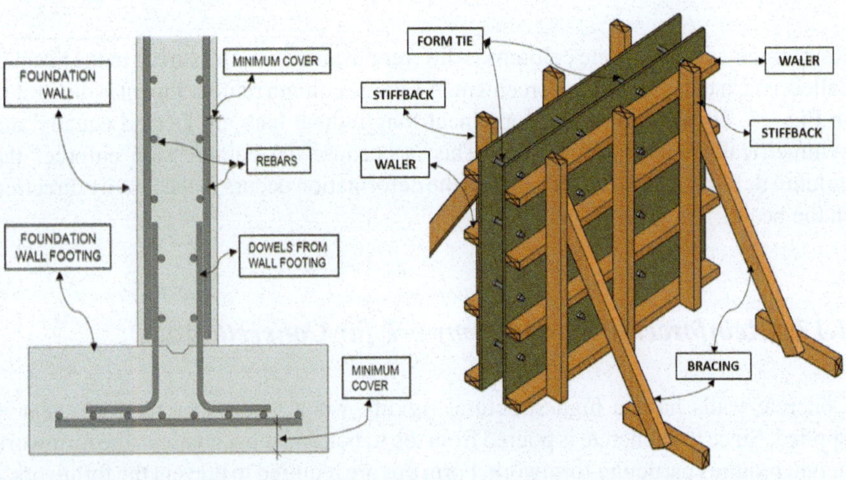

Fig. 14 Reinforcement of a foundation wall with footing and formwork of a shear wall

1.18 Types of Concrete Slabs

Concrete slabs are divided into two categories: ground-supported slabs and elevated slabs:

- **Ground-Supported Slab:** Bear directly on compacted ground (grade).
- **Elevated Slabs**: Rest on and are part of the structural frame of the building. Therefore, they are called ***framed slabs*** or ***suspended slabs*** (Fig. 15).

1.18.1 Ground-Supported Slabs

Ground-supported concrete slab consists of the following layers at minimum from subfloor to soil:

- Concrete slab-on-ground
- Vapor retarder
- Sand or gravel sub-base as the drainage layer
- Soil compacting
- Earth

The ground-supported concrete slab design may be further enriched regarding the environmental and other conditions (Fig. 16).

1.18.2 Ground-Supported Stiffened Concrete Slab

The concrete slab is stiffened with perimeter and interior beams (ribs) in both directions, forming a ribbed concrete slab. Two available designs for stiffened slab-on-grade slabs are:

- Ribbed reinforced concrete slab.
- Ribbed post-tensioned concrete slab (Fig. 17).

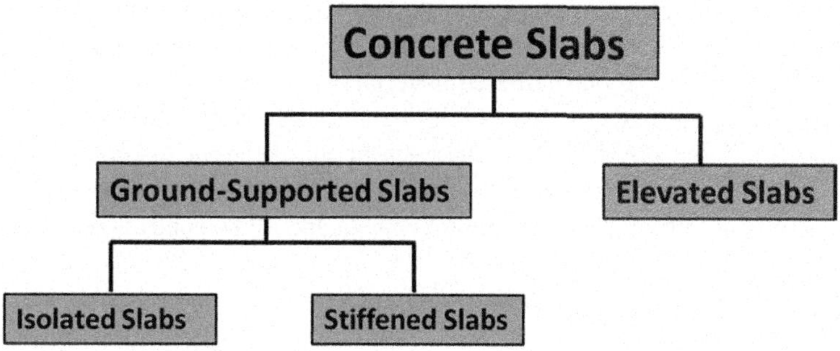

Fig. 15 Classification of concrete slabs

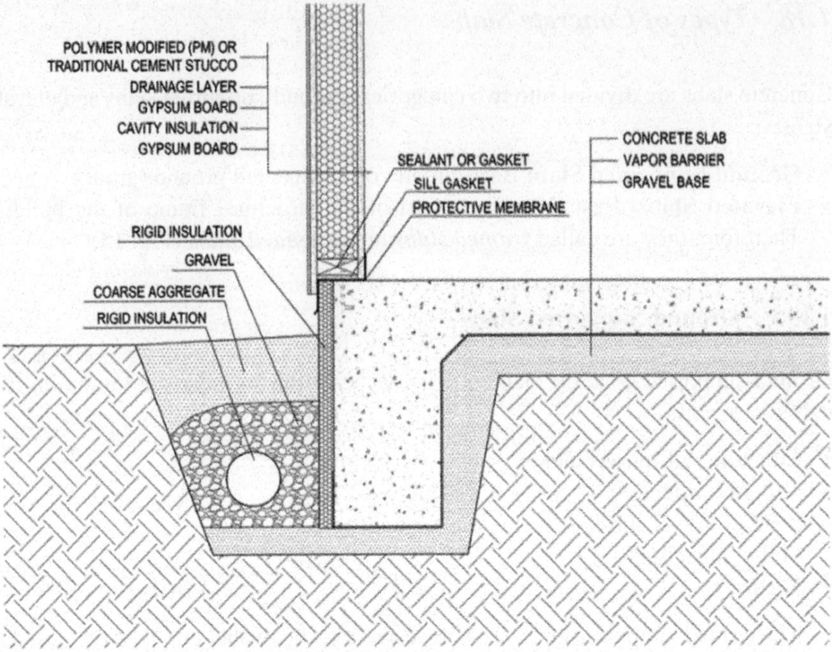

Fig. 16 An example of a slab-on-ground foundation

1.18.3 Control, Isolation and Expansion Joint

Cementous materials lose their moisture during their life cycle, and in order to prevent cracks due to the moisture lost, control joints are provided. The slab will crack in a random, haphazard pattern without control joints. The sawing of the slab generally provides control joints at a depth of 0.25 times the thickness of the slab.

Unlike control joints, isolation joints in a concrete slab extend the entire thickness of the slab. They are typically 1 cm wide and are provided to ensure that the slab is isolated from the building's structural components so that their movement (creep, foundation settlement, etc.) is not transferred to the slab. The joint space is generally filled with a flexible and plastic-based material.

Expansion joints exist between different construction systems of the same building. Since the physical characteristics of the different materials are varied, expansion joints are presented to eliminate the potential damage of seasonal activities (heating and cooling) due to temperature change and allow controlled movement of the building blocks due to lateral loads (wind and earthquake).

Long or wide buildings are divided into blocks to eliminate the cracks and collapse of the structure due to lateral loads such as earthquakes. The block dimension can be changed regarding the earthquake zones or other local factors and described by the local authorities. In Turkey, the maximum dimension of a building block can not be more than 40 m in tall buildings and 50 m in long buildings (Figs. 18, 19, 20 and 21).

Fig. 17 Ribbed reinforced concrete slab. *Image Credit* Özsarı Çelik

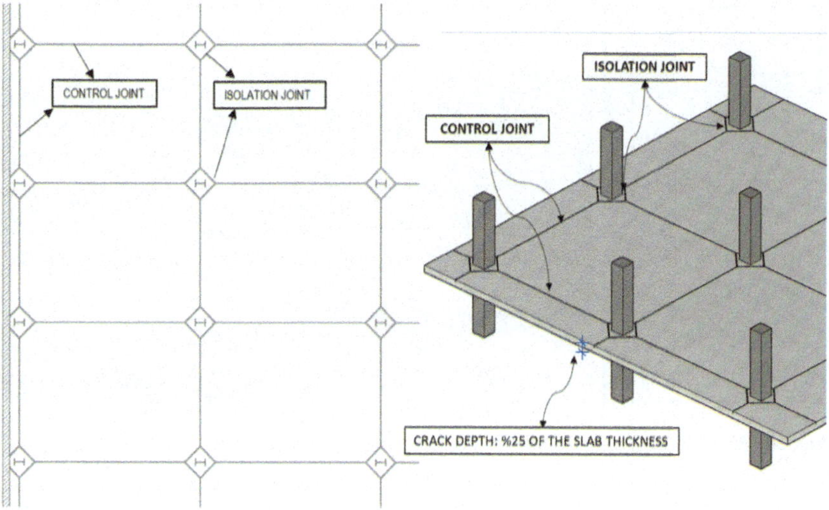

Fig. 18 Control and isolation joints in a concrete slab

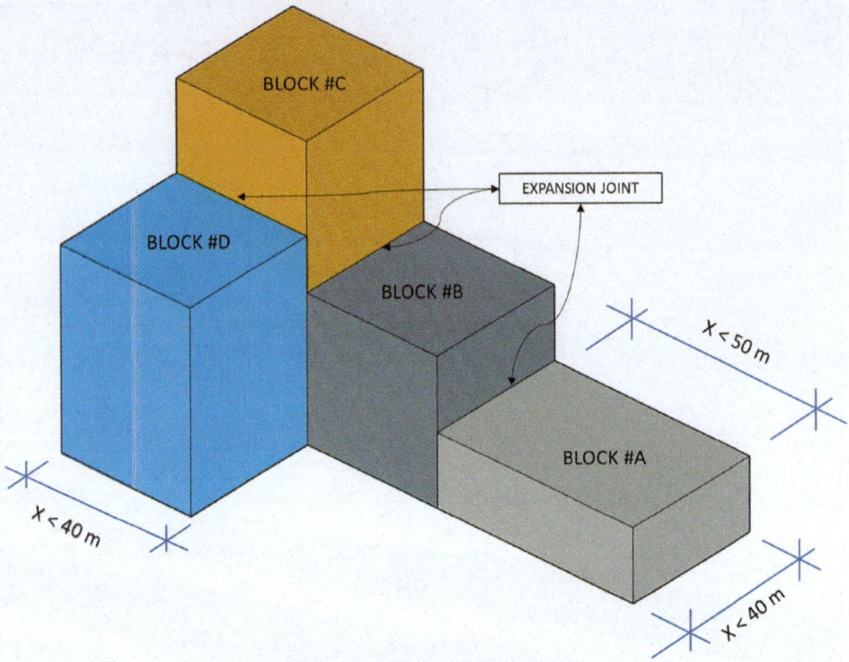

Fig. 19 Block division in a building for expansion joints

Fig. 20 An expansion joint is required for the safe expansion of the mixed structures

1.18.4 Types of Elevated Concrete Floor Systems

Site-cast reinforced concrete framing systems consist of horizontal elements (elevated floor, roof slabs, and beams) and vertical elements (columns and walls). Approximately 80% to 95% of the cost of materials and formwork of a concrete

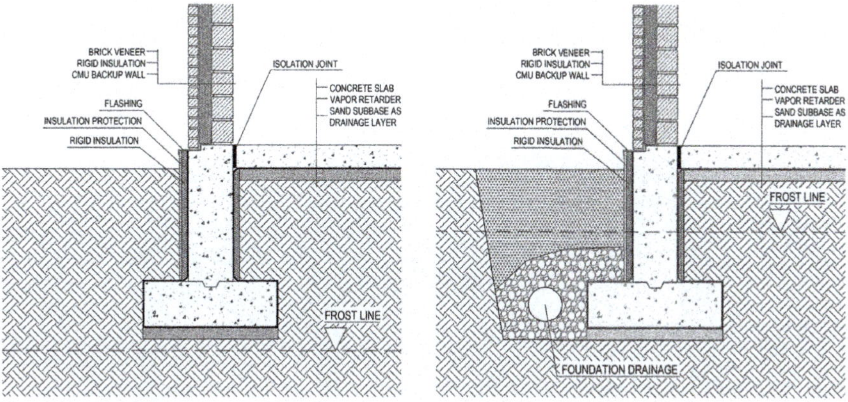

Fig. 21 An isolation joint illustration in a slab-on-ground foundation above the frost line (left) and below the frost line (right)

structural frame is in the horizontal framing elements of the frame. Elevated concrete floor systems can be classified as (i) Beam-Supported Floors and (ii) Beamless Floors (Fig. 22).

Beam–Supported Concrete Floors

A reinforced concrete floor slab with beams on all four sides can be one-way or two-way. They are also called one-way solid slabs or two-way solid slabs to distinguish them from one-way or two-way joists slabs.

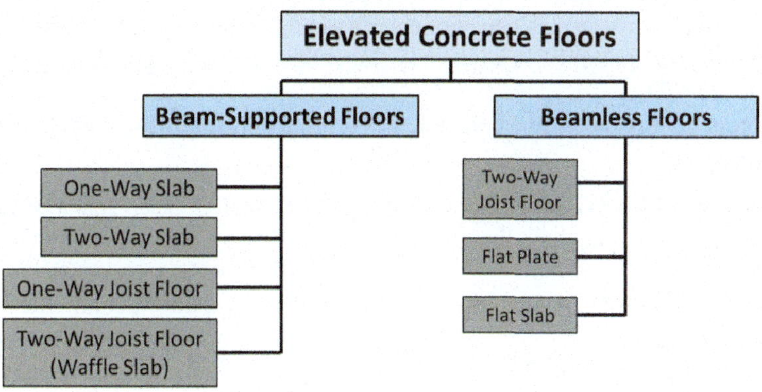

Fig. 22 Types of elevated concrete flooring systems

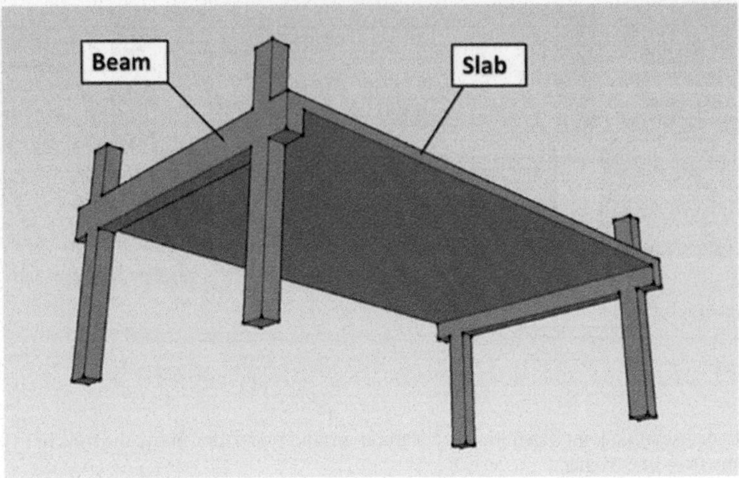

Fig. 23 One-way solid slab

- **One-Way Solid Slabs**

If the ratio of the long dimension to the short dimension of a four-side supported slab panel is greater than or equal to 2.0, most of the load on the slab is transferred to the long pair of beams; that is, the load path is along the short dimension of the slab panel (Fig. 23).

- **Two-Way Solid Slabs**

If the ratio of the long dimension to the short dimension of a four-side supported slab panel is shorter than 2.0, the slab is considered to behave as a two-way slab. However, real two-way slab behavior occurs when the ratio of the two dimensions is as close to 1.0 as possible. In a two-way slab, both directions participate in carrying the load (Fig. 24).

In order not to exceed 20 cm thicknesses in a slab, the following methods are used in solid slabs:

- **Beam and Floor Girder**: Floor girders carry the slab, placed sparsely between the floor beams, and their dimensions are bigger than joists but smaller than beams (Fig. 25).
- **Band Beam Floor**: A one-way slab floor with wide, shallow, continuous beams (Fig. 26).
- **One-Way Joist Floor**: Joists are aligned perpendicular to the beam and closely placed. The joists carry the floor slab, the beam carries the joists, and the columns carry the beams (Fig. 27).

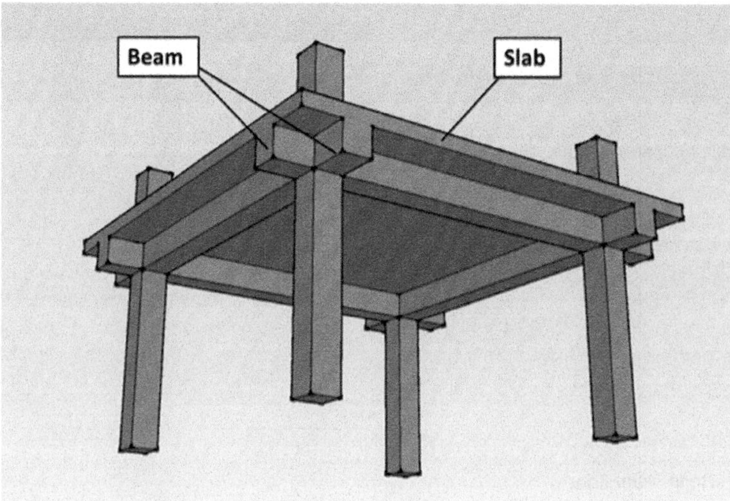

Fig. 24 Two-way solid slabs

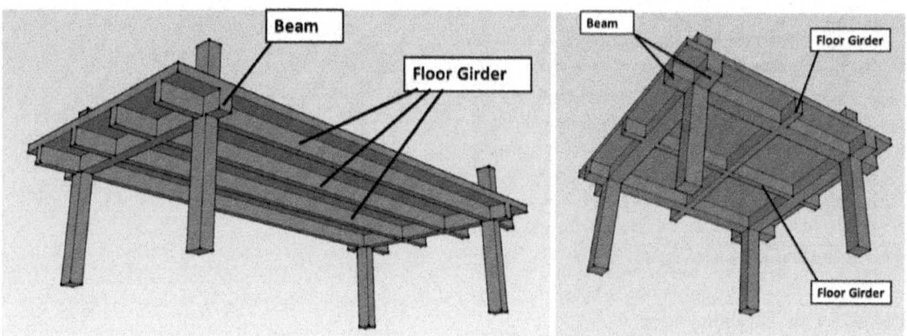

Fig. 25 Beam and floor girder

- **Two-Way Joist Floor**: The joists are placed closely in two dimensions. The resulting system allows wider column distances, allowing clear openings for public functions. The dimensions of the components, such as slab thickness, joist depth, and width, are particularly small, as illustrated in Fig. 28.

Beamless Concrete Floors

The beamless concrete flooring system is similar to the two-way joist flooring system. On the other hand, there is no beam used. Instead, column capital connects the column and slab to resist the shear forces (Figs. 29 and 30).

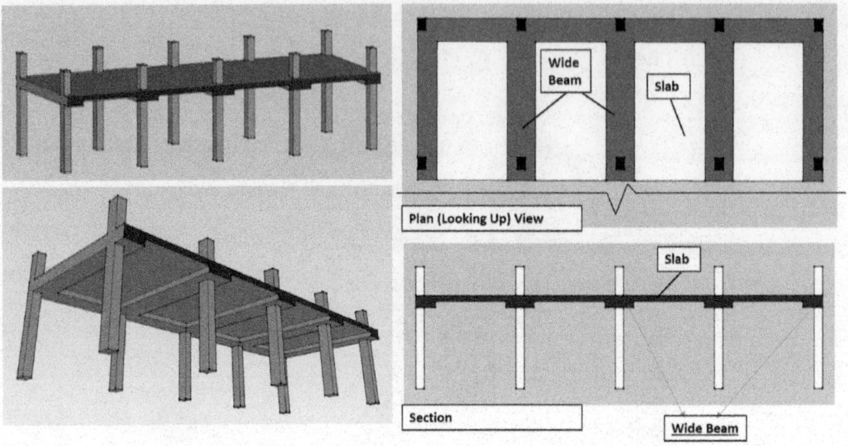

Fig. 26 Band beam floor

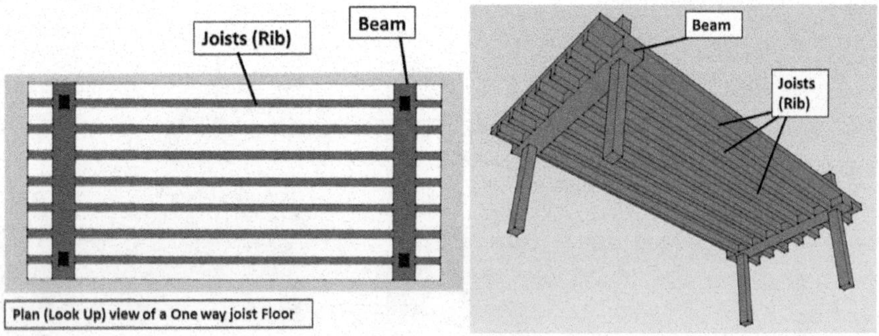

Fig. 27 One-way joist floor

- **Flat Plate**

A flat plate consists of a solid slab supported directly on columns. A flat plate is similar to a two-way banded slab, except that the beam bands in both directions are concealed within the thicknesses of the slab. Flat plates are suitable for occupancies with relatively light live loads, such as hotels, apartments, and hospitals, where small column-to-column spacing does not pose a major design constraint. A flat plate slab results in a low floor-to-floor height, and its formwork is economical (Fig. 31).

- **Flat Slab**

A flat slab is similar to a flat plate but has column heads, called drop panels. Drop panels provide shear resistance at columns where the shear maximizes. Flat slab is generally used with high live loads, such as parking garages, storage, etc. (Fig. 32).

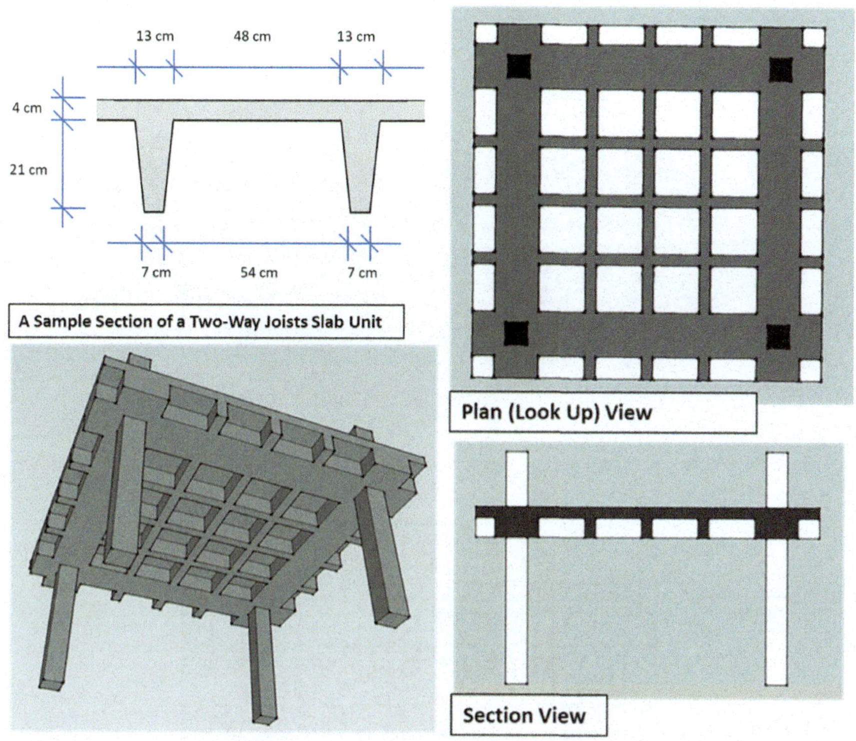

Fig. 28 Two-way joist floor

1.19 Precast Concrete

Precast concrete members are divided into two categories regarding their applications: (i) architectural precast and (ii) structural precast.

- Architectural Precast: Concrete elements that are used as nonstructural cladding elements.
- Structural Precast: Includes all elements of a building's structural frame (floor/roof slabs, columns, and walls).

1.19.1 Structural Precast Concrete Members

Structural precast concrete members are divided into horizontal-spanning elements and vertical-spanning elements.

- **Horizontal-Spanning Elements**: Generally, they are prestressed. Hollow-core slabs, solid planks, double-tee units, and inverted-tee-beams.
- **Vertical-Spanning Elements**: as columns and walls

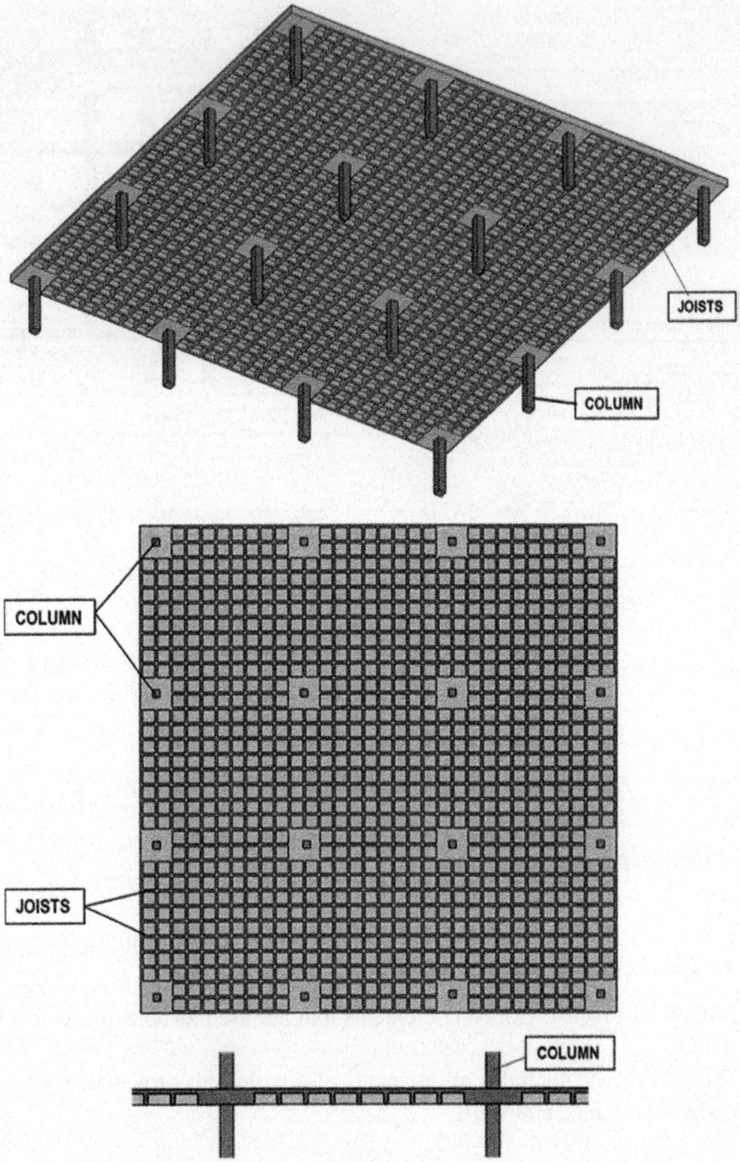

Fig. 29 Waffle slabs are commonly used as beamless slabs

1.19.2 Reinforced Concrete Bearing Wall Construction

Site-cast reinforced concrete bearing walls can be used instead of masonry bearing walls. When used in conjunction with site-cast reinforced concrete floor slabs, they provide a robust structure because of the inherent continuity between the vertical and

Fig. 30 A two-way joist flooring is used as a skylight

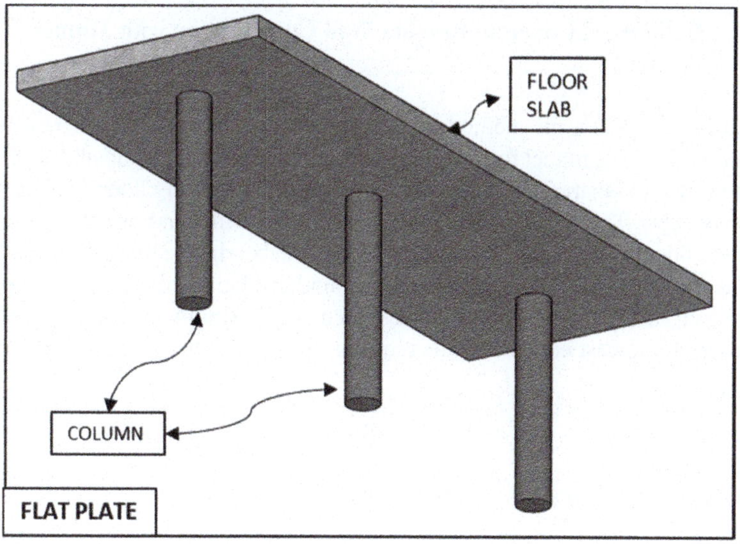

Fig. 31 Flat plate

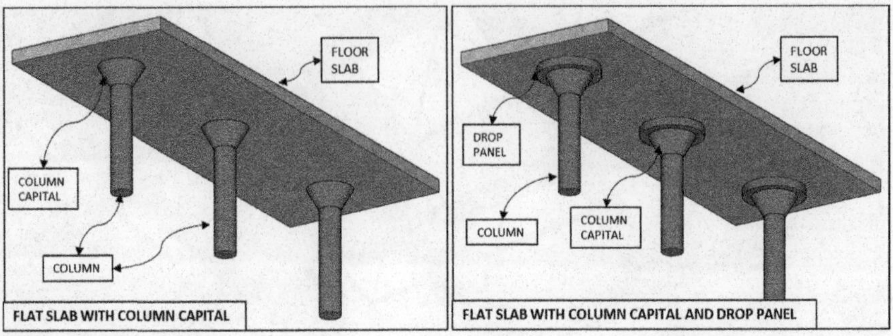

Fig. 32 The flat slab examples

horizontal elements of the building. In other words, the joints between the walls and the floor slabs of a site-cast reinforced-concrete bearing Wall structure are tougher than those in masonry Wall and precast-concrete hollow-core slab structures. Site-cast concrete-bearing Wall structures are better able to resist lateral loads. Site-cast reinforced concrete walls can seamlessly integrate with a site-cast reinforced concrete column-beam frame in the case of a hybrid (bearing Wall and frame) system.

1.19.3 Reinforced Concrete Bearing Wall Construction with Tunnel Form

Tunnel Form is a favorite method in high and mid-rise reinforced concrete-bearing wall structures. The tunnel form consists of prefabricated and collapsible steel forms (of an inverted L-shape) that allow the walls and the overlying floor slabs to be cast simultaneously. The system is feasible only with a modular and repetitive floor plan, such as residential buildings. A reinforced concrete bearing wall has the same advantages as a masonry bearing building: good sound insulation, high fire resistance, and a mold-free structure. The disadvantages, such as inflexibility in spatial organization and exterior site-cast concrete walls requiring water-resistive cladding, are also the same (Fig. 33).

1.20 Precast Concrete Connection Details

Precast concrete column and beam are connected by steel fasteners, as indicated in Fig. 34. The steel fasteners were placed in the formwork and attached to the reinforcement of the columns and beams.

Fig. 33 Tunnel form reinforced concrete load-bearing wall construction

1.21 Span and Interval for Concrete Structural Systems

See Table 1

1.22 Construction of a Concrete Frame Structure

In the following figures, various concrete frame structures are facilitated on each floor of a single building (Figs. 35, 36, 37, 38, 39, 40, 41, 42, 43, 44, 45 and 46).

Reference

Eren, Ö. (2014). Büyük Açıklıklı Çelik Yapılar. Arı Sanat.

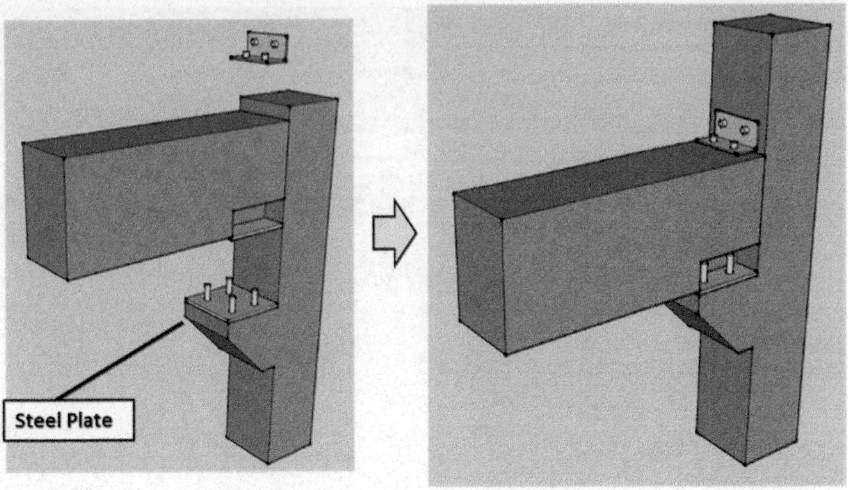

Fig. 34 Precast concrete beam and column connection details

Table 1 Span and interval dimensions for concrete structural systems

Floor type	Material	Span (m)
Flat plate	Concrete	4–8
	Pre-tensioned concrete	9–10
Flat plate with column draop	Concrete	5–10
	Pre-tensioned concrete	12–14
T od L beam used slab	Concrete	5–15
	Pre-tensioned concrete	9–24
One-way solid slab	Concrete	2–7
	Pre-tensioned concrete	5–9
Two-way solid slab	Concrete	6–11
One-way joist floor	Concrete	4–12
	Pre-tensioned concrete	10–18
Two-way joist floor (Waffle slab)	Concrete	9–15
	Pre-tensioned concrete	10–22

Source (Eren, 2014)

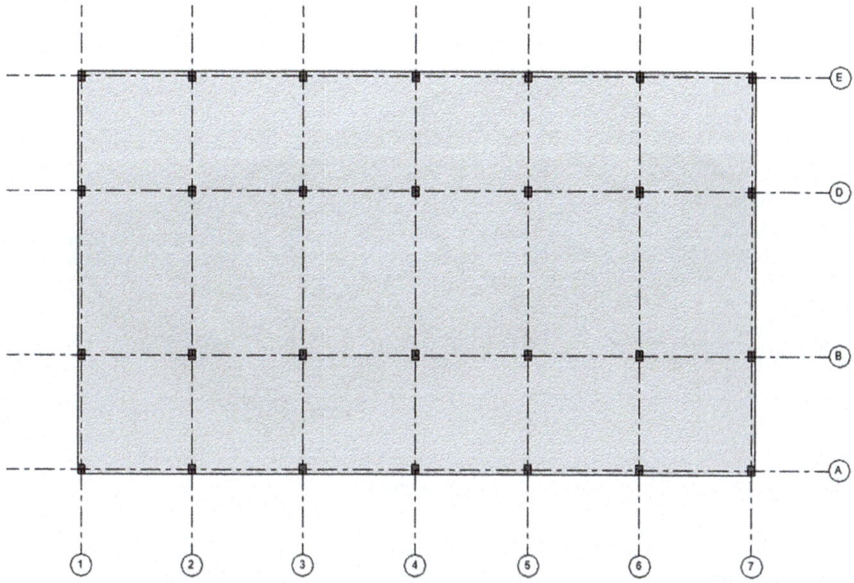

Fig. 35 Basement floor plan of a reinforced frame structure

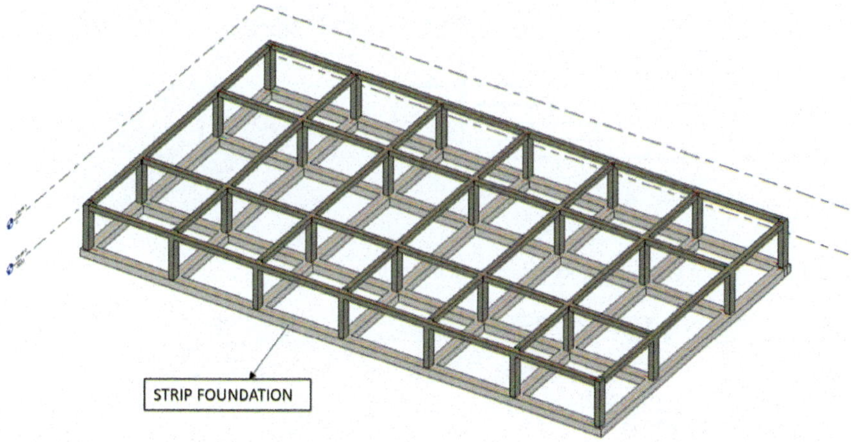

Fig. 36 3-Dimensional illustration of a strip foundation, column, and two-way solid floor systems at the basement floor

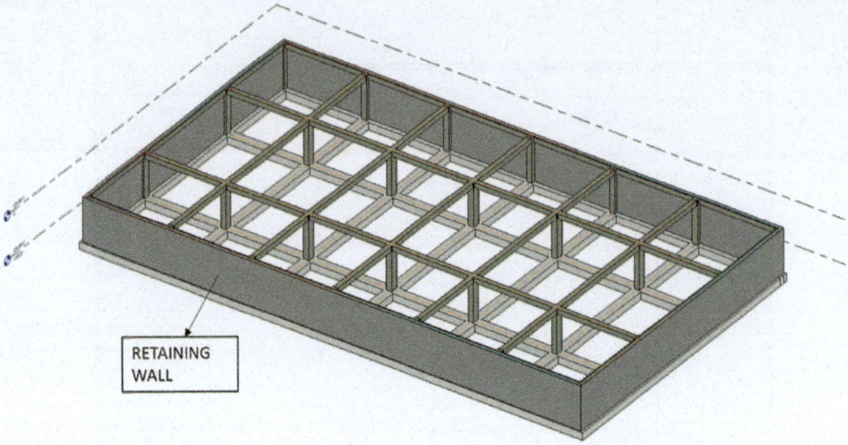

Fig. 37 Retaining wall enclosure at the basement floor

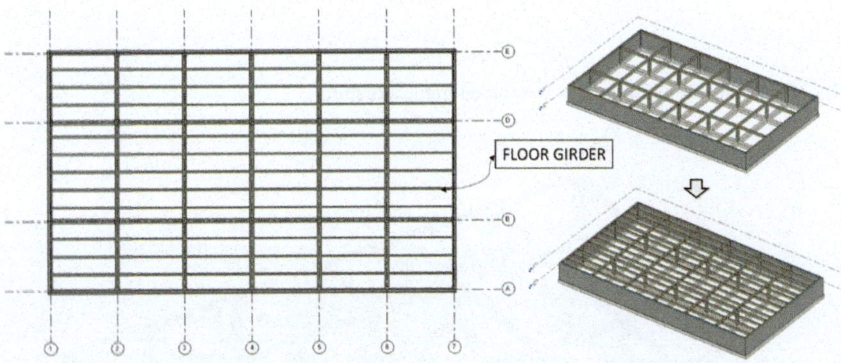

Fig. 38 Design of beam and floor girder system on the ground floor

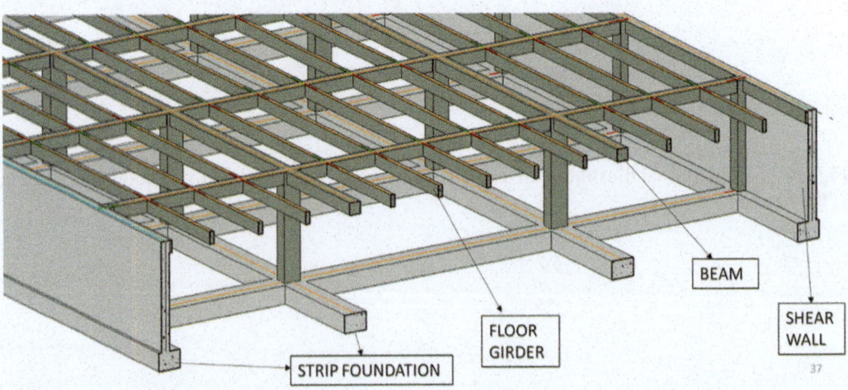

Fig. 39. 3-Dimensional illustration of beam and floor girder system on the ground floor

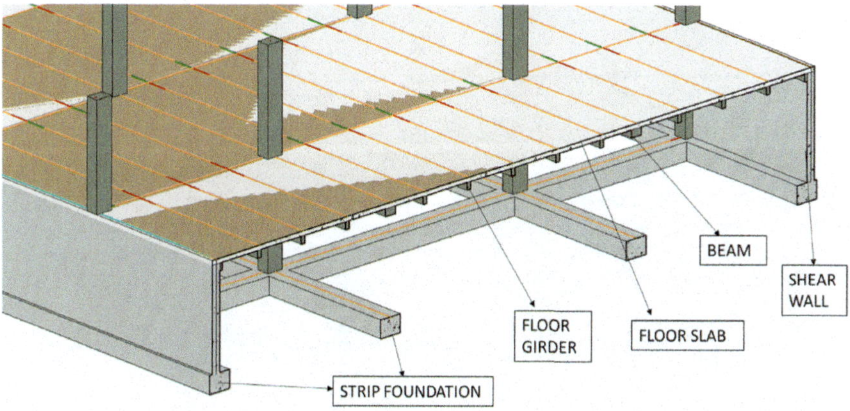

Fig. 40 Beam and floor girder system with floor slab

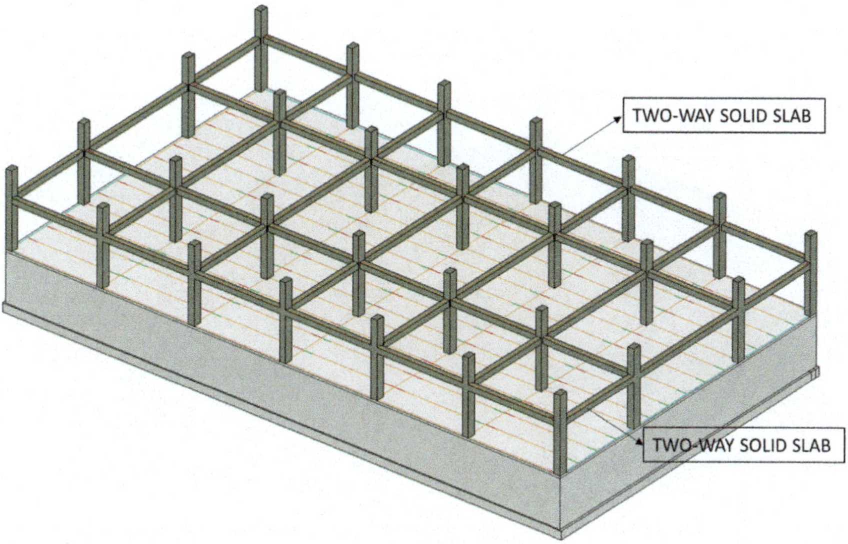

Fig. 41 Two-way solid frame system on the first floor

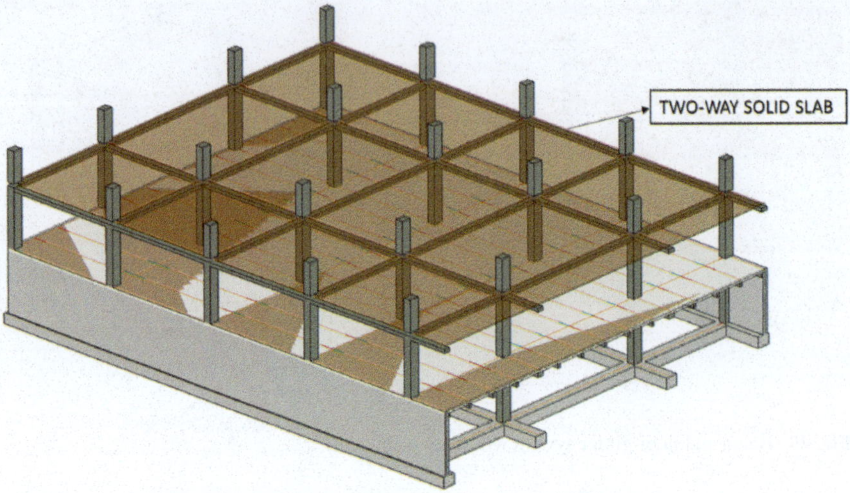

Fig. 42 Two-way solid slab system on the first floor

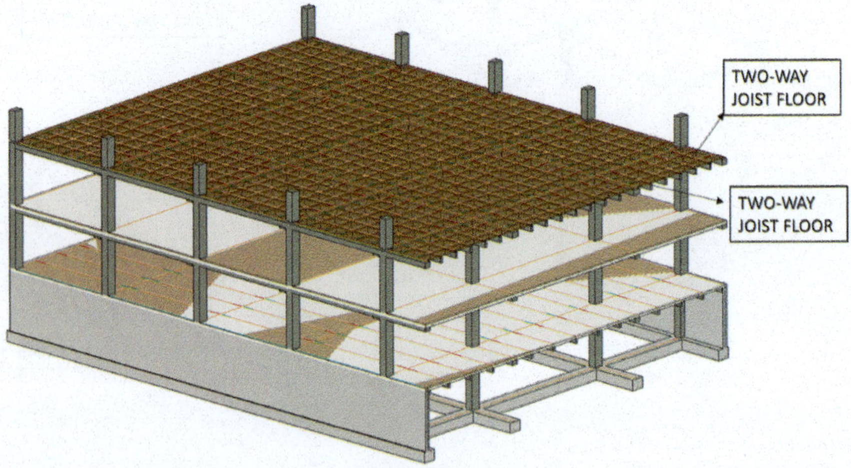

Fig. 43 Two-way joist floor system on the second floor

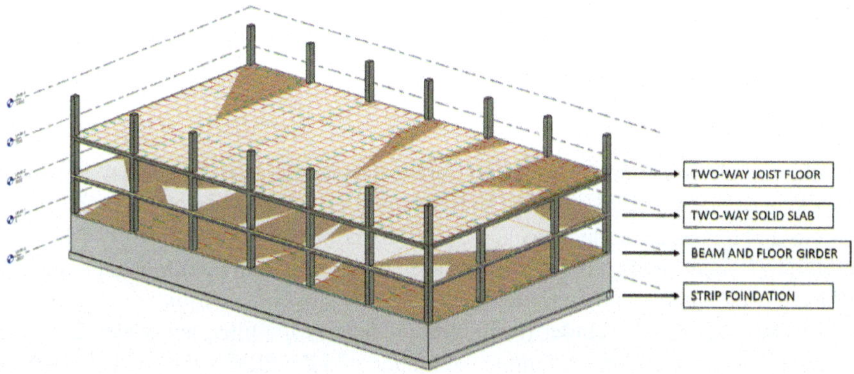

Fig. 44 Various concrete flooring systems are used in the basement, ground, and first-floor

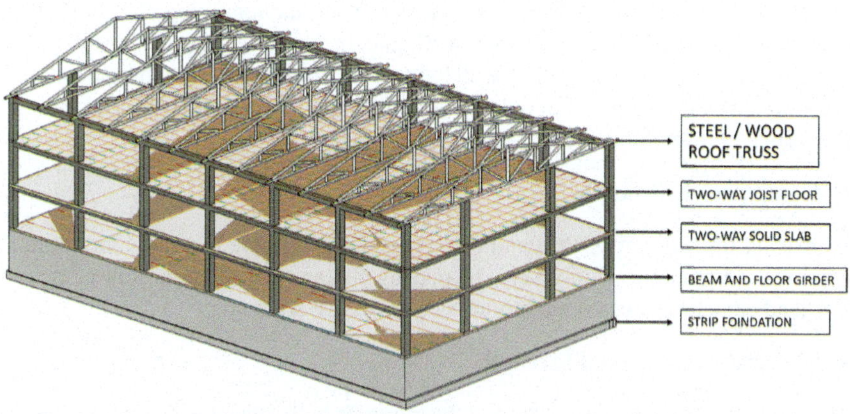

Fig. 45 Steel or wood truss system at the roof for enabling wide spans

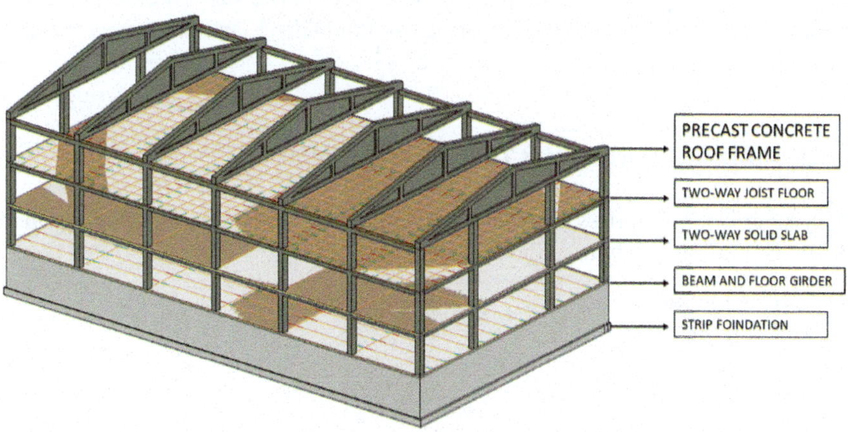

Fig. 46 Precast concrete roof frame at the roof for enabling wide spans

2 Masonry Wall Superstructures

2.1 Masonry Mortar

Bricks are generally made from clay; brick masonry is also called clay masonry. Block masonry is called concrete masonry due to being made of concrete block. Bricks and blocks are called clay and concrete masonry units, respectively. Masonry units are bonded with mortar to yield a composite building component—a Wall. Thus, mortar is the common ingredient in all masonry construction.

Mortar consists of a binder (cementitious material), a filler, and water. Portland cement and hydrated lime comprise the binder, and the filler is the sand. Mortar, as a binder of units, not only helps seal the Wall against water and air infiltration but also provides a cushion between the units. Lime inside the mortar in its plastic state improves its workability and water retentively. A mortar consisting of only Portland cement without lime is coarse and less workable. In the hardened state, lime improves the water resistance of the Wall. A wall built with Portland cement and lime mortar is more watertight than one with only Portland cement mortar. There are two strength properties of mortar in general:

- **Compressive Strength**

Increasing the amount of Portland cement concerning lime increases the mortar's compressive strength. Conversely, increasing the amount of lime with respect to Portland cement decreases the mortar's compressive strength.

- **Flexural Tensile Bond Strength**

The bond between a masonry unit and the mortar is both a chemical and a mechanical bond. Bond strength of masonry related to types of units, surface roughness of units, quality (such as the pressure applied between the units at the time of mortaring), curing conditions (air temperature, wind, and humidity), and amount of water in mortar.

2.1.1 Masonry Mortar Joint Thickness and Profiles

A masonry wall comprises horizontal and vertical mortar joints called ***bed*** and ***head joints.*** Some commonly used mortar joint profiles:

- **Concave joint**
- **Raked joint**

- **Flush joint**
- **Weathered joint**
- **Struck joint.**

2.2 *Manufacture of Bricks*

Conceptually, brick manufacturing consists of the following six operations:

- **Mining clay from the ground**
- **Grinding and sieving clay to a fine powder**
- **Mixing water with sieved clay**
- **Forming wet clay into the desired brick shape (green bricks)**
- **Drying green bricks**
- **Firing dried bricks in a kiln.**

Brick shapes can be formed by one of the following two methods:

- **Extrusion of wet clay through *extruded bricks***
- **Molding wet clay—*molded bricks.***

2.2.1 Dimension of Masonry Units

Both brick and concrete block masonry units have three types of dimensions:

- ***Specified Dimensions***: Finished dimension that the specifier has requested and the manufacturer desires to achieve.
- ***Actual Dimensions***: Because the manufacturing process is not perfect, the actual dimensions of a unit are different from the specified dimensions. These differences between the specified and actual dimensions must lie within the *dimensional tolerance* established by the industry for the product.
- ***Nominal Dimensions***: Specified dimension plus one mortar joint thickness equals nominal dimension (Fig. 47).

2.2.2 Hollow Versus Solid Masonry Unit

Both brick and concrete blocks generally contain voids. A solid masonry unit is less than 25% hollow. A hollow masonry unit is 25% or more hollow. A brick without cores must be specified as 100% solid. A 100% solid brick is generally used for paving or as coping at the top of walls. Solid brick walls generally have structural characteristics, while hollow bricks do not (Fig. 48).

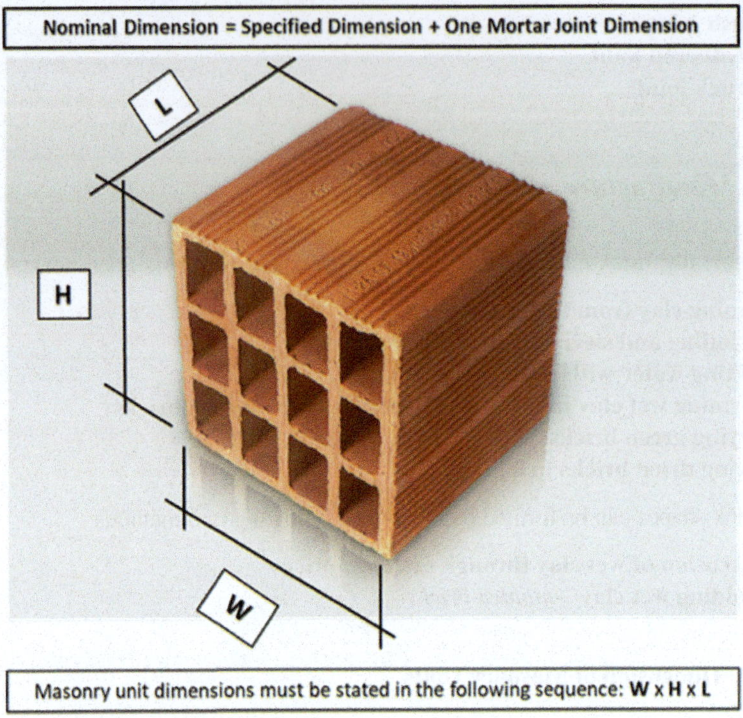

Fig. 47 Dimensions of a brick masonry unit

2.2.3 Bond Patterns in Masonry Wall

Bricks can be assembled in a wall in several patterns, called bond patterns or simply as bonds. Functionally, the bond is meant to stagger the units so that an increasing number of underlying units shares the load on one unit. One-wythe masonry wall (a wall whose thickness equals the width) can have a ***Stack Bond and running Bond***. The brick's exposed face is called a stretcher face or a stretcher. Brick can have six different face orientations on a wall elevation: ***stretcher, header***, ***rowlock***, ***soldier***, ***shiner***, ***and sailor.*** Figures 52, 53, and 54 illustrate the widely used bond types as American, English and Flemish (Figs. 49, 50 and 51).

2.3 Concrete Masonry Units (CMUs) Properties and Construction

Concrete masonry units (CMUs), called concrete blocks, are larger than brick ones. A CMU also contains particularly large voids than brick (Fig. 55).

Fig. 48 Hollow and solid masonry units

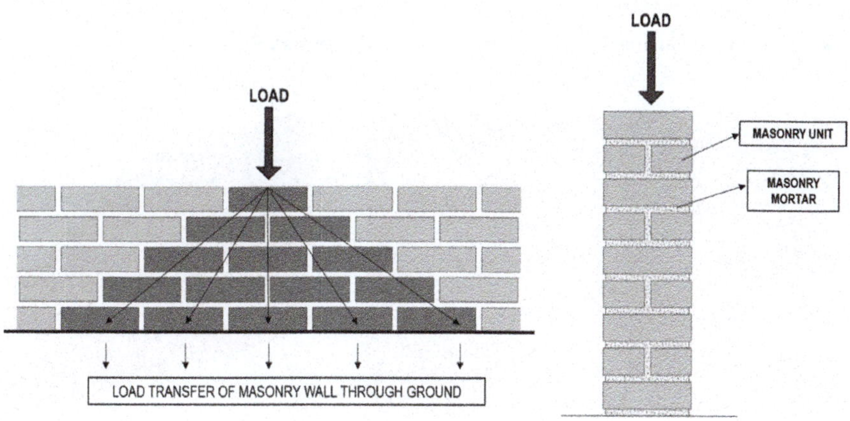

Fig. 49 Load transfer of a masonry wall

2.3.1 Lintel Unit

Lintels are used to present a structural underlayment for the door and window opening consisting of a reinforcing bar and concrete infill, as depicted in Fig. 56. The external dimensions of the lintel unit are the same as the normal CMU unit, so it would be the same as the normal CMU unit within the CMU wall.

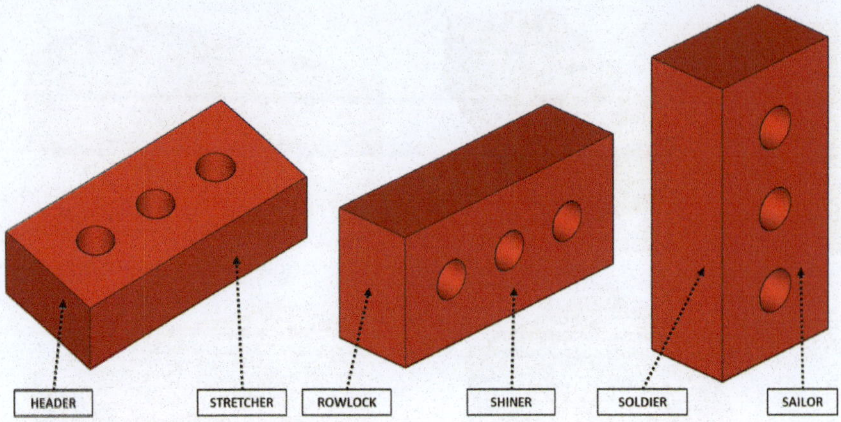

Fig. 50 Naming of brick unit surfaces

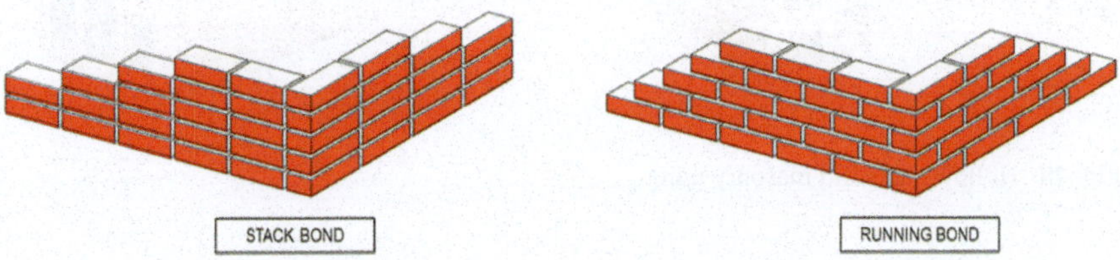

Fig. 51 Stack and running bonds in a single wythe brick masonry wall

Fig. 52 American (common) bond brick wall

2.3.2 Reinforcement in Cmu Wall

CMU blocks are heavier than brick but have bigger sizes. Therefore, the larger size of the CMU enables faster construction. Unlike bricks, whose bed and joints are fully mortared, a CMU is generally mortared only on its exterior periphery, referred to as ***face-shell mortaring*** . A typical CMU wall contains joint reinforcement for

Fig. 53 English bond brick wall

Fig. 54 Flemish bond brick wall

shrinkage control, which is not required in brick walls other than a stack-bonded brick wall. A reinforced CMU wall is generally partially grouted because only cells that contain reinforcement need the grout. A reinforced brick wall, on the other hand, must be fully grouted in all cases (Fig. 57).

2.3.3 Shrinkage Control in Cmu Walls

Clay bricks expand during service, whereas CMUs shrink. Shrinkage of CMU continues for several months after its manufacture. There are two ways to control the shrinkage of CMU walls:

- **Horizontal Reinforcement**: To resist tensile stress in the wall caused by the shrinkage of units. Horizontal reinforcement cannot neither prevent the formation of shrinkage cracks nor can it reduce the total width of cracks in a given length of a wall. However, reinforcement distributes the cracks in the wall by increasing their number and reducing the width of individual cracks—the total width of cracks remains unchanged. Thus, instead of one or two large cracks, numerous small

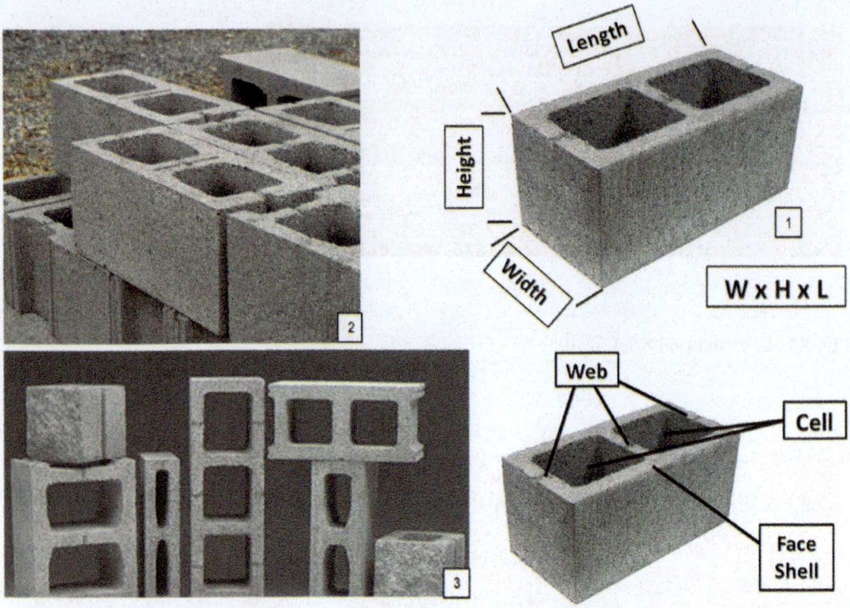

Fig. 55 Concrete Masonry Unit (CMU)

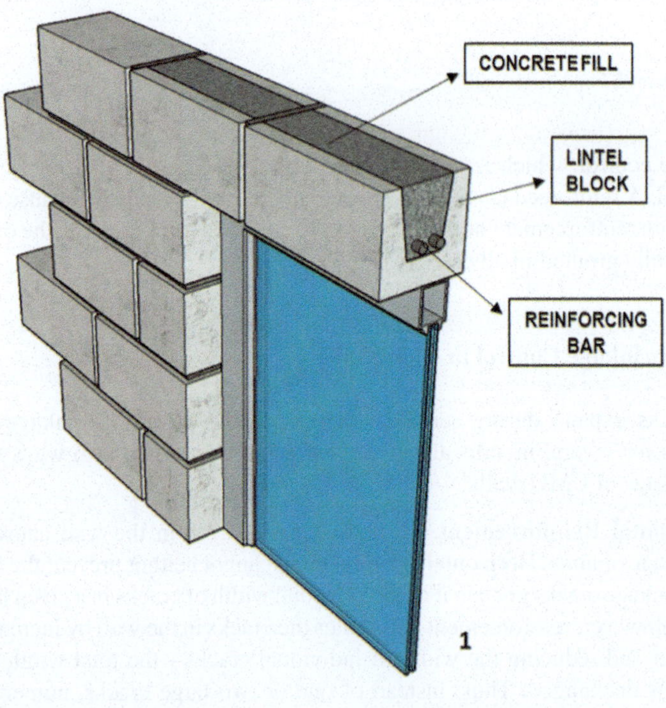

Fig. 56 Lintel CMU block

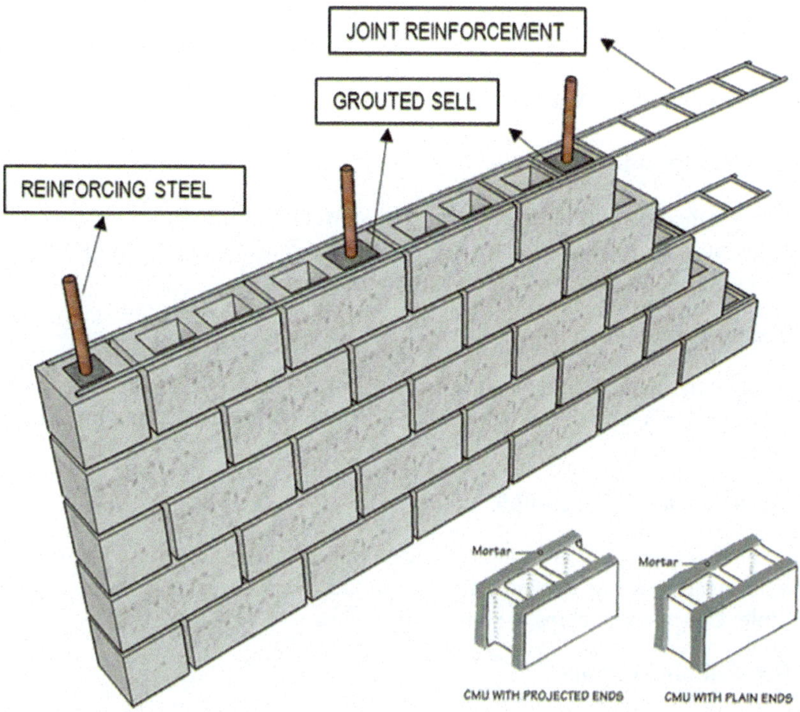

Fig. 57 Vertical (reinforcing steel) and horizontal (joint reinforcement) in CMU wall

cracks are formed in the wall, which are more water resistant, heal more easily, and can be sealed with a coating.

- **Control (shrinkage) Joints**: to enable the division of a long CMU wall into smaller segments.

2.3.4 Grout

The purpose of masonry grout is to fill the voids in masonry walls so that the grout, the masonry units, and the reinforcement are integrated into a composite whole. Grout is a cementitious mix, in many ways, similar to concrete. Concrete durability is important, but it is not a concern for grout because the units protect it. A small amount of lime is permitted in grout but not in concrete. Grout is placed in masonry voids; thus, it contains more water than concrete. In other words, grout is a little bit of a fluent mix to allow it to flow down the voids, which are relatively small in size. There are two types of grout:

- ***Fine Grout*** consists of sand only as aggregate.
- ***Coarse Grout*** consists of sand and coarse aggregate as aggregate.

2.4 *Stone Masonry Wall*

2.4.1 Rubble Stone Masonry Wall

- **Random Rubble**

It is the roughest and the cheapest form of stonewalling. Since stones are not uniform in shape and size, they are arranged carefully to distribute pressure over the maximum area and simultaneously avoid long vertical joints.

- **Square Rubble**

The stones are made roughly square and used in construction. The facing stones are provided with a hammer-dressed finish. Larger stones are used as quoins. Chips are not used as bedding.

- **Miscellaneous Type Rubble**

The stones for masonry are roughly shaped into irregular polygons. The stones are then arranged to avoid vertical joints in the face work. Break the joints as soon as possible. Use of stone chips to support the stones.

- **Dry Rubble Masonry**

The mortar is not used in the joints; therefore, it is the cheapest, but a more skilled workforce is required in construction. The dry rubble masonry is used for non-load-bearing walls like compound walls (Fig. 58).

2.4.2 Ashlar Masonry

In ashlar masonry, square or rectangular blocks are dressed and have extremely fine bed and end joints.

- **Broken Ashlar Masonry**

In broken ashlar, a particular course is not followed. The horizontal joint does not exceed four feet. The stones are of dissimilar sizes but still have their apt positions. It is left natural, showcasing the rock-faced texture or, at times, dressed in tools.

- **Coursed Ashlar Masonry**

It is immediately between ashlar and rubble masonry. The faces of each stone are hammer-dressed, but the vertical joints are not as straight and fine as in ashlar masonry.

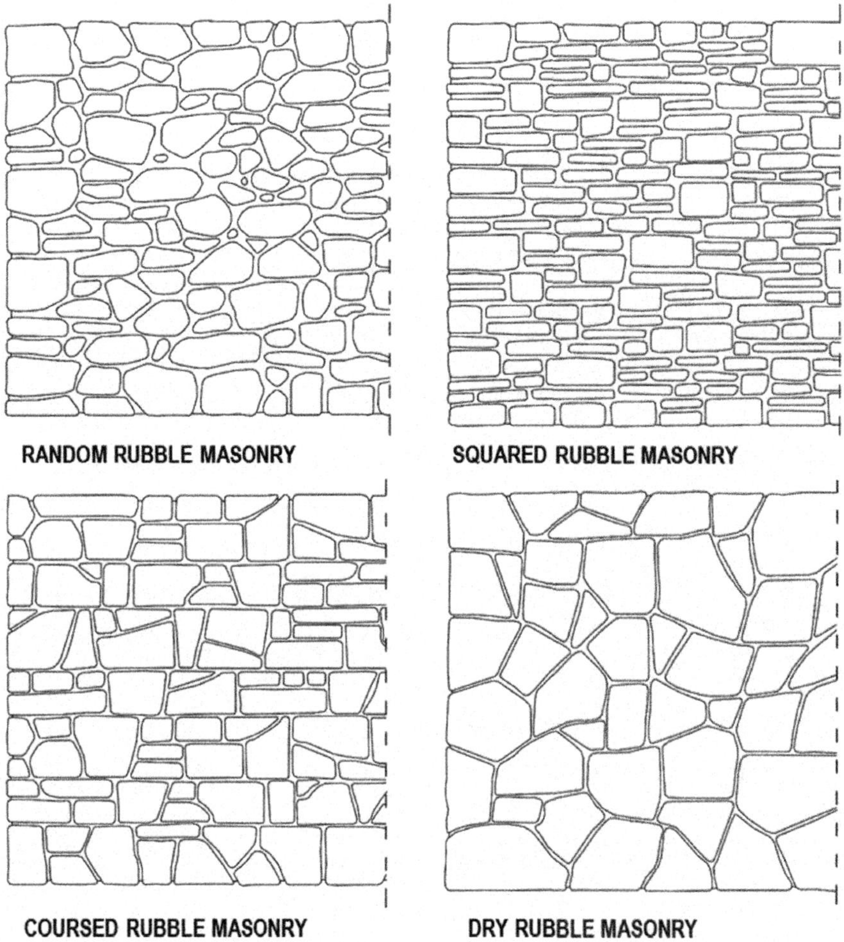

Fig. 58 Rubble stone masonry wall types

- **Random Ashlar Masonry**

Stone blocks are of varying height and length. Also, their arrangement has no specific rule. It has a general course of alignment; however, the vertical joints need not be one over the other. Although it looks very random, it has an in-built harmony. The variations in height and length have to be proportional to the overall structure for strong bonding; thus, the positioning must be mindfully crafted.

- **Ashlar Chamfered**

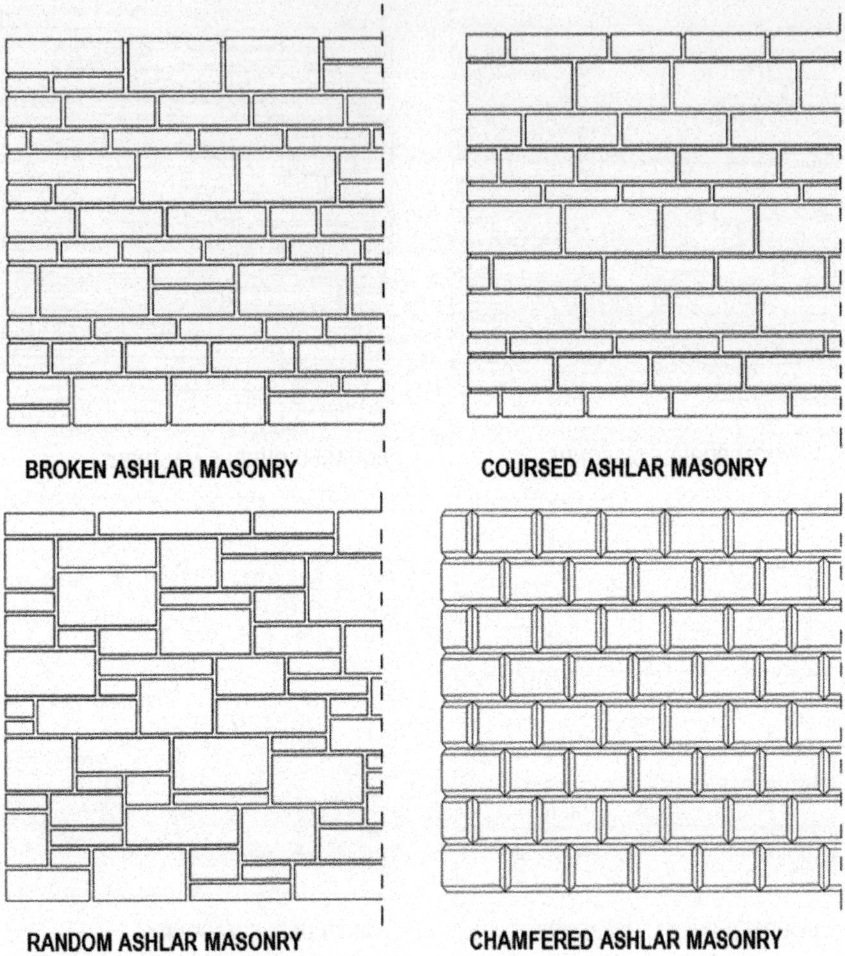

Fig. 59 Ashlar masonry wall types

It is a special type of ashlar rock-faced in which the strip provided around the perimeter of the exposed face is chamfered at an angle of 45° to a depth of 25 mm (Fig. 59).

2.5 Masonry Wall Construction

Regarding its structural usage, masonry walls are divided into two categories: load-bearing and non-load-bearing. Load-bearing walls can resist lateral loads where columns, such as shear and retaining walls, cannot represent the same performance.

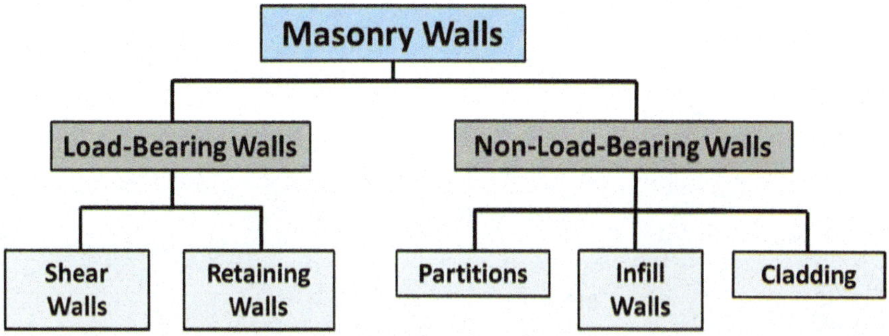

Fig. 60 Various applications of masonry walls

Thus, depending on their application area, these wall types are appropriate for reducing the hazardous impact of earthquakes, wind, and soil pressure. Non-load bearing walls, on the other hand, are used as enclosures to create a space (infill walls) to separate two spaces from each other (partitions) or for covering purposes (cladding) to protect the building superstructure from environmental effects such as sun, rain, snow, and wind. Due to not having any load-bearing capacity, non-load-bearing walls have limitations when the wall height increases by 2.5–3 m. Bond beams become a necessity when the height of the wall increases by 3 m, and it is used for bearing the floor structures (Fig. 60).

2.5.1 Load Bearing Wall

A load-bearing masonry structure is also referred to as a bearing wall structure. Load-bearing walls can also enclose and divide the spaces, which is unavailable in steel or concrete frame structures. Reinforced masonry Wall has several advantages over plain masonry at little additional cost, such as increasing the flexural and shear strengths. Low-rise, plain masonry-bearing Wall structures have excellent performance records not subjected to earthquakes or extreme winds.

Before the frame structure's invention, the masonry Wall system was the only system to construct major buildings. Although several large and tall masonry structures were built, their design was based on arbitrary rules formulated intuitively rather than scientifically. This intuitive understanding occurs with building codes requiring that the thicknesses of exterior walls increase progressively toward the lower levels. As the skeleton (iron and later steel) frame became popular, the use of masonry-bearing Wall systems went out. However, after World War II, masonry load-bearing wall construction again became popular during the reconstruction of Europe (Fig. 61).

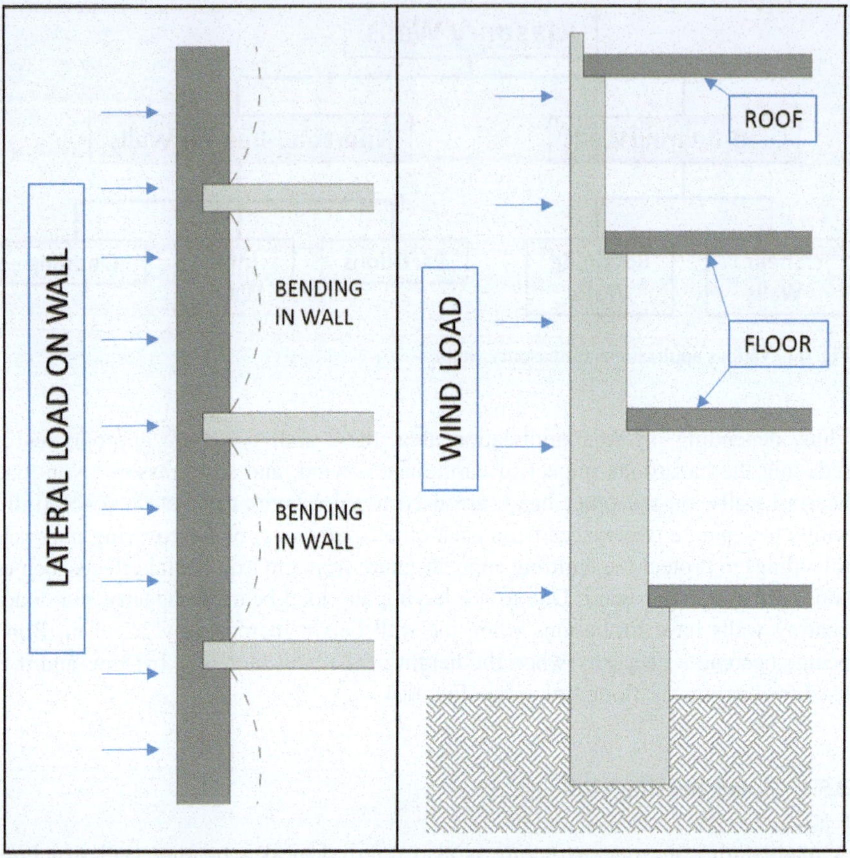

Fig. 61 Lateral load resistance in masonry wall (left and middle image)

2.5.2 Bond Beams

Bond beams provide continuity at the wall and stabilize the level. They are located below the floor levels to bear the load transferred from floor and roof structures. This is because bond beams distribute the load to the below masonry structure.

The repetitiveness of floor layout requires that all load-bearing or shear walls are continuous to the foundation. This is particularly important in high-rise bearing Wall structures, where there is a temptation to provide larger uninterrupted spaces at the first floor or basement level. The discontinuity of walls at foundations, referred to as a soft story, is structurally feasible but requires a heavy transfer structure. The transfer structure carries the super-imposed gravity loads and provides the required lateral load resistance. A soft story is particularly problematic in seismic zones (Figs. 62 and 63).

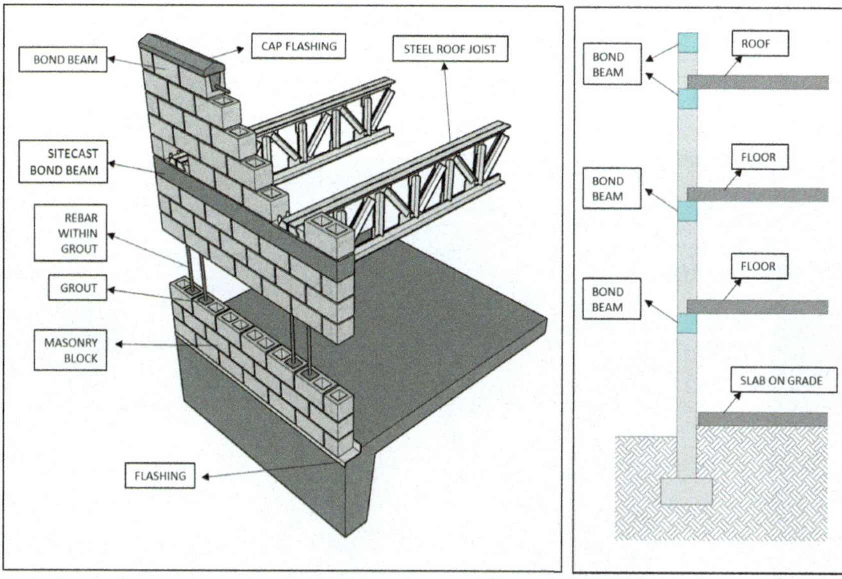

Fig. 62 Bond beam in CMU walls

2.5.3 Wall Layout in Masonry Structures

For a multi-floor bearing wall structure, a cellular-type structure is best achieved when the walls on an upper floor are at the same location as the ones on a lower floor, resulting in repetitive floor plans. Repetitive floor plans are structurally more efficient and more easily constructible. A multi-floor bearing Wall structure is ideal for occupancies in which the economy is a major consideration. Such occupancies are generally residential—apartment buildings, hotels, motels, student dormitories, correction facilities, hospital wards, etc. An asymmetrical Wall layout leads to rotation of the building (torsion) under lateral loads. The greater the asymmetry, the greater the torsion created, which increases the cost of the structure.

Floor and roof decks can either be one-way or two-way decks. In a one-way deck, the deck spans between two opposite walls; that is, the gravity load on the deck is carried in one direction across the two opposite walls. In a two-way deck, the load is transferred to all four walls. Two-way decks are generally of site-cast concrete. One-way decks may be made of steel, precast concrete, or wood. The use of one-way decks yields the following two types of bearing 2all plans:

- **Cross-bearing Wall plan**: The bearing walls are transverse to the main axis of the building.
- **Longitudinal-bearing Wall plan:** Provides larger, unobstructed interior spaces. It is used where interior walls are architecturally undesirable (Figs. 64, 65, 66 and 67).

Fig. 63 The bond beam application in a brick wall

2.6 Construction of a Building with Masonry Wall

The following images illustrate a two-story building made of masonry walls and wood floor decks (Figs. 68, 69, 70, 71, 72, 73, 74, 75 and 76).

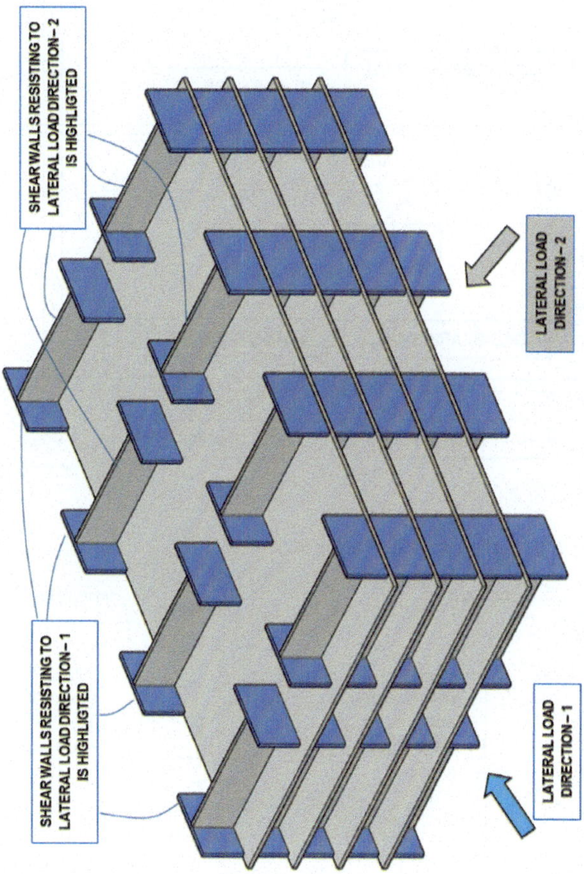

Fig. 64 Lateral load directions and necessary load-bearing wall direction set for resistance to the lateral loads

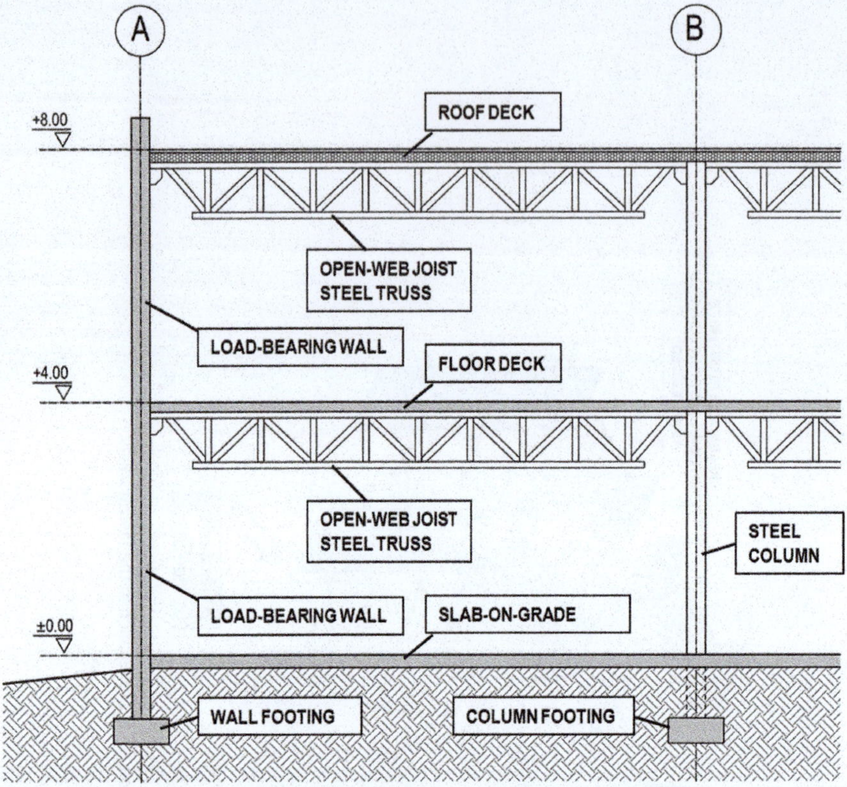

Fig. 65 Floor and roof deck connected to bearing wall

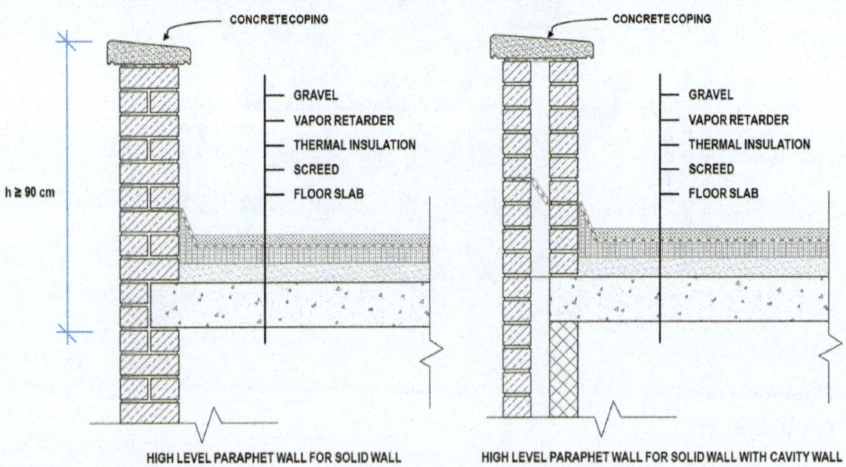

Fig. 66 Parapet details of brick masonry wall having more than 90 cm parapet height

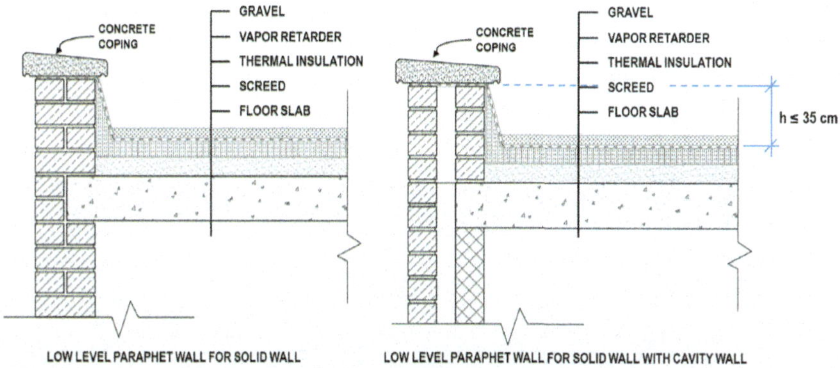

Fig. 67 Parapet details of brick masonry wall having less than 90 cm parapet height

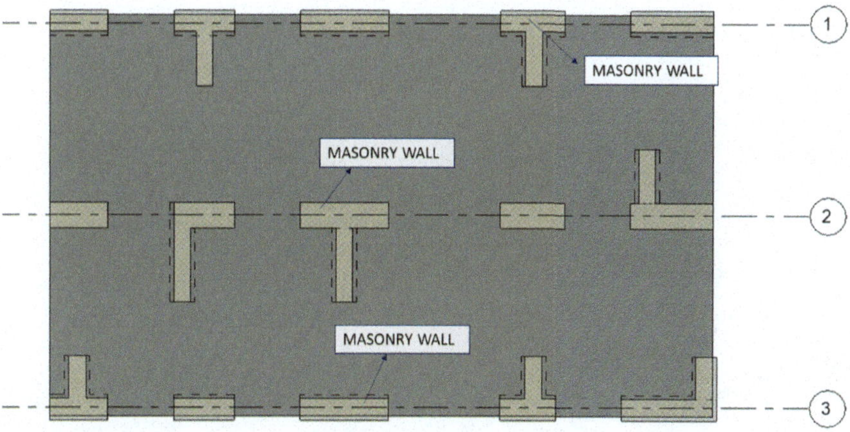

Fig. 68 A ground floor plan with masonry walls

3 Wood Construction Systems and Methods

3.1 Materials for Wood Construction: Engineered Wood Products, Fasteners, and Connectors

Two types of manufactured wood products are available in the construction industry:

- **Engineered Wood Products:** Wood products that are engineered for **structural applications.**
  - **Glue-Laminated Wood**
  - **Structural Composite Lumber**
  - **Wood I-Joists**

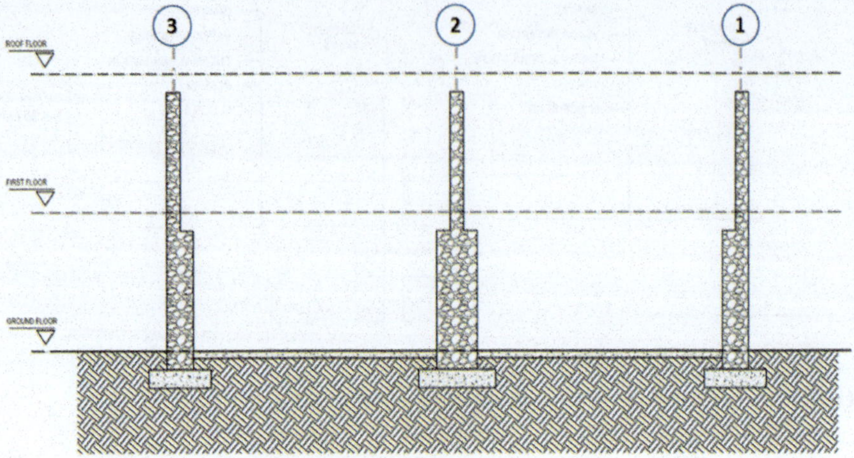

Fig. 69 Section of a two-story masonry wall building

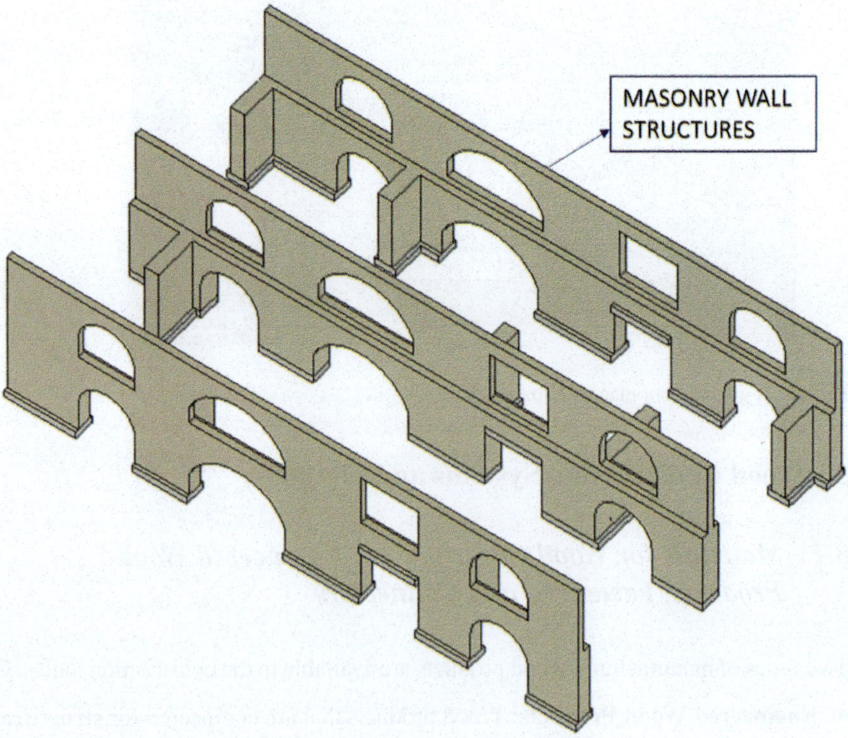

Fig. 70 3-Dimensional illustration of a two-story masonry building

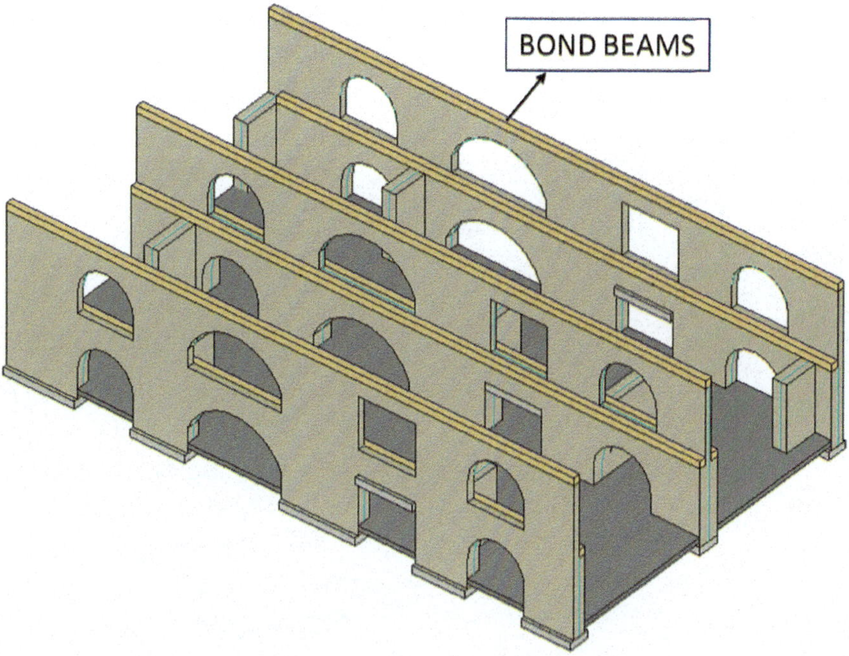

Fig. 71 Bond beams are needed to be added before floor and roof construction

  - **Plywood**
  - **Oriented Strand Board (OSB)**
  - **Wood Trusses**

- **Industrial Wood Products**
  - **Particle Board:** Cabinet Work, Furniture, Heavy-Duty Shipping Containers.
  - **Medium-Density Fiberboard (MDF):** MDF is produced to replace solid lumber due to having a smooth surface, allowing precise machining to form complex and intricate moldings.
  - **High-Density Fiberboard (HDF)—*Hardboard:*** Floor Underlayment

3.1.1 Engineered Wood Products

Glue-Laminated Wood (Glulam)

The wooden laminations are glued side-by-side and face-to-face to form a **Glue-Laminated Wood**—*Glulam* (Figs. 77 and 78).

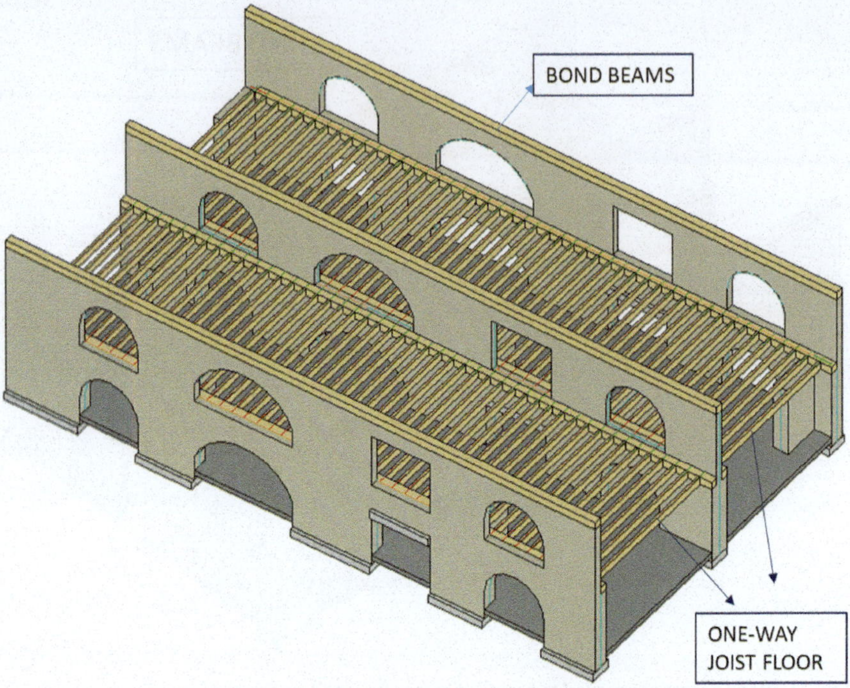

Fig. 72 If a wood or steel floor deck will be constructed, a one-way joist floor is the only option

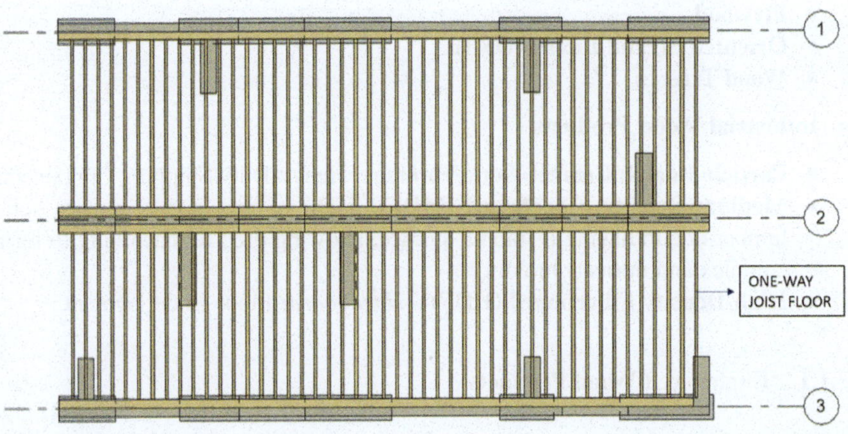

Fig. 73 Plan view of one-way joist floor deck

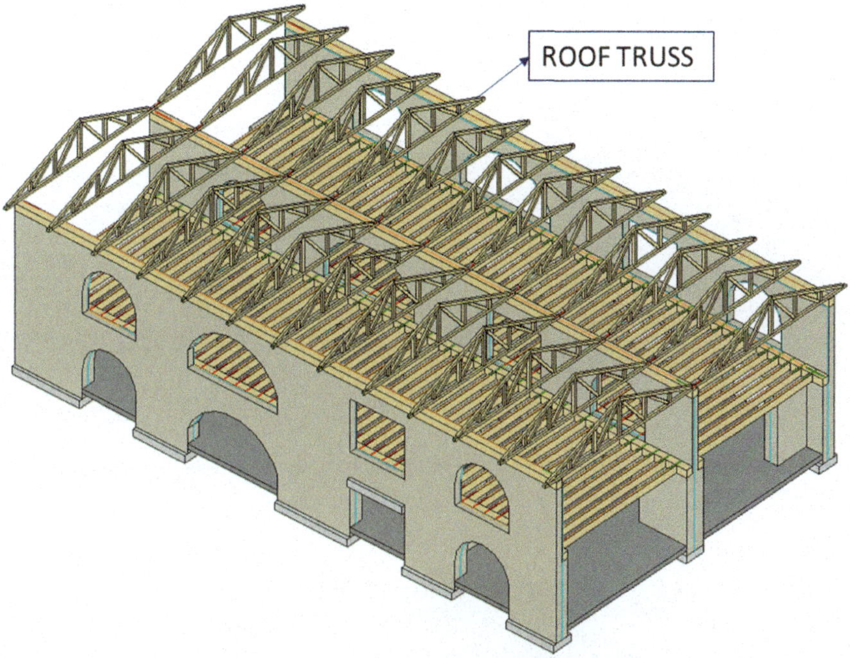

Fig. 74 Wood trusses can be used on the roof floor

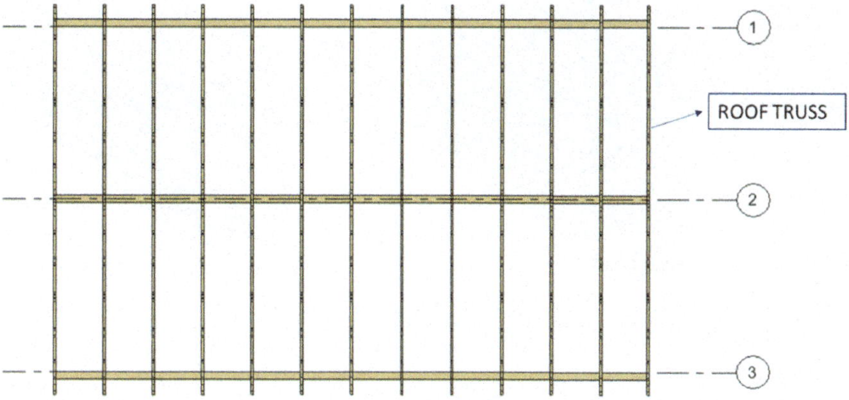

Fig. 75 The sequence of the roof trusses is not as dense as the joists at the floor deck

Structural Composite Lumber (Lvl)

Laminated Veneer Lumber (LVL) is produced by gluing together dried wood veneers. The wood grain in all veneers runs in the same direction, unlike the grain in plywood, where the veneers are cross-grained between laminations. Thus, LVL is stronger

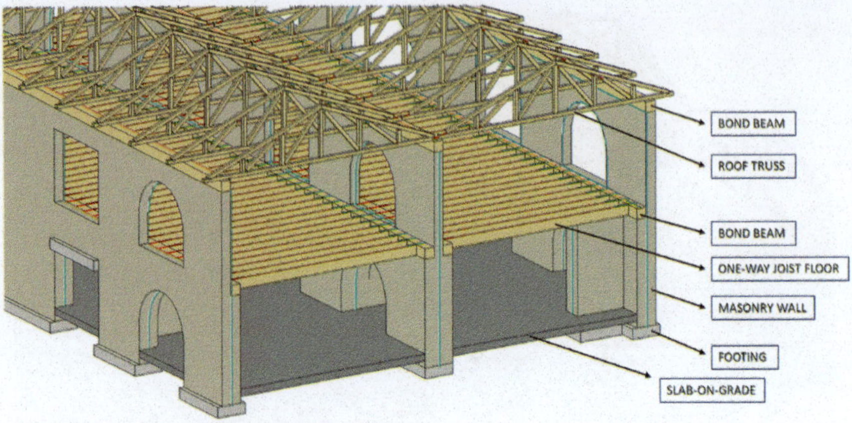

Fig. 76 3-Dimensional illustration of a two-story building structure made of masonry wall and wood floor deck

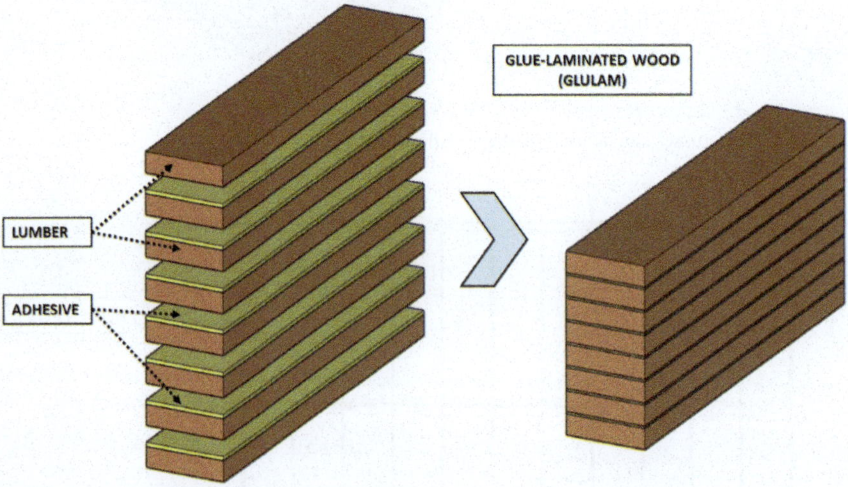

Fig. 77 Glue-laminated wood

along the grain and weaker across the grain. LVL is generally used as floor joists and rafters (Fig. 79).

Structural Composite Lumber: Psl

Parallel Strand Lumber (PSL) is a variation of LVL. PSL is used for Beam and Header applications (Fig. 80).

Fig. 78 Various applications with glue-laminated wood. *Source* Naswood (2023)

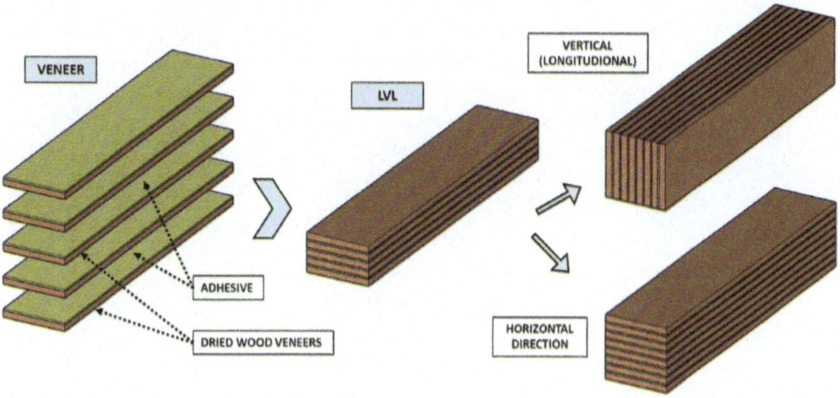

Fig. 79 Manufacture of laminated veneer lumber (LVL)

Wood I Joists

Wood I-Joists are made by gluing wood flanges to a wood web. I-Joists are commonly used as floor joists and roof rafters (Fig. 81).

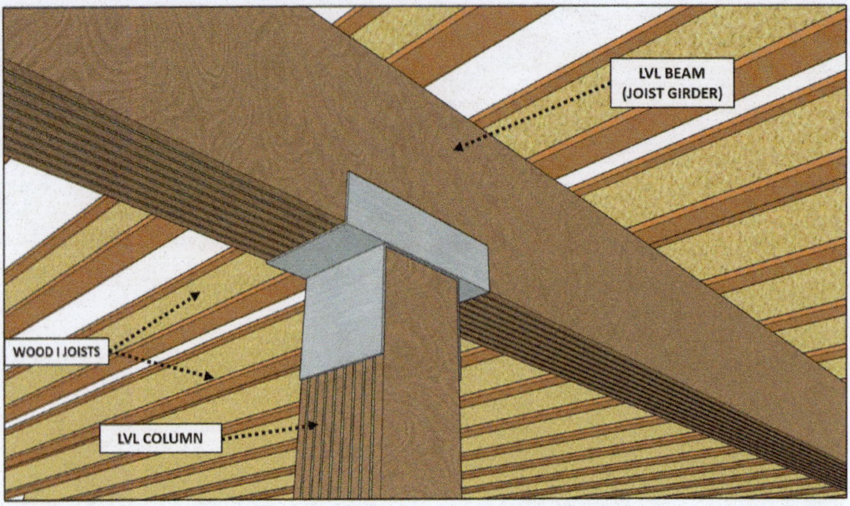

Fig. 80 Usage of parallel strand lumber (PSL) as beam and column

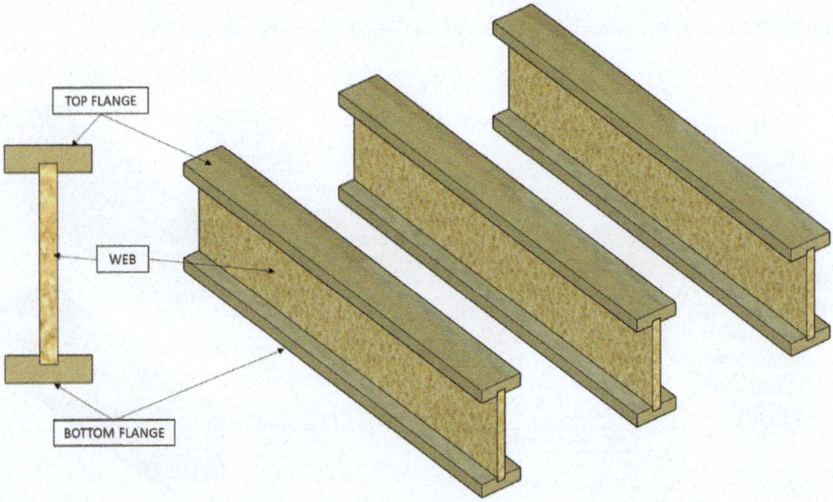

Fig. 81 A Wood—I joist example

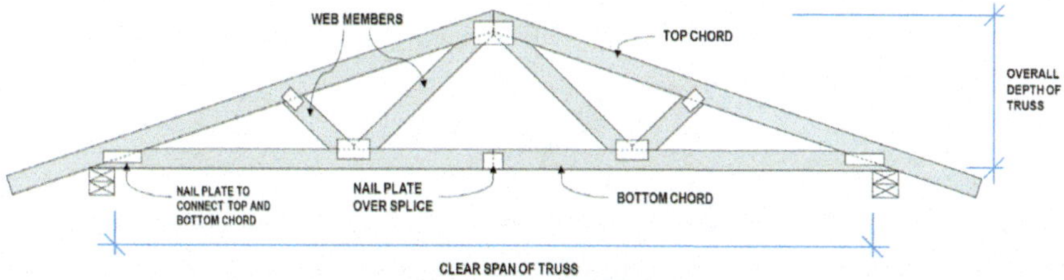

Fig. 82 Terminology of a wood roof truss

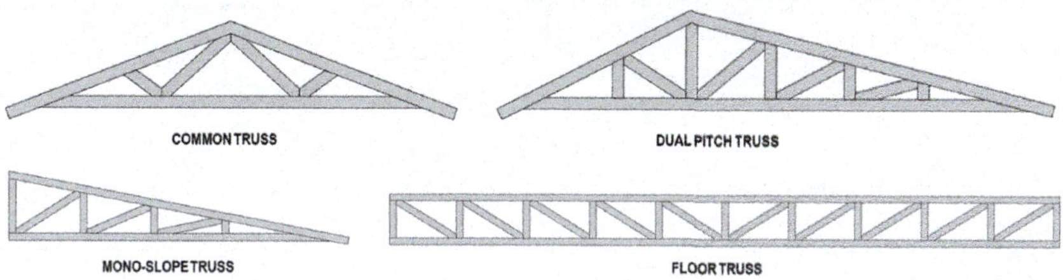

Fig. 83 Various wood roof truss types

Wood Trusses

A truss consists of individual members that are joined together to form an array of interconnected triangular frames. A truss is more rigid than a beam with the same amount of material because a triangle is a naturally rigid geometric shape that resists being distorted when loaded from any direction (Figs. 82 and 83).

- Roof Trusses (Figs. 84 and 85)
- Floor Trusses (Fig. 86)

Wood Panels: Plywood and Osb

Plywood panels are made by gluing wood veneers under heat and pressure. An oriented Strand Board (OSB) panel is a more efficient technology than a Plywood panel. The shredded wafer-thin wood strands are compressed and glued to create an OSB panel. OSB is cheaper than Plywood. The plywood panel surface is smoother than the OSB. OSB panels are more prone to edge swelling if they remain wet.

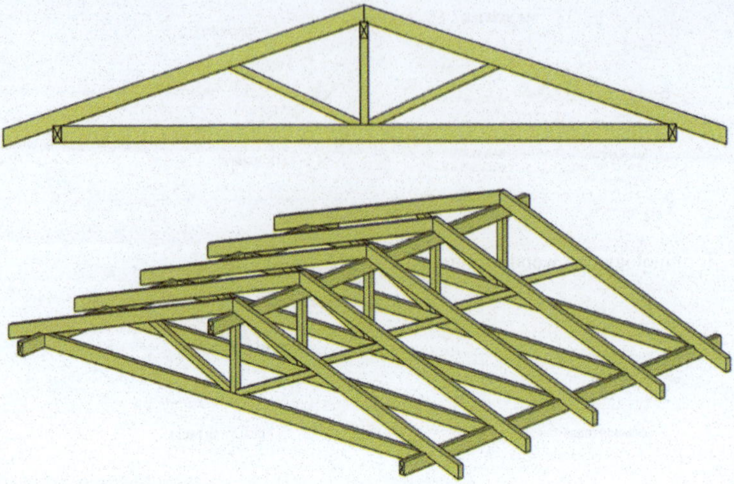

Fig. 84 A Sample roof truss

3.2 *Fasteners for Connecting Wood Members*

3.2.1 Mortise and Tenon Joint

See Fig. 87.

3.2.2 Housed—Mortise and Tenon Joint

See Fig. 88.

3.2.3 Dovetail Joint

See Fig. 89.

3.2.4 Nailed Connections

- Face—Nailed Connections
- End—Nailed Connections
- Toe—Nailed Connections (Fig. 90)

Fig. 85 Various applications of roof trusses. *Source* Naswood (2023)

3.2.5 Sheet Metal Connectors in Wood

- Face—Mounted Hanger (Fig. 91)
- Top—Mounted Hanger (Figs. 92 and 93)

3.3 Wood Frame Structures

3.3.1 Interlocking Members System

See Figs. 94 and 95.

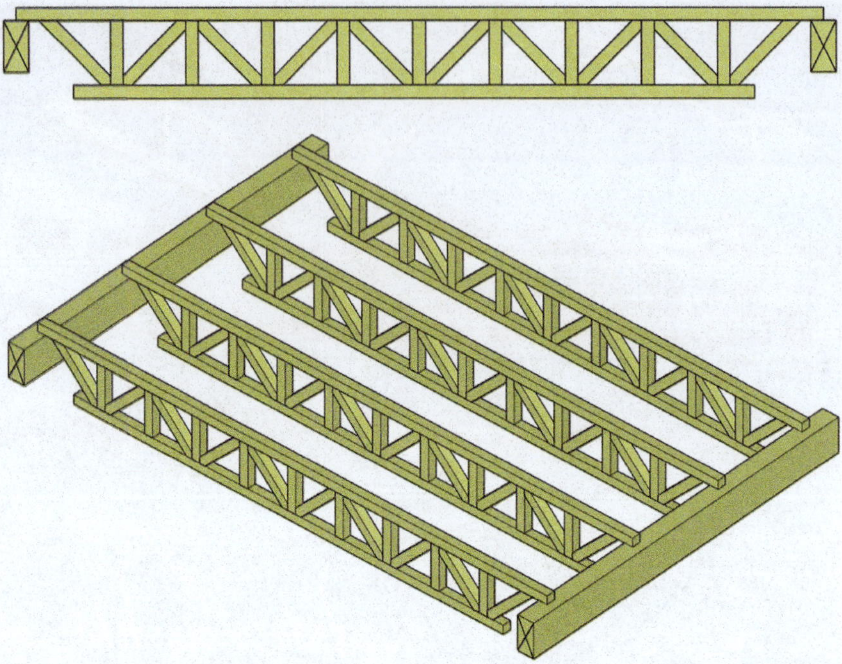

Fig. 86 Sample wood floor truss

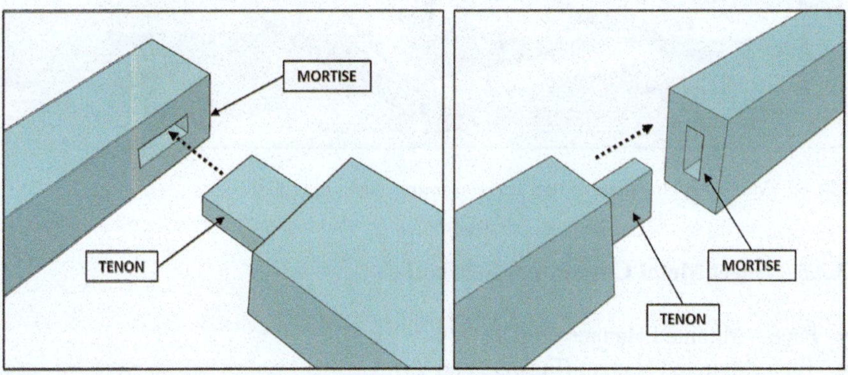

Fig. 87 Mortise and tenon joint

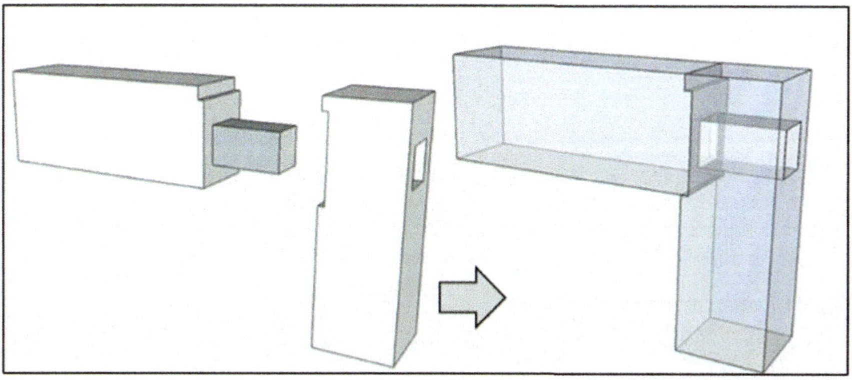

Fig. 88 Housed mortise and tenon joint

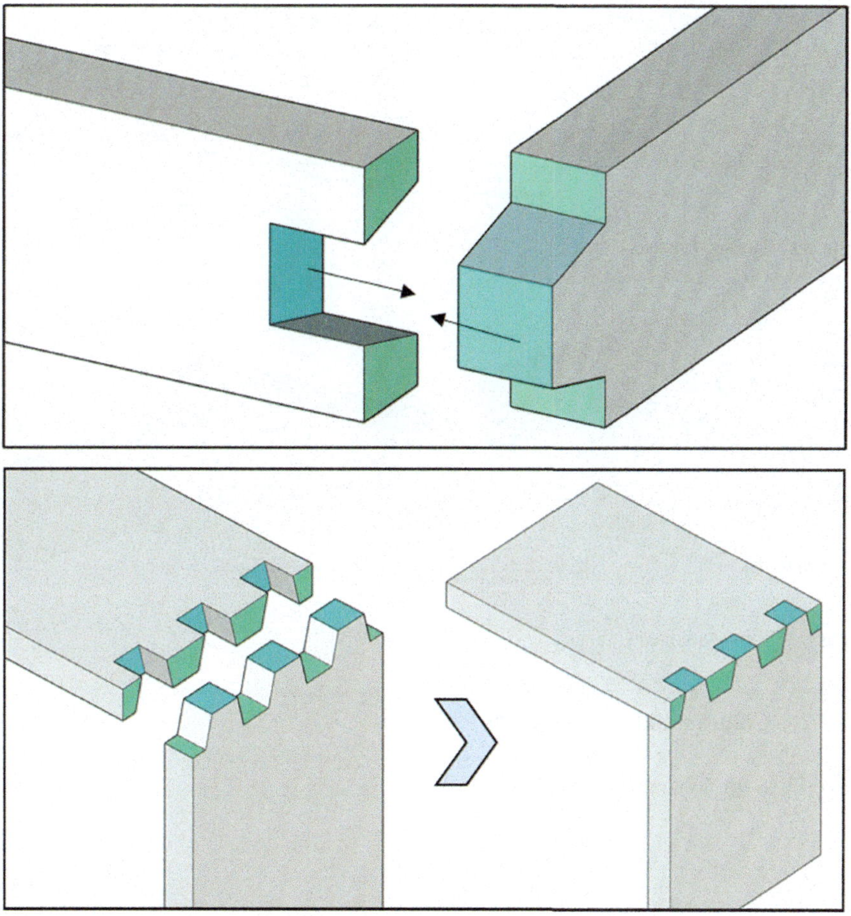

Fig. 89 Dovetail joint

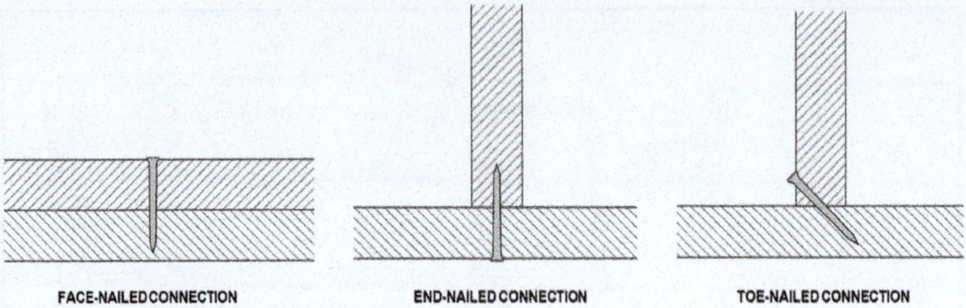

Fig. 90 Various nailed connection types

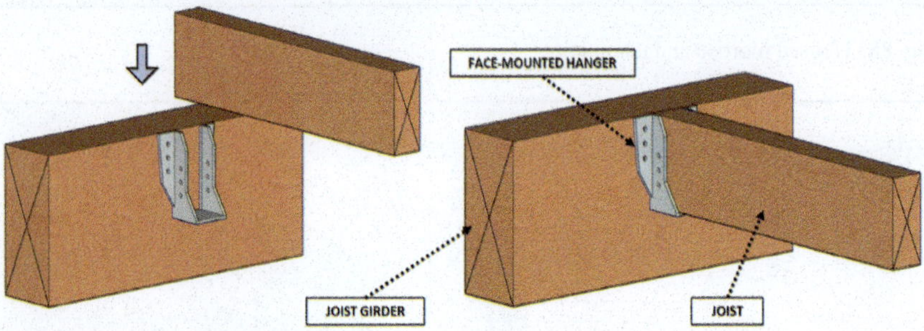

Fig. 91 Face—Mounted Hanger

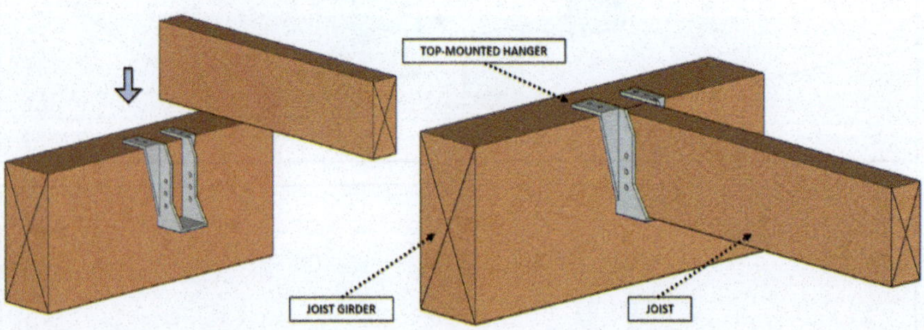

Fig. 92 Top—Mounted Hanger

3.3.2 Column and Beam System

See Figs. 96, 97 and 98.

Fig. 93 Various applications of face-mounted hangers. *Source* Naswood (2023)

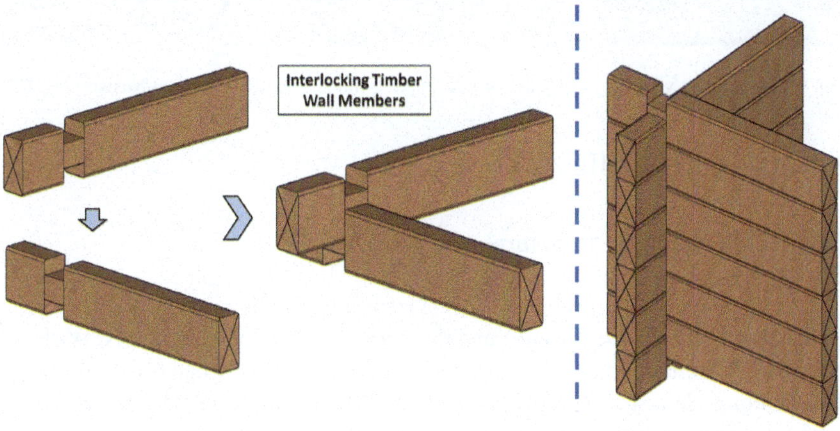

Fig. 94 Wood wall structure by interlocking timber members

Fig. 95 A wooden villa project by use of interlocking timber members. *Source* Naswood (2023)

3.4 Light-Frame Wood Construction

3.4.1 Balloon Frame Structures

Before the Wood Light-Frame (WLF) system, most buildings had thick masonry walls, heavy timber posts, beams, and thick wood plank floors and roofs. WLF was invented in traditional heavy construction materials and is called a balloon frame. The building elements of WFL are so light that two or three workers can easily work. The connections between members are made of simple nails. The Studs run the full height of the building, from the sill plate at the bottom to the top plate under the rafters. The continuity of studs is the major limitation in balloon frames due to the lack of long straight members, making them more expensive. Thus, the balloon frame was modified over time into the platform frame, in which the individual studs are only one story high (Figs. 99 and 100).

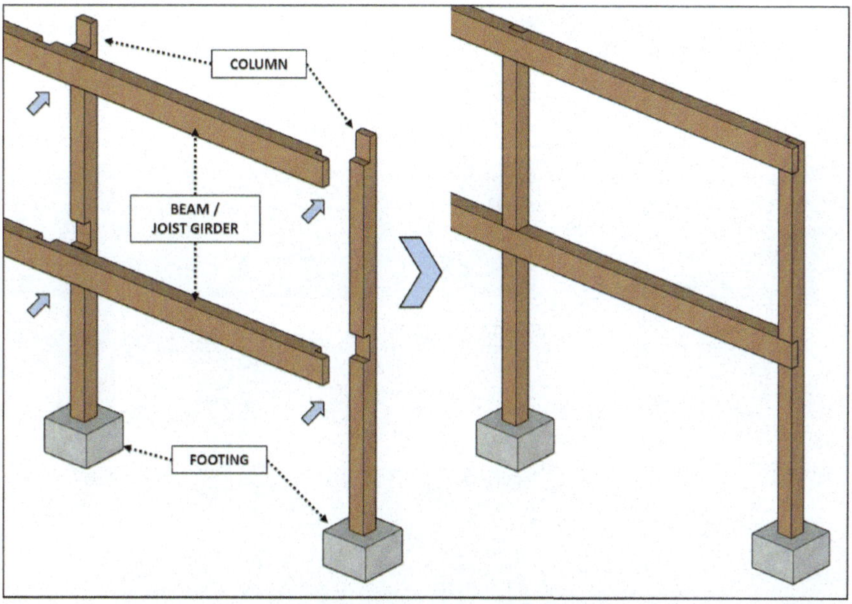

Fig. 96 Column and beam framing system for wood construction

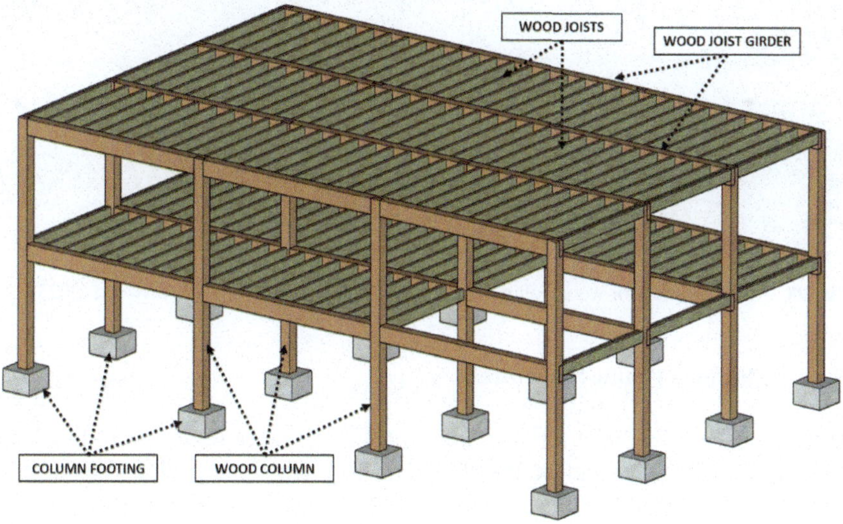

Fig. 97 Wood column and beam framing system

Fig. 98 An application for wood column and beam framing system. *Source* Naswood (2023)

3.4.2 Platform Frame Structures

In the balloon frame, wall framing, floor framing, and roof framing are completed for the entire building before the structural floor (subfloor or floor sheeting) is constructed. In the platform frame, the subfloor at the first level is completed soon after laying floor joists at that level. This subfloor provides a platform for the workers to stand on and build the following story. Due to providing a platform to workers, the platform frame increases the safety of the workers (Figs. 101 and 102).

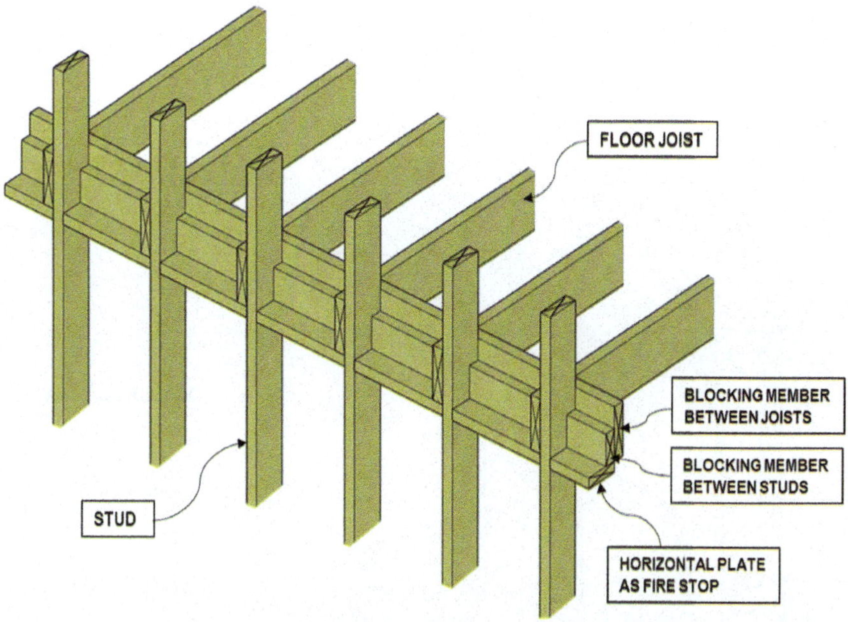

Fig. 99 Floor and wall assemblies of balloon frame structure

3.4.3 Comparison of Balloon Frame and Platform Frame Construction

Balloon frame is more stable than platform: Continuous studs in balloon frame increase the stability of the structure while floor-by-floor construction makes the construction process much easier for platform frame when compared with balloon frame structure. Studs are in building height in balloon frame, limiting the building height. Furthermore, the availability of long studs is another challenge. This is because studs are made of timber logs cut from trees. Trees are naturally growing, and it is hard to control and manage the length of the trees. However, there is no such deficiency in platform frame structures. Balloon frame structure made of studs having building height requiring less connection than platform frame. Each connection causes a shake of the building in case of impact by lateral loads. This situation brings less shake of building structures for balloon frame structures when compared with platform frame structures (Fig. 103).

3.4.4 Frame Configuration in Light Frame Walls

Bottom Plate

See Fig. 104.

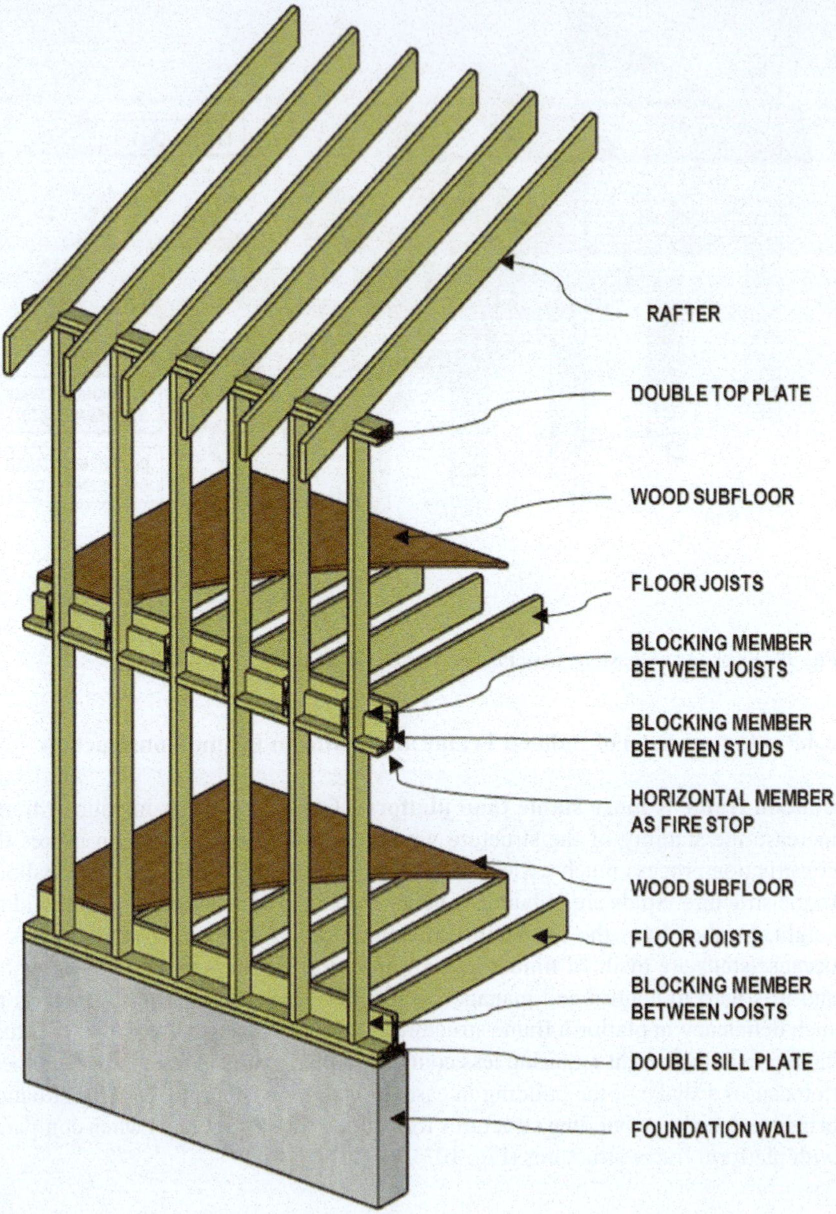

Fig. 100 Balloon frame structure

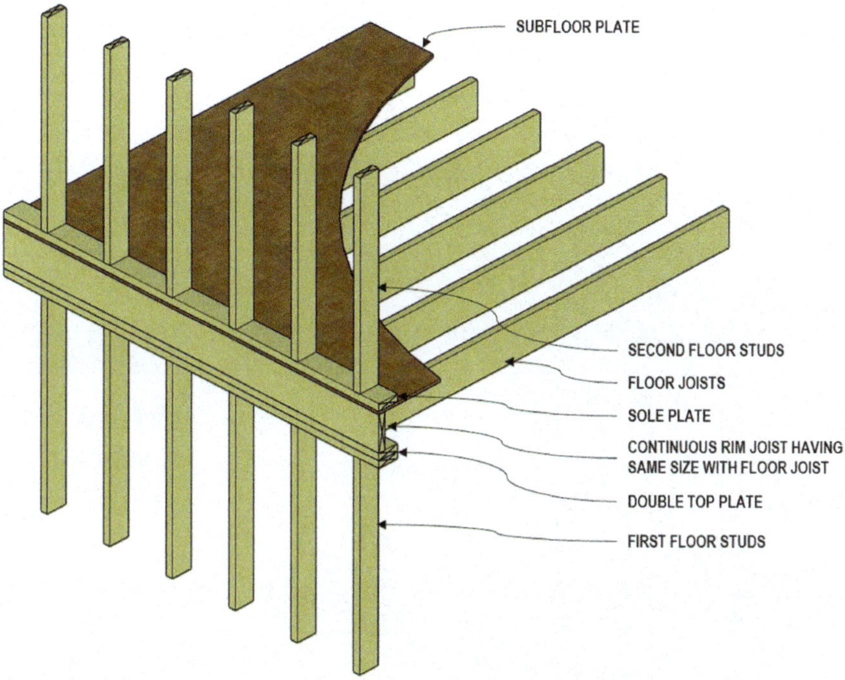

Fig. 101 Floor and wall assemblies of platform frame structure

Corner Arrangement of Studs

See Figs. 105, 106.

Framing Around the Wall Opening

See Fig. 107.

Floor Framing

See Fig. 108.

Installing a Load-Bearing and a Non-Load-Bearing Wall in the Wood Floor Deck

See Fig. 109.

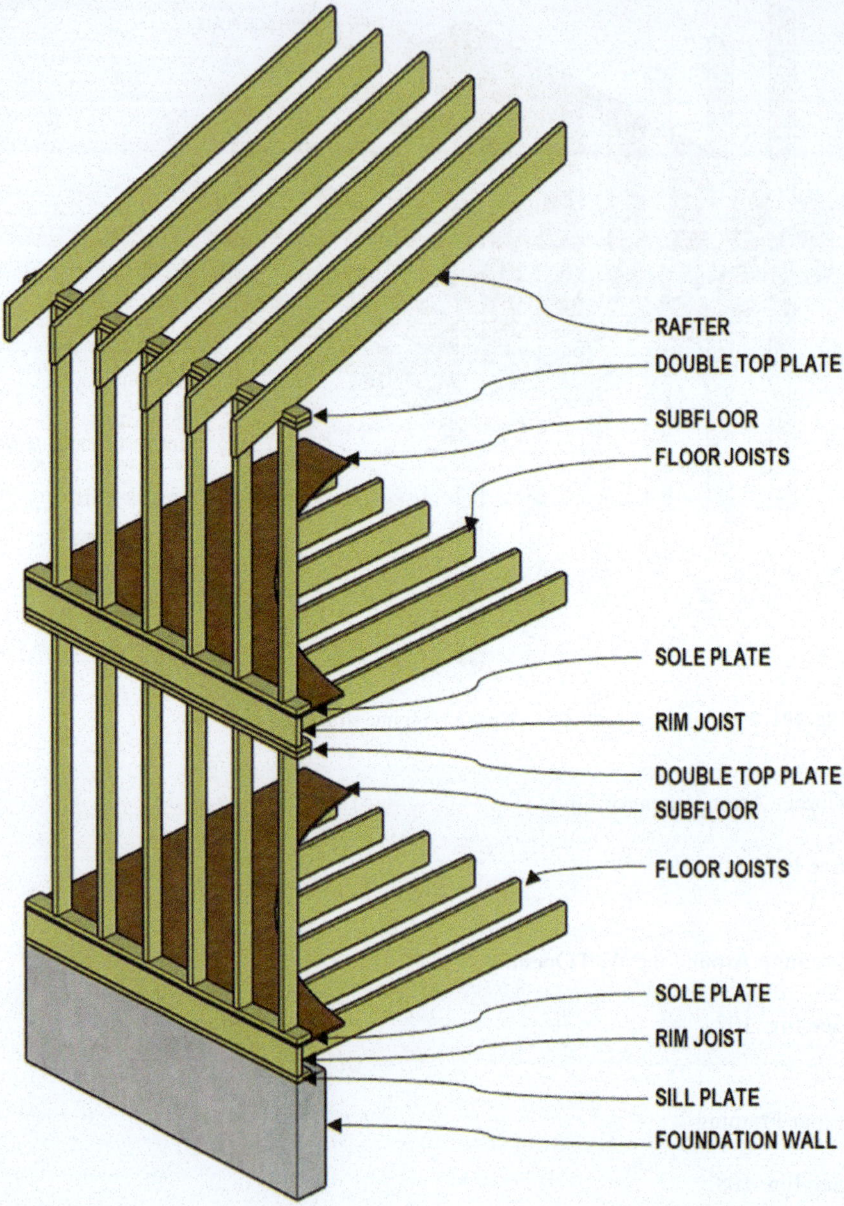

Fig. 102 Platform frame structure

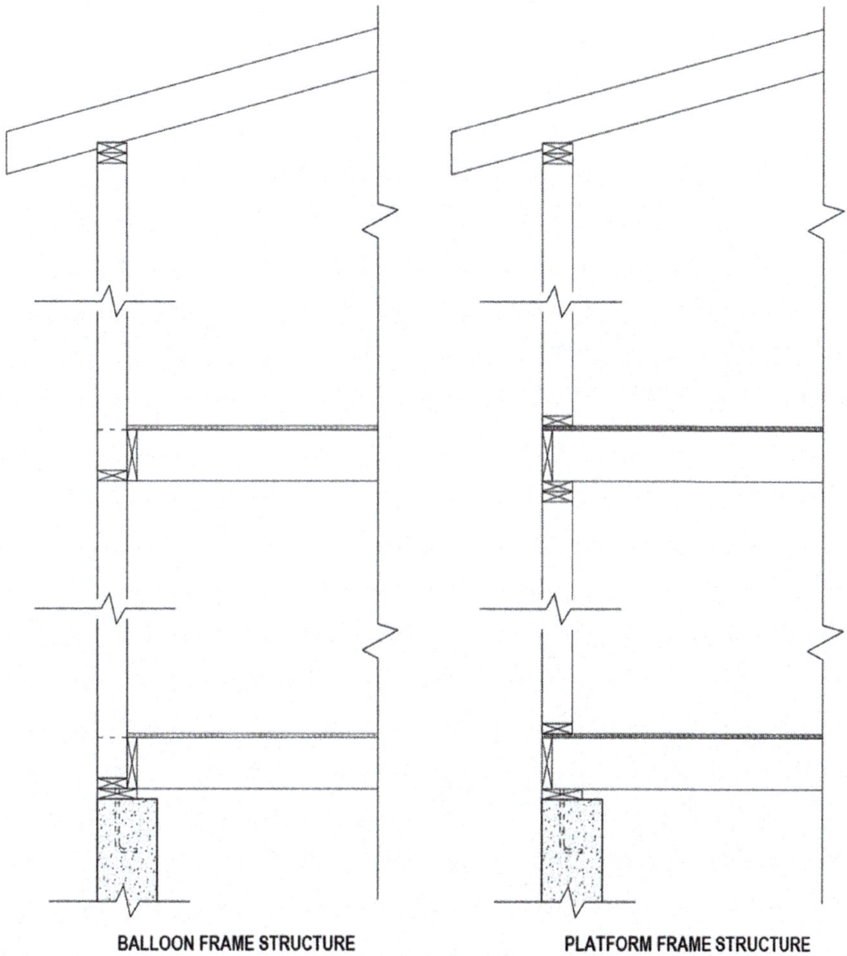

Fig. 103 Comparison of balloon frame and platform frame structure

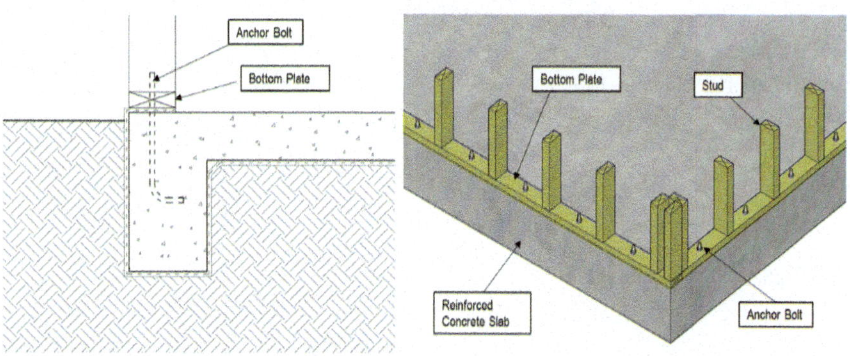

Fig. 104 Illustration of the bottom plate and its connection to the slab-on-grade foundation

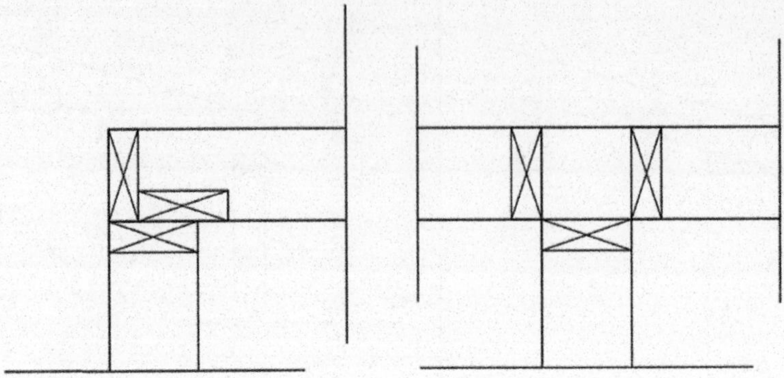

Fig. 105 Arrangement of studs at corners and connections

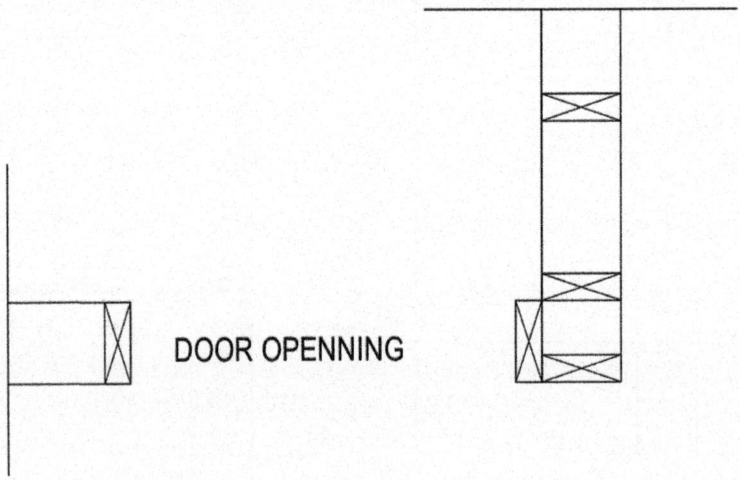

Fig. 106 Arrangement of studs at door openings

3.4.5 Roof Types and Framing

See Figs. 110, 111 and 112.

3.5 Some Applications

See Figs. 113, 114 and 115.

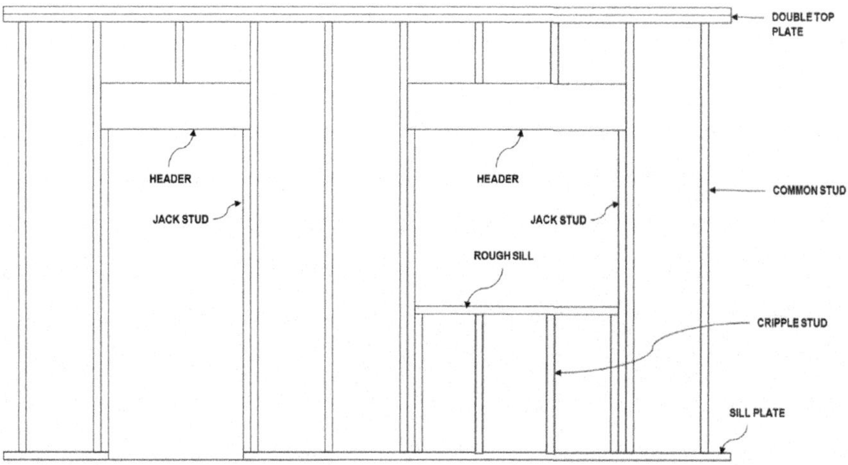

Fig. 107 Framing around door and window openings

Reference

Naswood: *Naswood Wood Company, located in Antalya, all images are retrieved from the below link in 2023 by the permission of the company*: https://www.naswood.com.tr/projelerimiz/

4 Steel Construction Systems and Methods

4.1 Production of Modern Steel

4.1.1 Evolution of Modern Steel

Steel is an alloy of various elements, mainly carbon and iron. Steel provides high tensile strength with minimum weight compared to other construction materials. The wrought iron and cast iron are the predecessors of steel (Fig. 116).

4.1.2 Cold Rolled and Hot Rolled Steel

If the last form of the steel component is shaped at a high temperature near the melting point, this process is called "Hot Rolling," and the product is called *"Hot Rolled Steel."* If the last form of the steel component is shaped without implementing any heat to the steel by simply bending or rolling the material, this process is called "Cold Rolling," and the end product is called *"Cold Rolled Steel."* (Fig. 117).

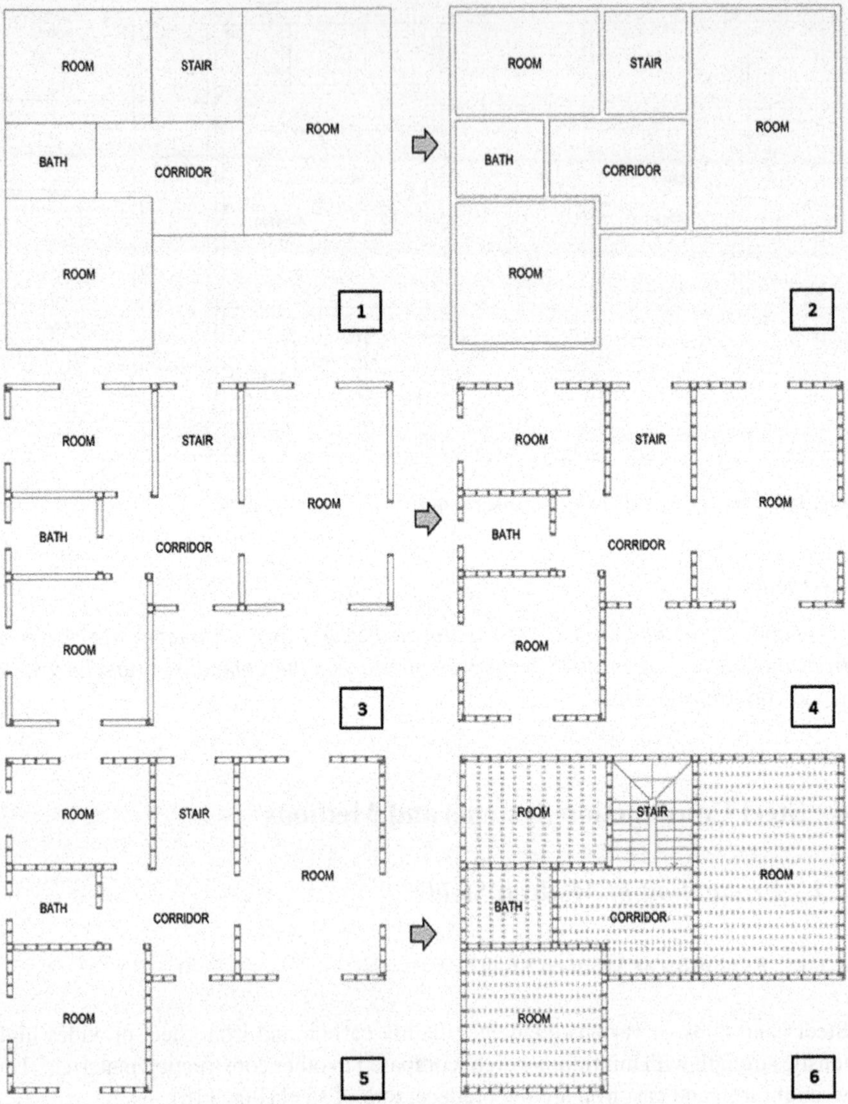

Fig. 108 Arrangement of wall studs and floor framing for a simple plan layout

4.2 *Classification of Steel Components*

Application-Based Classification

- **Structural Steel**: Including Steel Cross Sections, such as **I** –Sections, **H** Sections, **T**-Sections, **C**-Sections (Channels), **L**-Sections (Angles), Plates, Pipes and Rectangular Tubes

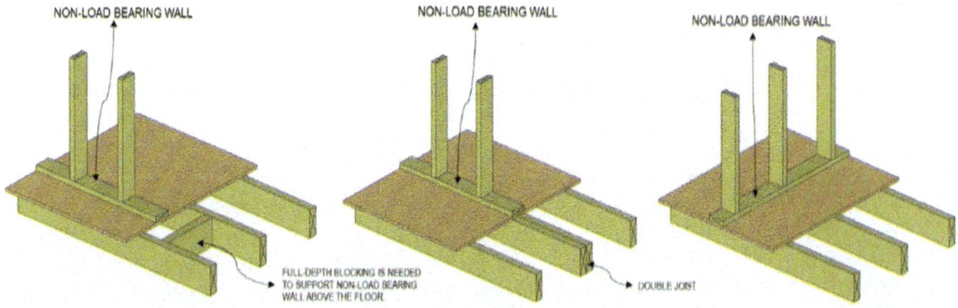

Fig. 109 Installment of a non-load-bearing wall in wood floors

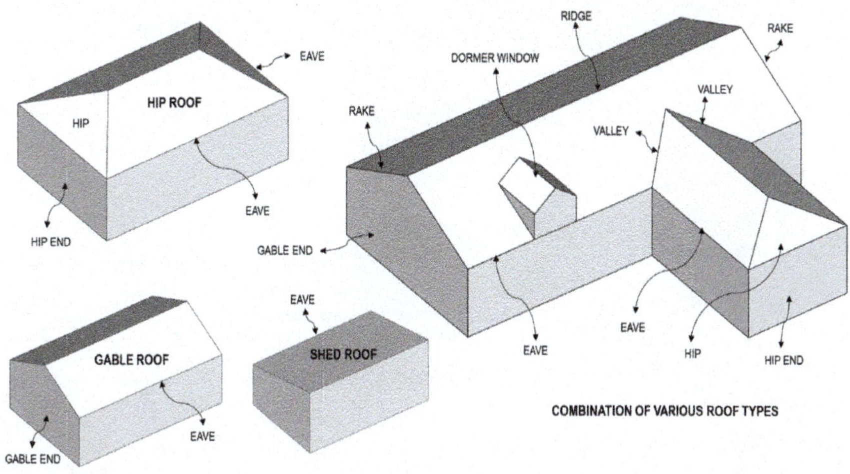

Fig. 110 Illustration of various roof types and terminologies

- **Cold-Formed Steel**: Cold-Formed Steel (Light-Gauge Steel) members are produced by bending sheets to produce various corrugated profiles at room temperatures.
- **Reinforcing Steel**: Deformed round steel bars (rebars) used in concrete slabs, beams, and columns.
- **Pre-stressing (post-tensioning) Steel**: These are used in precast or post-tensioned concrete members to replace reinforcing steel.

Strength-based Classification:

- The content of elements (iron, carbon, chromium, magnesium, oxygen, etc.) inside a steel alloy affects the yield strength of steel products.
- Thus, the content inside the steel alloy is standardized, and steel products are classified depending on the yield strength.

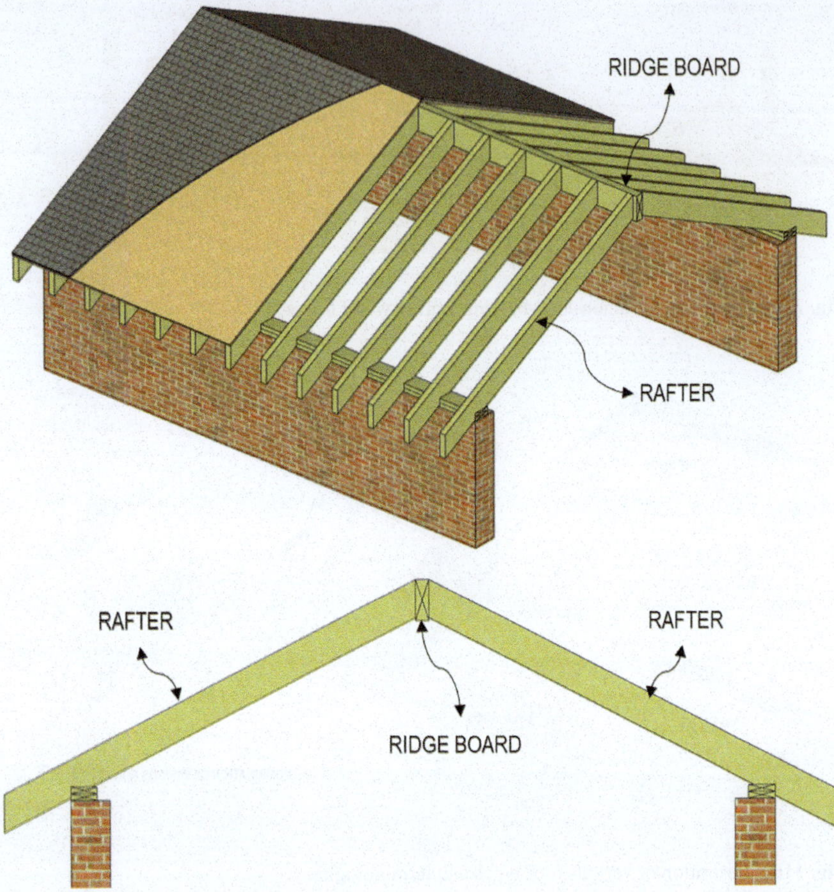

Fig. 111 Terminology for gable and hip roof framings

- This classification is important for structural calculations.

Metallurgy-Based Classification:

- **Carbon Steel**: Steel that amount of carbon is between 0.1% and 0.3% with other metals in a minimum percentage.
- **Alloy Steel**: Stainless Steel (the main alloying element is Chromium) and Weathering Steel.

4.2.1 Structural Steel

Hot-Rolled Structural Steel Sections

- **I-Sections**

Fig. 112 A timber steep roof application upon a concrete flat roof

- **W**-Shapes
- **S**-Shapes
- **H**-Shapes
- **M**-Shapes

Properties of flanges and webs are changing in each of these shapes.

- **C**-Shapes
- **T**-Shapes
- **L**-Shapes
- **Pipes**: Has circular hollow
- **Tubes**—Hollow Structural Section (HSS)
- **Bars**: There is no hollow

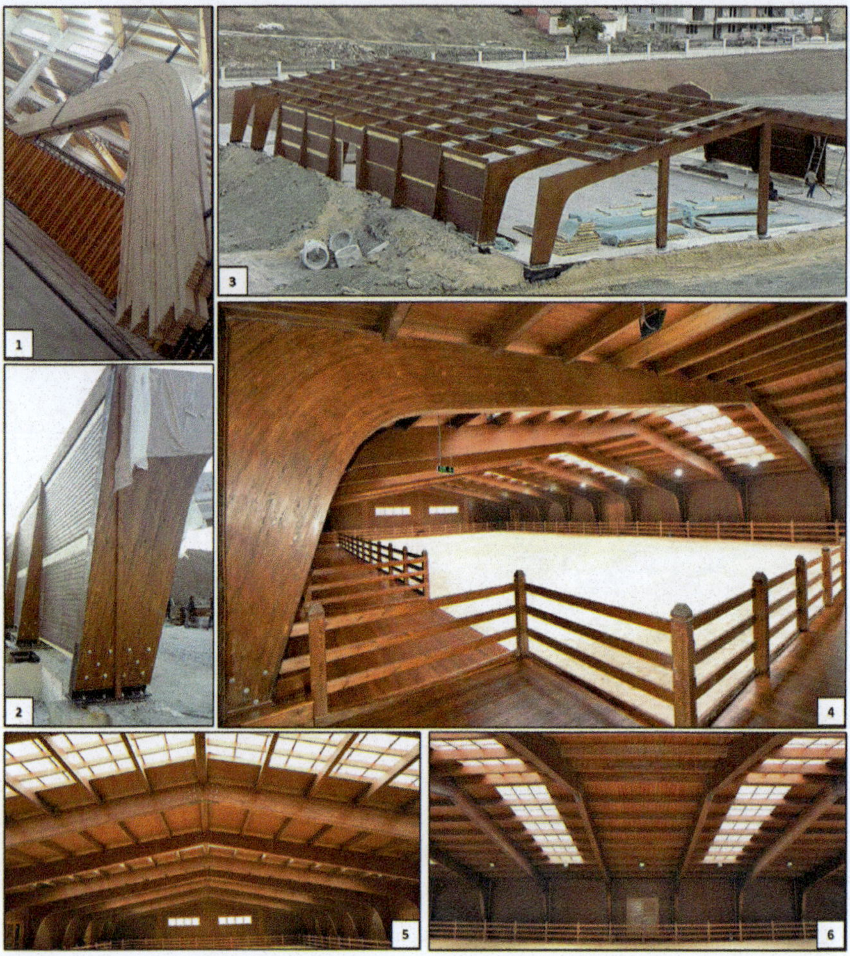

Fig. 113 A LVL timber framing application is used for wide-span covering. *Source* Naswood (2023)

- **Plates**: Thick sheets
- **Bent-Plate Sections**
- **Built-Up Sections**

4.3 Design of Steel Skeleton and Frame Structures

Steel frame structures' design resemble concrete and wood frame structures in form and utility. However, regarding its characteristics, steel behavior, function, and structural capability are differentiated from wood and concrete structures.

Fig. 114 A LVL timber framing application in curvilinear form. *Source* Naswood (2023)

4.3.1 Steel Framing

The design of steel framing is similar to concrete framing structures. Column-beam can be applied with bracing to the structures to resist lateral loads (earthquake, wind). The design of steel framing starts with the decision of framing members. Compared with columns, beam options are higher (Fig. 118).

Available steel columns can be categorized as follows:

- Universal Structural Steel Sections: I Sections, C Shapes, Pipes, Tubes (Hollow Steel Section—HSS),
- Built-up Sections: Plate built-up sections and universal structural steel section-oriented built-up sections

Universal Structural Steel Columns

In the following images, sample steel columns are provided (Fig. 119).

Fig. 115 A LVL timber framing application is used to build a pedestrian bridge. *Source* Naswood (2023)

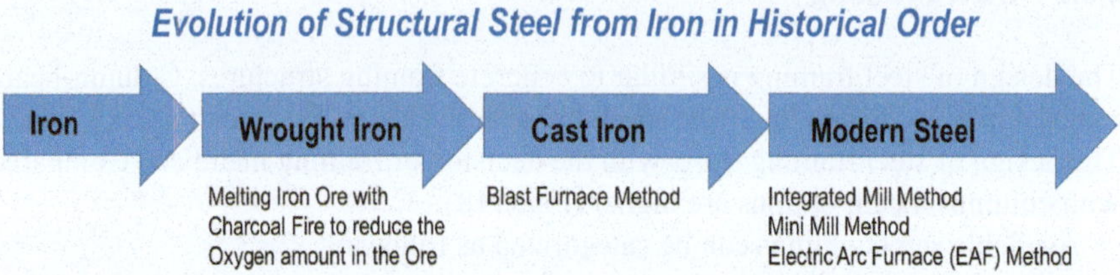

Fig. 116 Evolution of structural steel from iron in historical order

Plate-Based Built-Up Sections

See Fig. 120.

Universal Structural Steel Section-Based Built-Up Columns

- **Direct Connection** (Fig. 121)

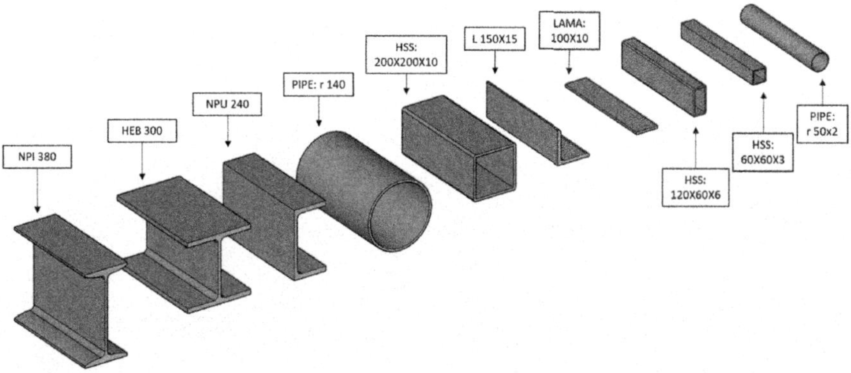

Fig. 117 Some hot rolled and cold rolled steel sections

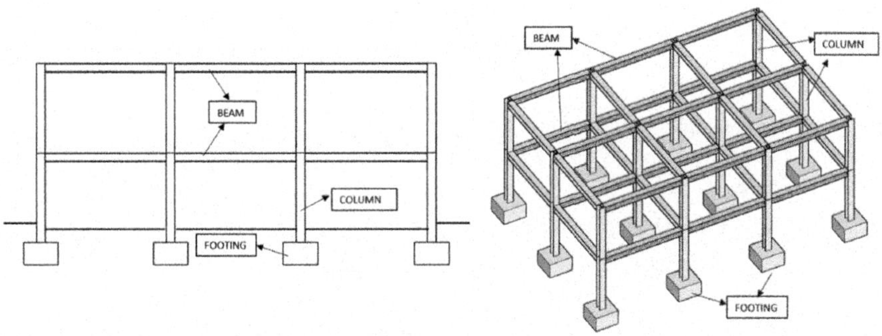

Fig. 118 Steel framing and its terminologies

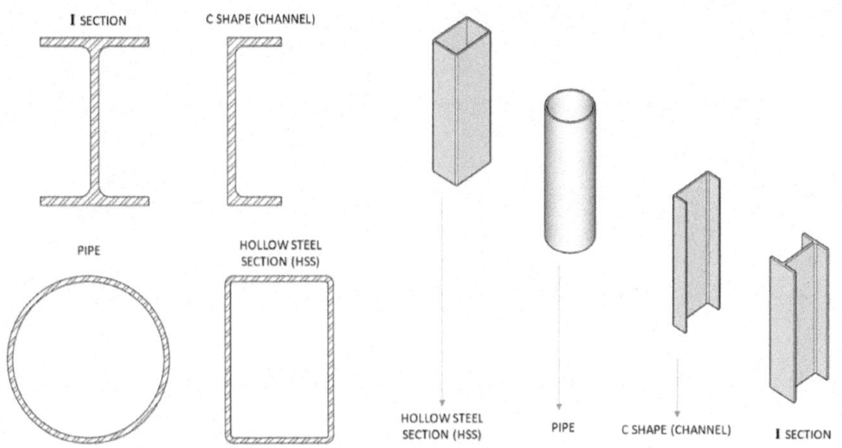

Fig. 119 Some steel sections can be used as unique columns

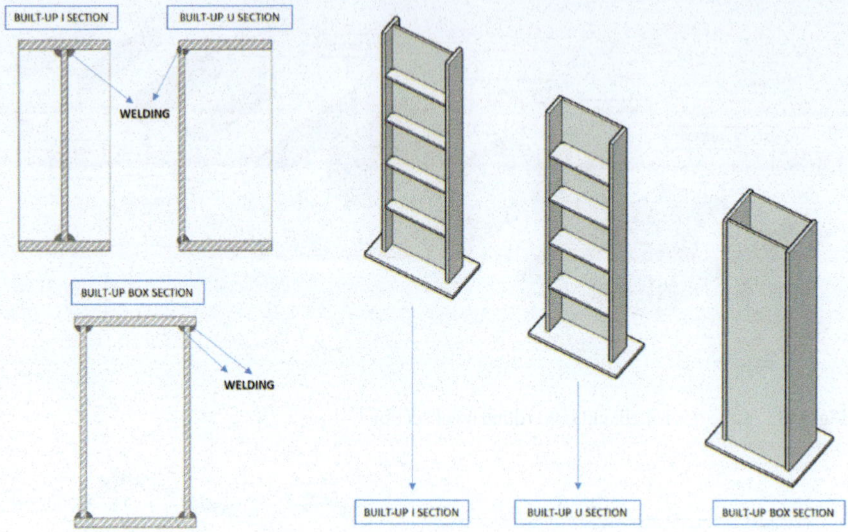

Fig. 120 Some samples for plate-based built-up sections

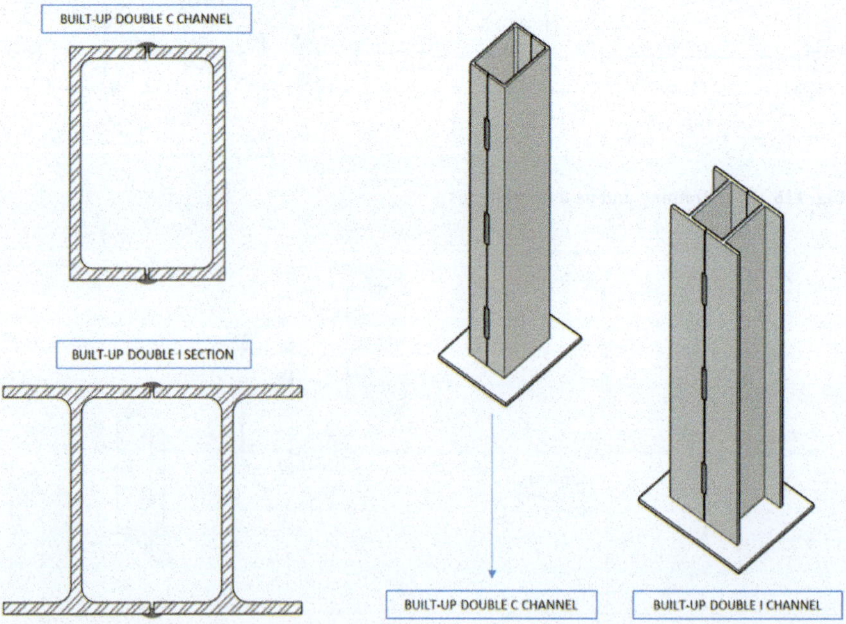

Fig. 121 Built-up columns by direct connection of universal structural steel sections

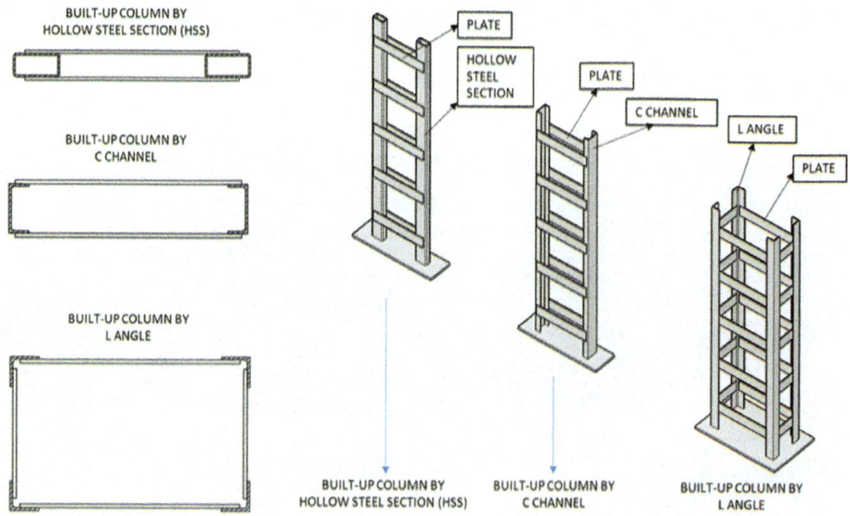

Fig. 122 Built-up columns connected with steel plates

- **Connection with Steel Plates** (Fig. 122)

4.4 *Steel Floor and Roof Structures*

4.4.1 Steel Beam, Joist Girder, and Joist

See Figs. 123, 124, 125, 126 and 127.

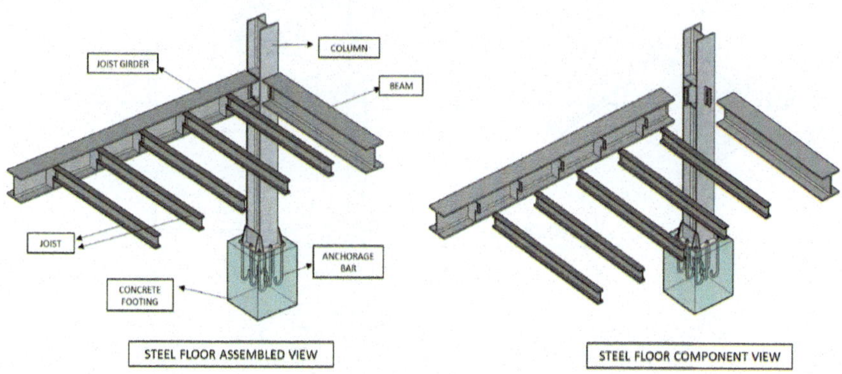

Fig. 123 The floor system consists of I section joist girder and I section joist

Fig. 124 Application of I section floor joist girder and I section floor joist. *Source* Özsarı Çelik (2023)

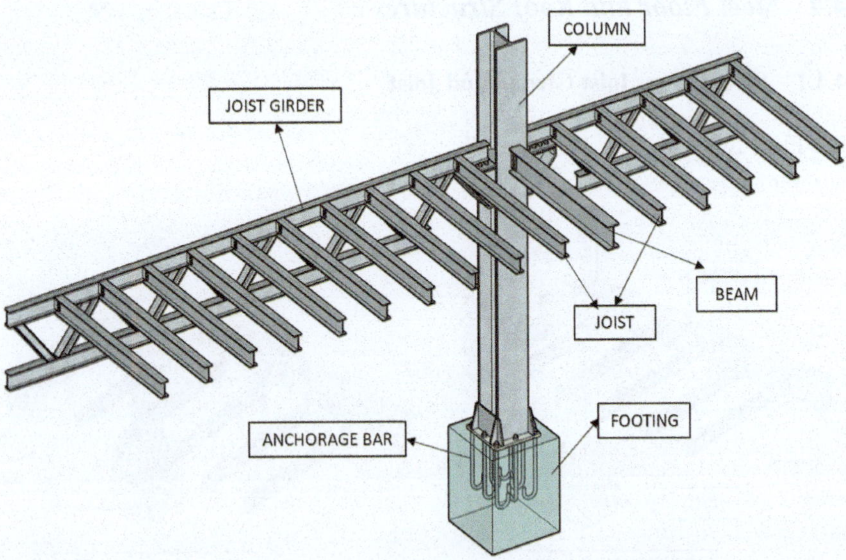

Fig. 125 The floor system consists of a truss floor joist girder and I section joist

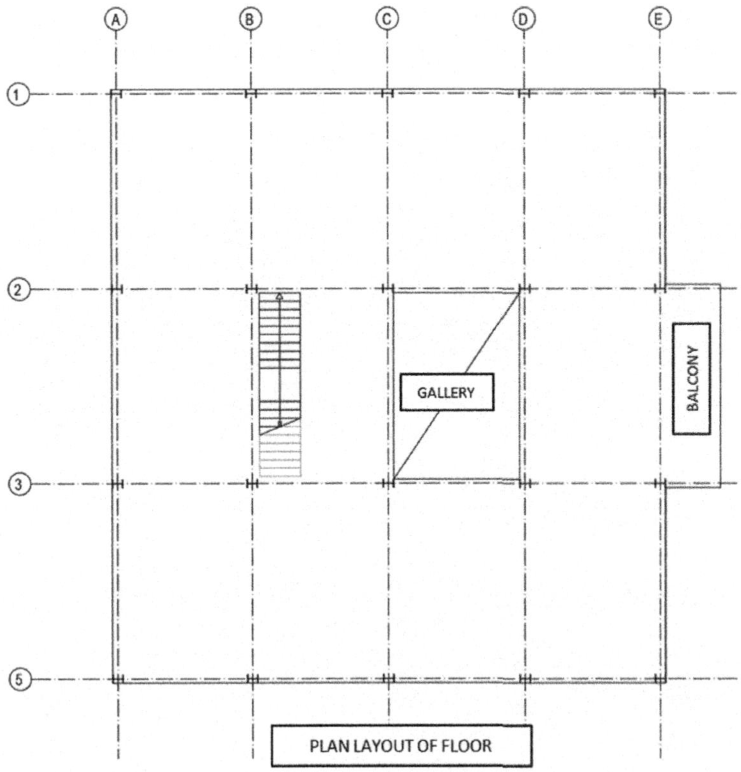

Fig. 126 Sample floor plan

4.4.2 Steel Deck

The steel deck is the flooring option for steel structures acting compositely with a concrete slab, forming a lightweight, cost-effective floor construction that does not require traditional formwork with props. Steel decking has a high strength-to-weight ratio, which decreases erection and material handling costs. It is an ideal product for multi-story buildings.

Figures 128 and 129 illustrate a sample steel deck. A galvanized corrugated metal sheet is installed above the joist and beam frame of the structure. The floor joists include shear studs and intertwine with the corrugated sheet. Welded Wire Mesh (WWM) is placed as a reinforcement of the slab. Then, the concrete is poured into the composition. At each beam and joist girder, a pour stop is placed. The pour stop presents the required flexibility to concrete. Steel is more flexible regarding vibration, shaking, and other movements in a floor slab, while concrete is not. When the pour stops are not provided, the concrete starts to crack due to the movements in the deck. Thus, the pour stops provide control joints (Fig. 130).

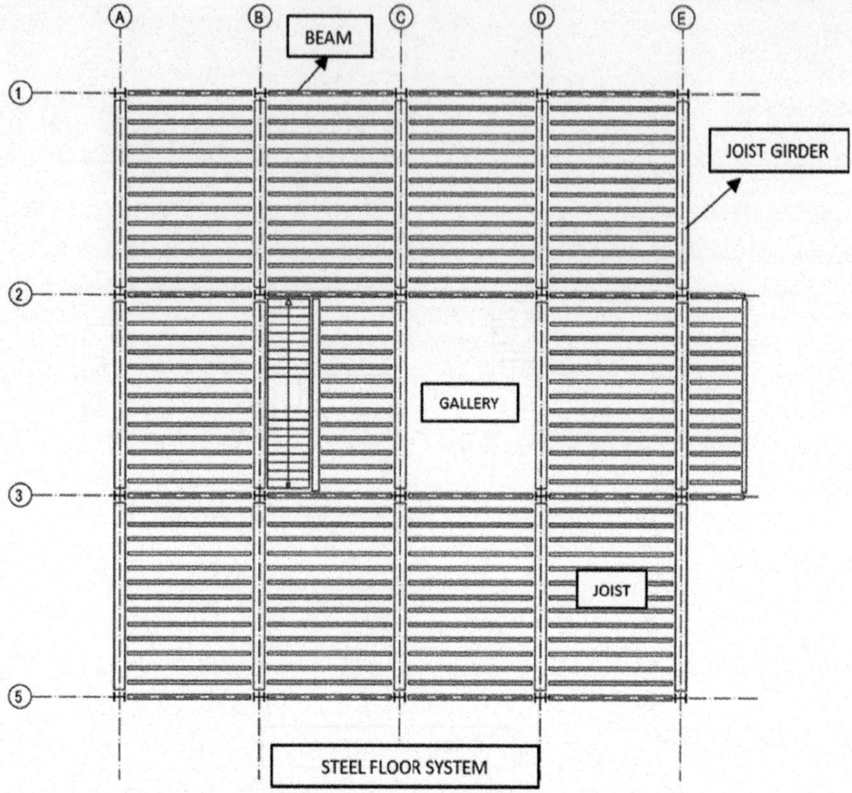

Fig. 127 Sample steel flooring system proposed for floor plan

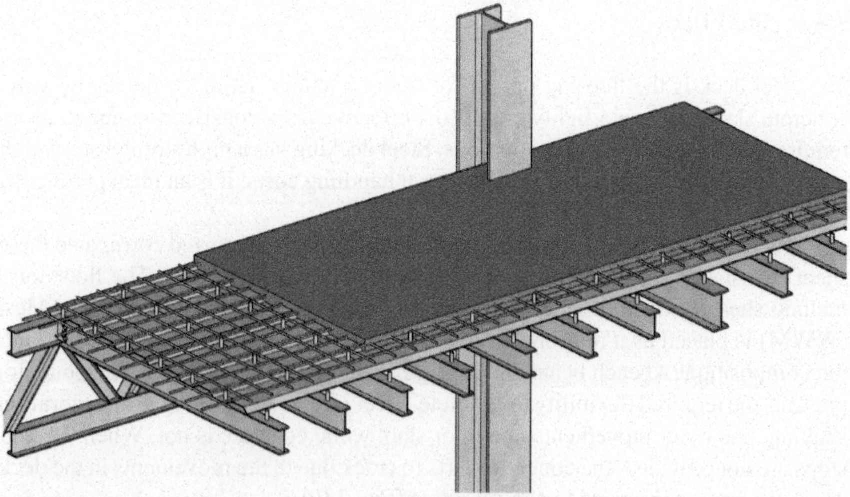

Fig. 128 Steel floor deck

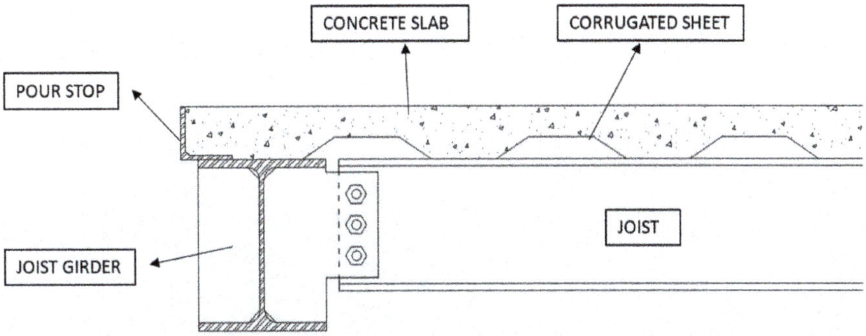

Fig. 129 Use of pour stop at the edge of the floor slab

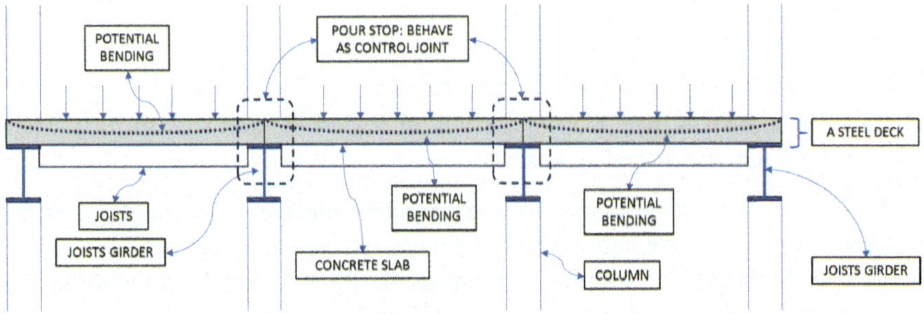

Fig. 130 The pour stop is used as a control joint in the steel decks

4.4.3 Steel Truss

Commonly used steel trusses are illustrated in Fig. 131:

4.5 Corrosion and Fire Protection of Steel

The following methods can do protective coatings to prevent corrosion of steel:

- **Acrylics**
- **Epoxies**
- **Polyurethanes**
- **Zinc (Galvanizing)**.

The fire endurance of steel is poor. Steel's yield strength and modulus of elasticity drop to nearly 60% of their original values at about 600 C—a temperature well below the temperature used in a standard fire test. The shear strength of steel is reduced at high temperatures. For beams, girders, and girders joists, this may cause disastrous

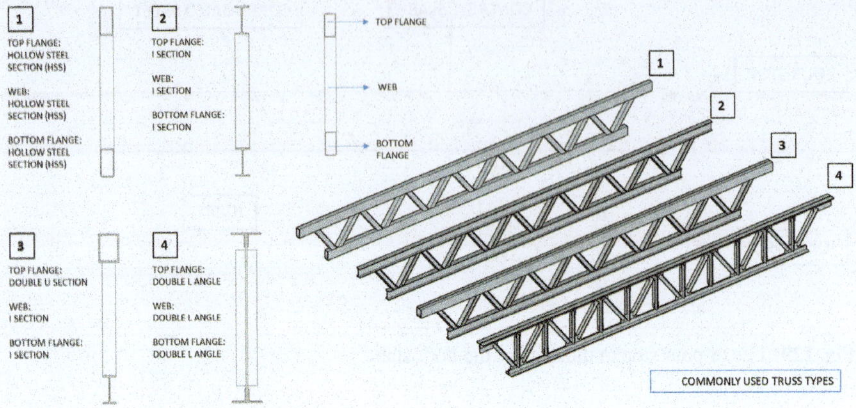

Fig. 131 Commonly used steel truss types

consequences, such as the collapse of the Twin Towers in New York in 2001. There are two ways to protect steel against fire:

- Insulate the steel component with noncombustible thermal insulation. Spray-applied fire protection and intumescent paints fall in this category.
- Encase the steel component with a non-combustible material with high thermal capacity, such as concrete gypsum board (Fig. 132).

- **Spray-Applied Fire Protection**

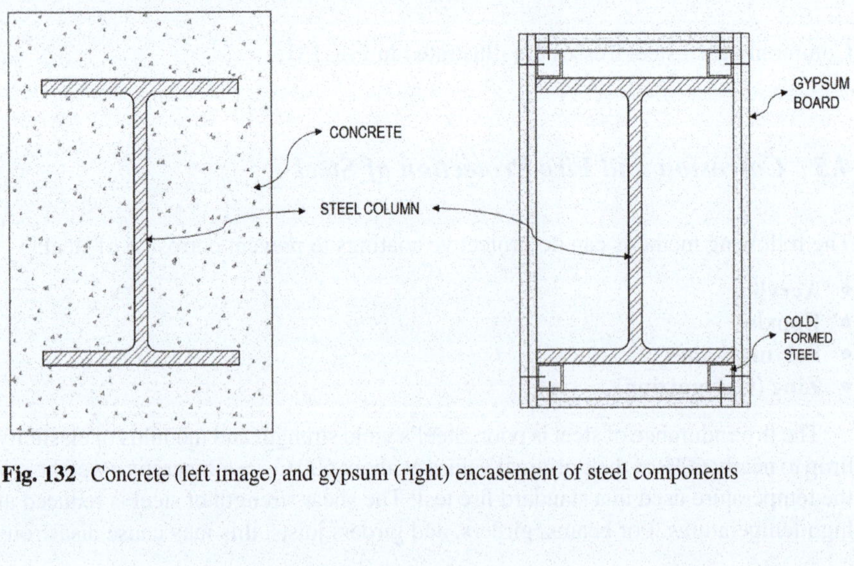

Fig. 132 Concrete (left image) and gypsum (right) encasement of steel components

Mineral fiber and binder, usually fiberglass and Portland cement. A cementitious mixture of Portland cement mixed with a lightweight aggregate such as expanded perlite or vermiculite.

- **Intumescent Paint**: It is typically 0.5 to 1.3 mm thick. It is beneficial for fires that are not longer than 2 h.
- **Suspended Ceiling**: Fire-proof materials in the suspended ceiling prevent the fire from accessing structural steel members.

4.6 Bolts and Welds

Comparison of Bolting versus Welding as connection methods are presented as follows:

- Bolts necessitate extra connection gusset plates.
- Welding has a much larger range of applicability than bolding.
- Welding requires controlled conditions of a shop.
- Welding is not preferable in site conditions; therefore, a bolt connection is applied.
- Welding requires a dry, non-dirty, and non-greased surface.
- Bolting is rapid in action and involves less skilled labor.
- Bolting enables prefabrication and montage of steel components manufactured in ateliers and then montage in construction sites.

4.7 Connections Between Framing Members

4.7.1 Column to Beam Connections

See Figs. 133 and 134, and 135.

4.7.2 Beam-To-Beam Connection

4.7.3 Joist Girder to Joist Connection

See Figs. 136, 137, 138, 139, 140, 141, 142 and 143.

4.7.4 Column to Column Connection

See Fig. 144.

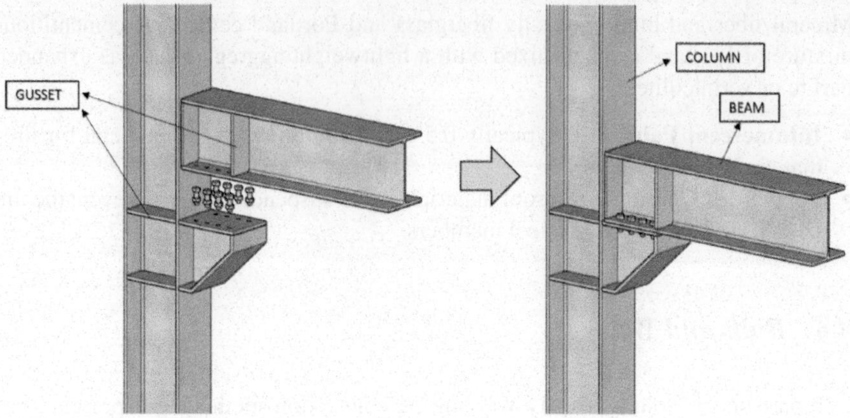

Fig. 133 The beam is horizontally connected to the column

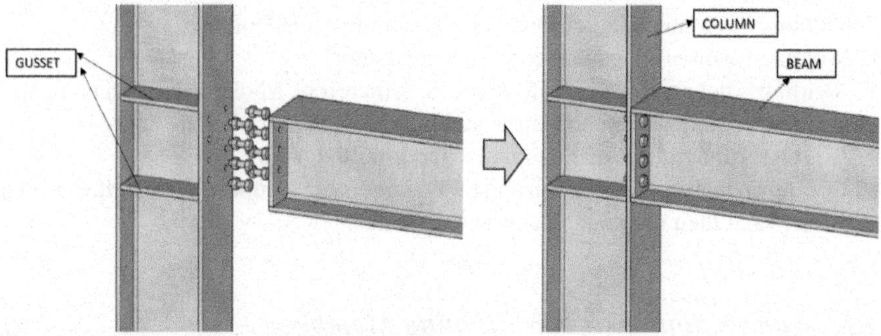

Fig. 134 The beam is vertically connected to the column

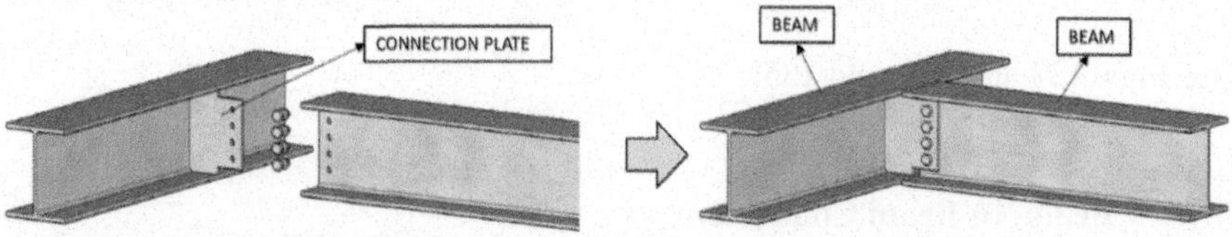

Fig. 135 Beam-to-beam connections

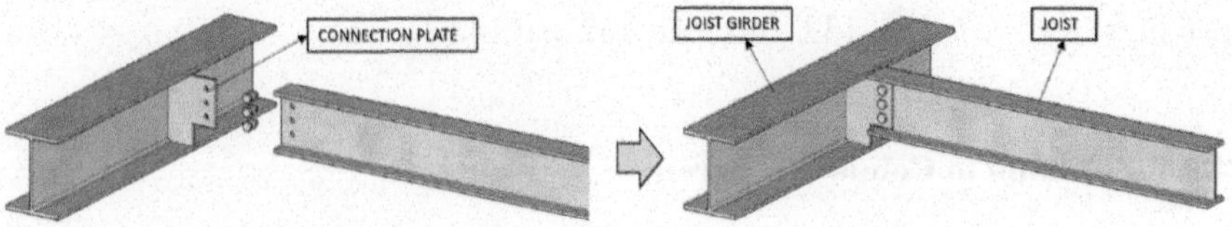

Fig. 136 Floor joist girder to joist connection

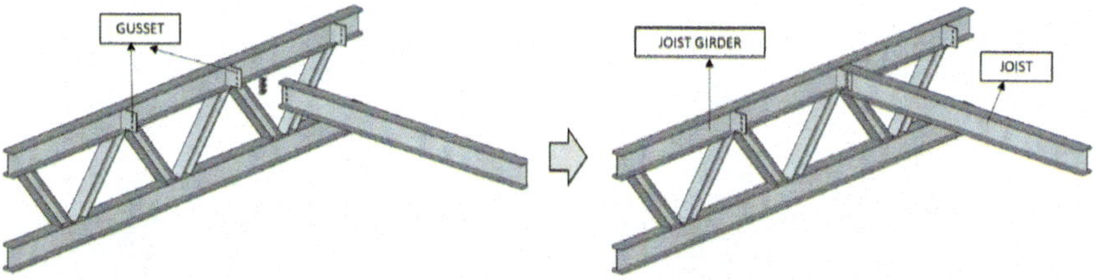

Fig. 137 Truss type floor joist girder to joist connection

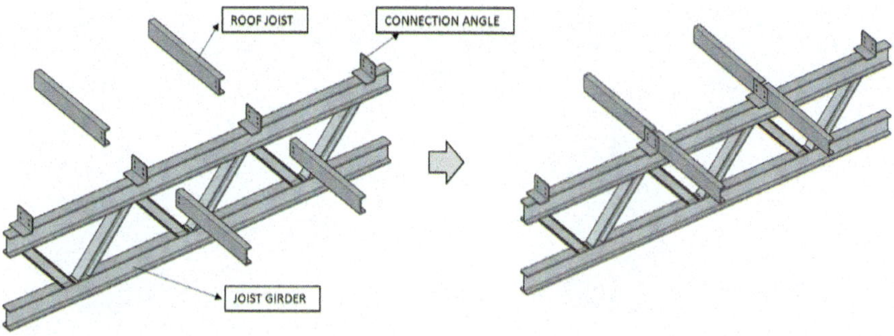

Fig. 138 Truss type roof joist girder to joist connection

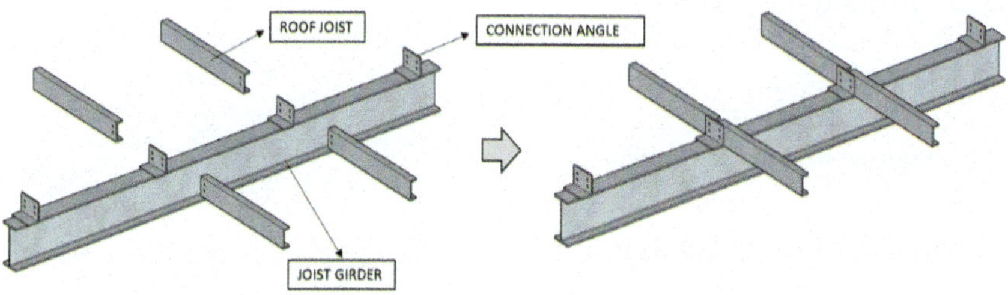

Fig. 139 Roof joist girder to joist connection

4.8 Bracing

See Figs. 145, 146 and 147.

4.8.1 Bracing in Tall Buildings

See Fig. 148.

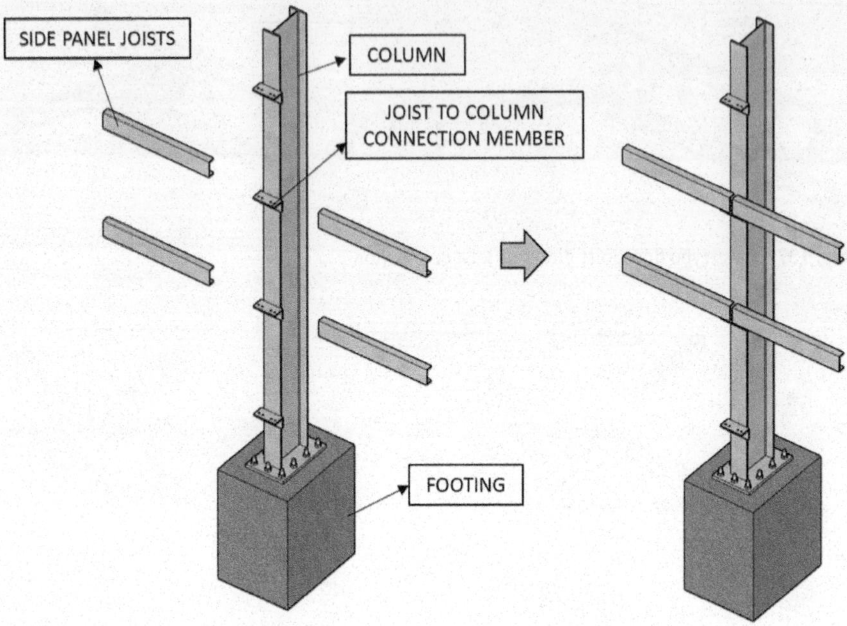

Fig. 140 Wall joist to column connection

4.8.2 Bracing Connections

L Angle

- Column-to-Bracing (Fig. 149)
- Bracing-to-Bracing (Fig. 150)

Pipe and Hollow Steel Section (Hss)

- **Column-to-Bracing** (Fig. 151)
- **Bracing-to-Bracing** (Fig. 152)

Fig. 141 Roof joists, wall joists, and rainwater gutter application. *Source* Özsarı Çelik (2023)

4.9 Structural Design Considerations of Steel

4.9.1 Span and Interval

See Fig. 153.

Fig. 142 Steel roof components

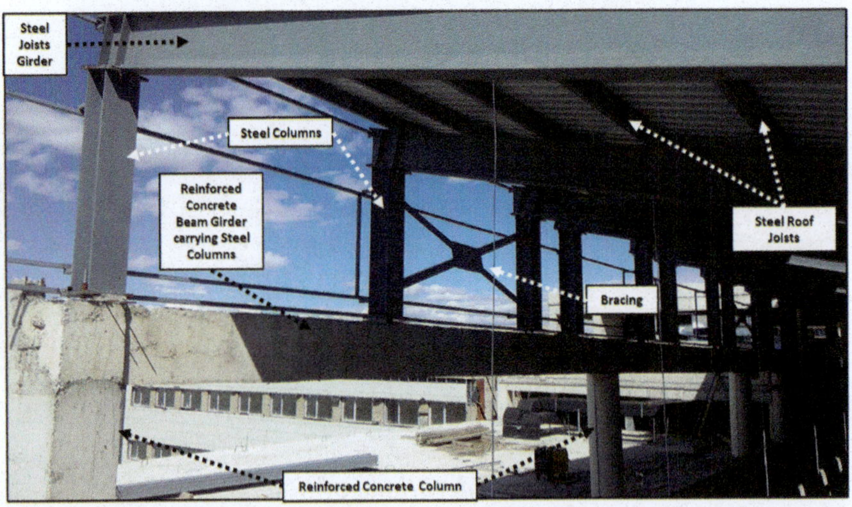

Fig. 143 Since steel is lighter than concrete, multiple steel columns could be carried by a single reinforced concrete beam

4.9.2 Span and Interval Limits for Structural Steel Members

See Figs. 154, 155, 156 and 157

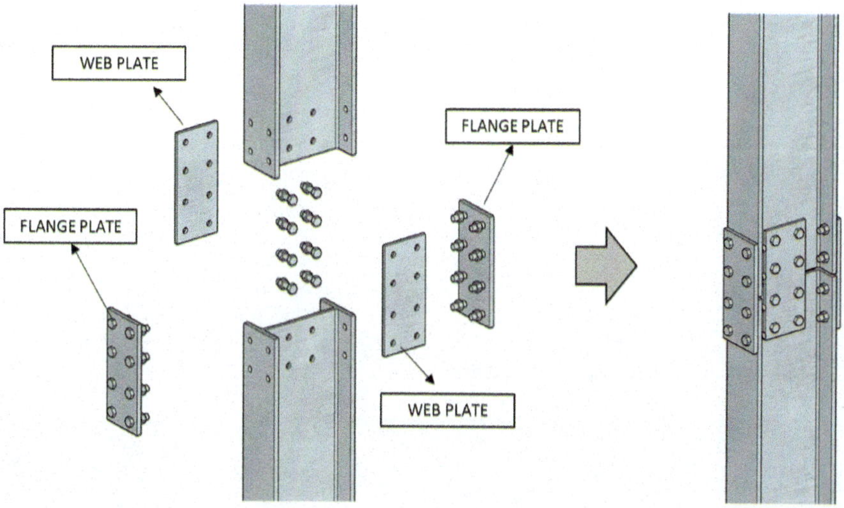

Fig. 144 Column-to-column connection

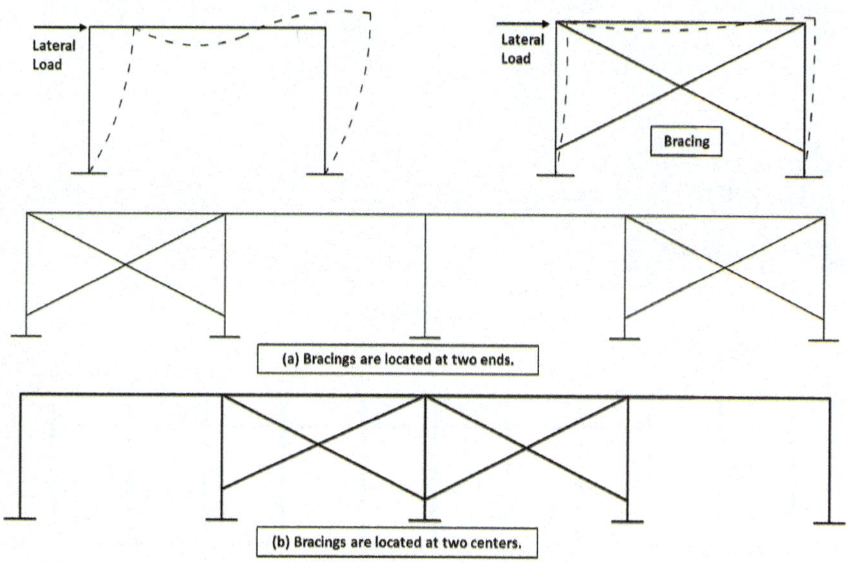

Fig. 145 Bracing allocation formations

References

Eren, Ö. (2014). Büyük Açıklıklı Çelik Yapılar. Arı Sanat.

Özsarı Çelik, located in Antalya, all images were retrieved in 2023 with the company's founder's permission.

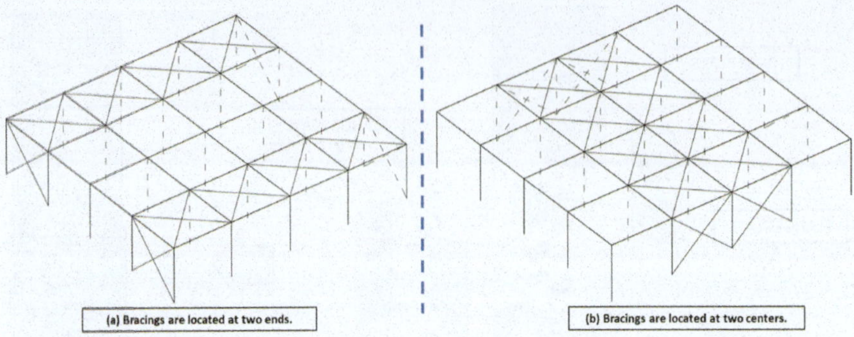

Fig. 146 Bracing for low-rise steel buildings

Fig. 147 Bracing application. *Source* Özsarı Çelik (2023)

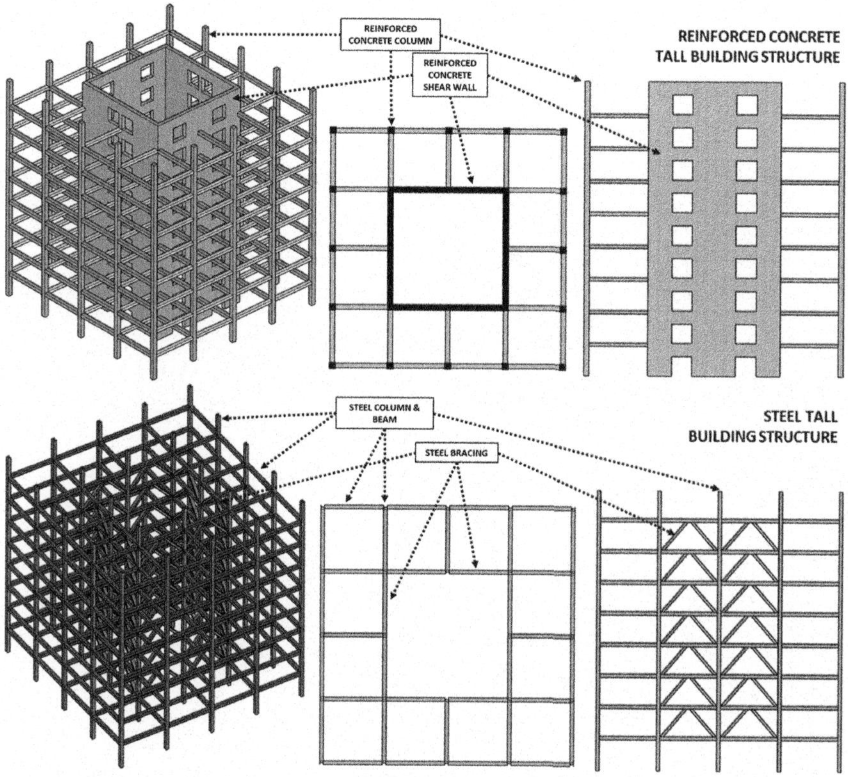

Fig. 148 Steel bracing application in the structural core behaves as a reinforced concrete shear wall for tall buildings

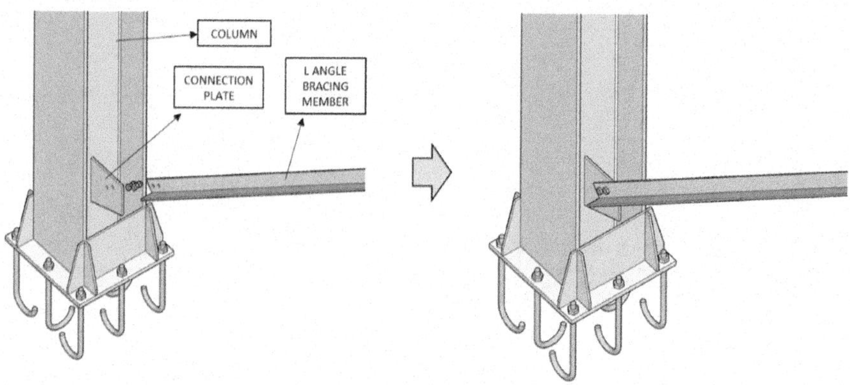

Fig. 149 Column to L angle bracing connection detail

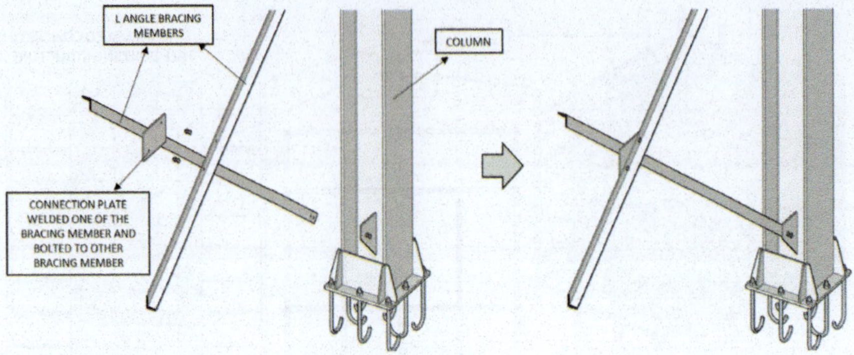

Fig. 150 Bracing-to-bracing connection

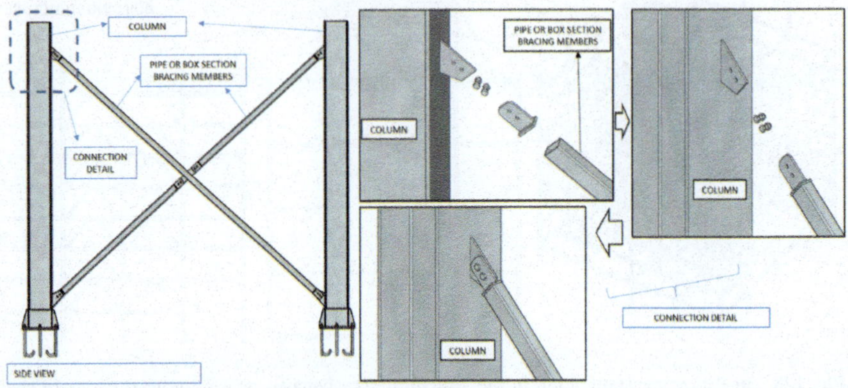

Fig. 151 Column to pipe or hollow steel section connection

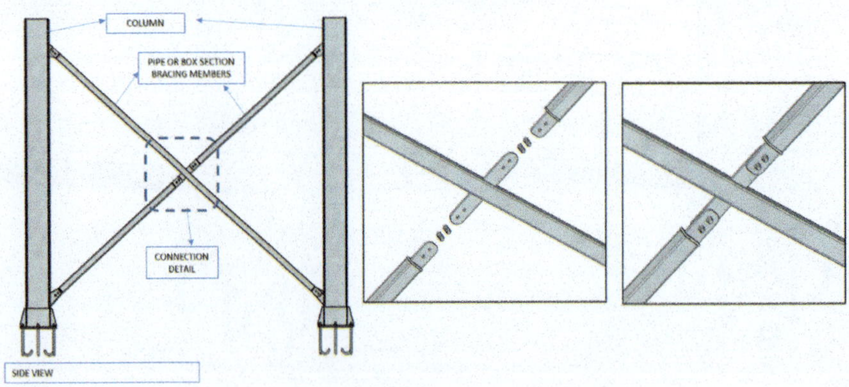

Fig. 152 Bracing to bracing connection for pipe or hollow steel section

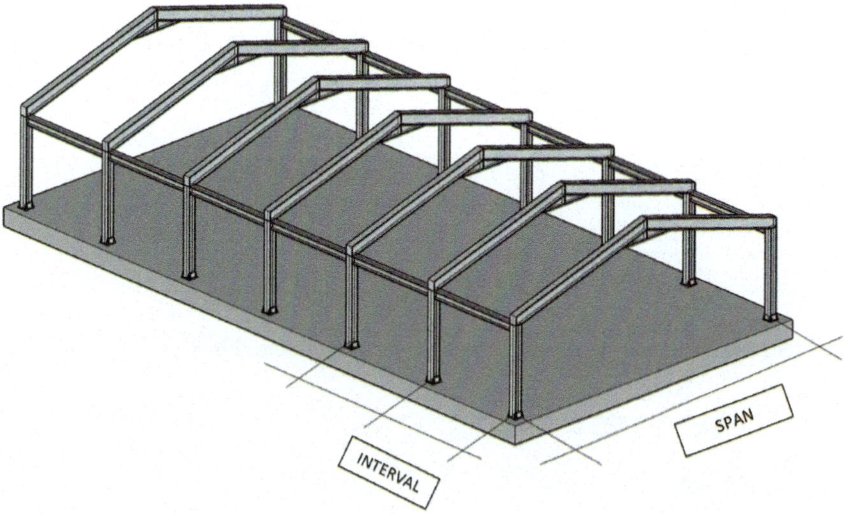

Fig. 153 Span and interval for low-rise steel building

MEMBER	HEIGHT (m)	SPAN (ratio)*
Hot Rolled Steel Sections	2-8: Single Floor 2-4: Multiple Floor	20-25 7-18
Built-Up Columns	4-10	20-25
Steel and Concrete Mix Columns	2-4	6-15

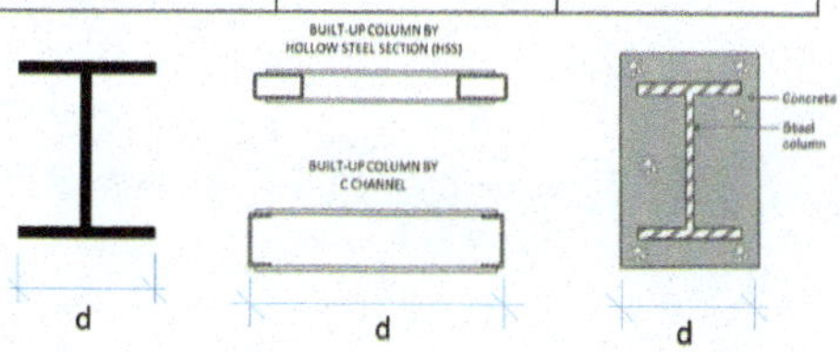

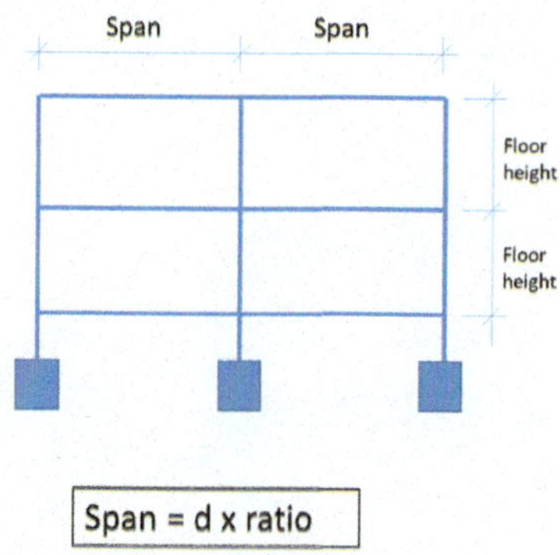

Fig. 154 Span limits for steel frame systems. *Source* Eren (2014)

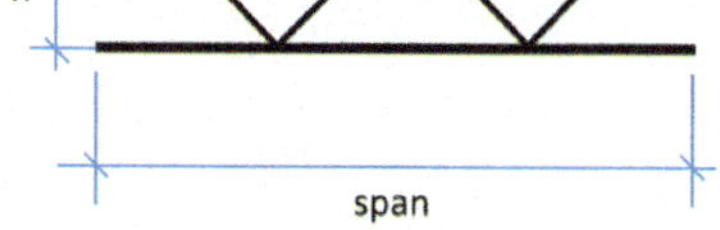

SPAN (m)	HEIGHT OF THE TRUSS (mm)	
	STRAIGHT TRUSS	SPACE TRUSS
10	1000	1000
15	1200	1200
20	1400	1400
30	1800	1600
40	2500	2200
50	3000	2800
60	4000	3800
70	5000	4800

Fig. 155 Span and height limits for steel truss systems. *Source* Eren (2014)

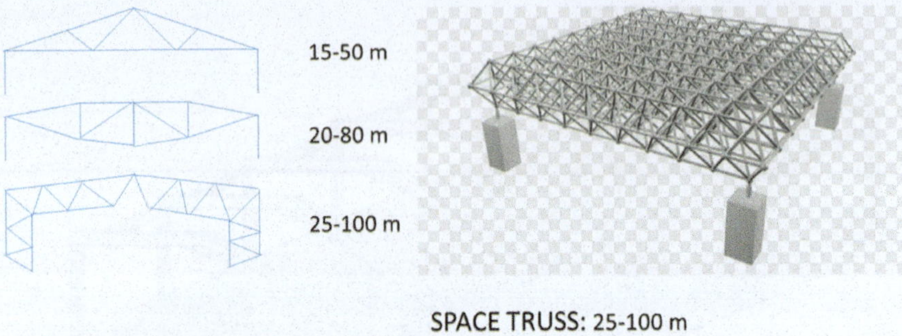

Fig. 156 Span limits for special types of roof trusses. *Source* Eren (2014)

Fig. 157 A space truss application in a construction site

5 Pedestrian Building Circulation Systems

5.1 *Stairs*

5.1.1 Stair Fundamentals

- **Tread**: Horizontal surface on which one walks
- **Riser**: Vertical component separates one tread from another
- **Stair Width**: Length of tread at stair (Fig. 158)

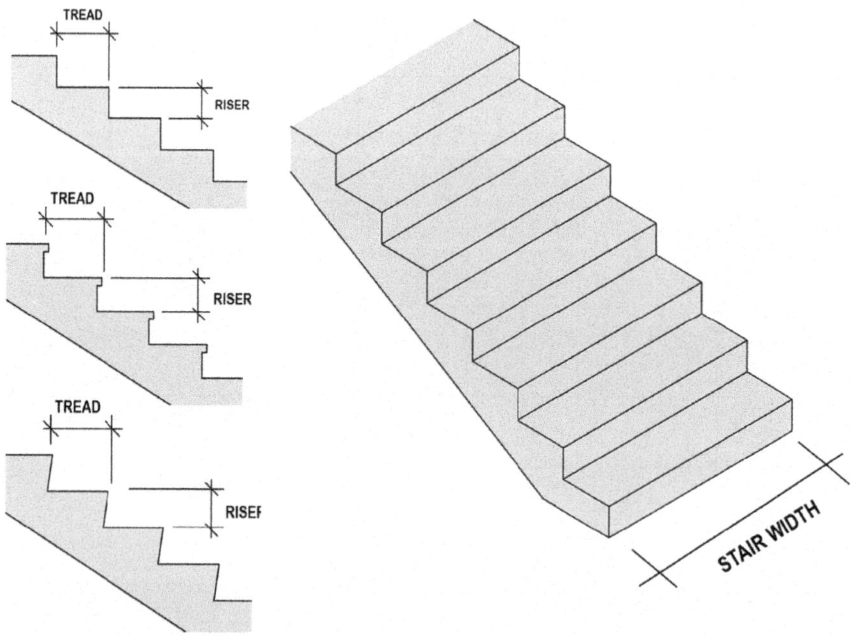

Fig. 158 Stair terminology

5.1.2 Stair Types

See Fig. 159.

5.1.3 Stair Design

See Figs. 160, 161, 162, 163, 164, 165, 166, 167, 168, 169, 170, 171, 172, 173, 174, 175, 176, 177, 178, 179, 180, 181, 182, 183, 184, 185, 186, 187 and 188.

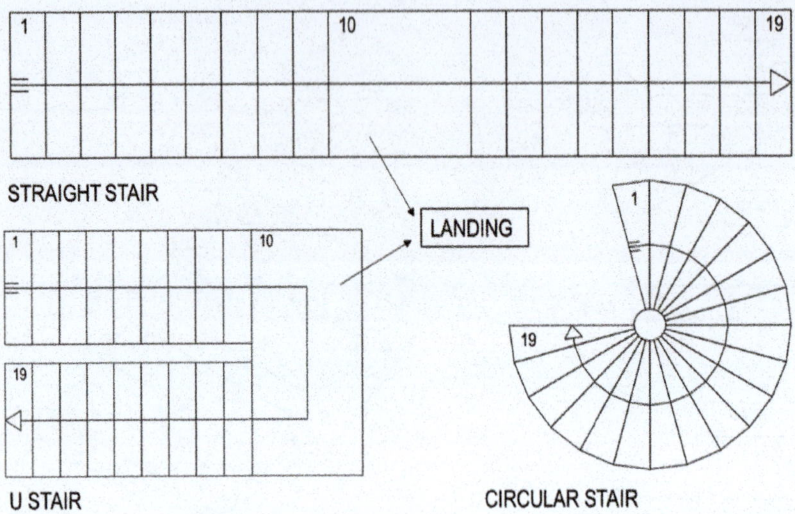

Fig. 159 Plan view of straight, U, and circular stair

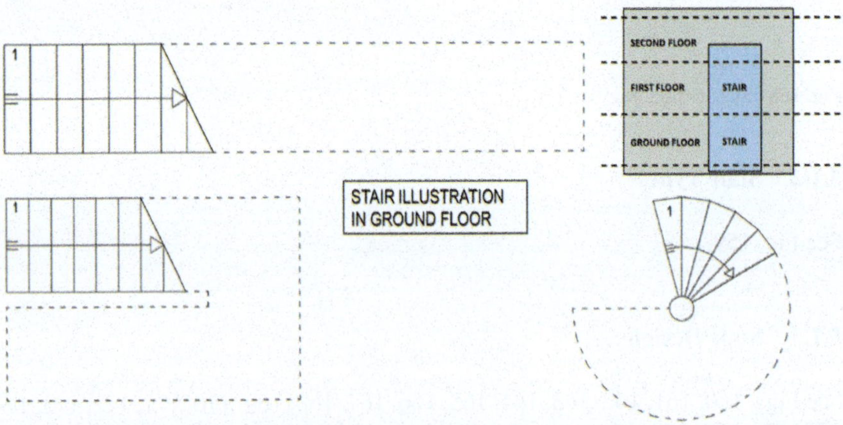

Fig. 160 Stair illustration at the ground (starting) floor

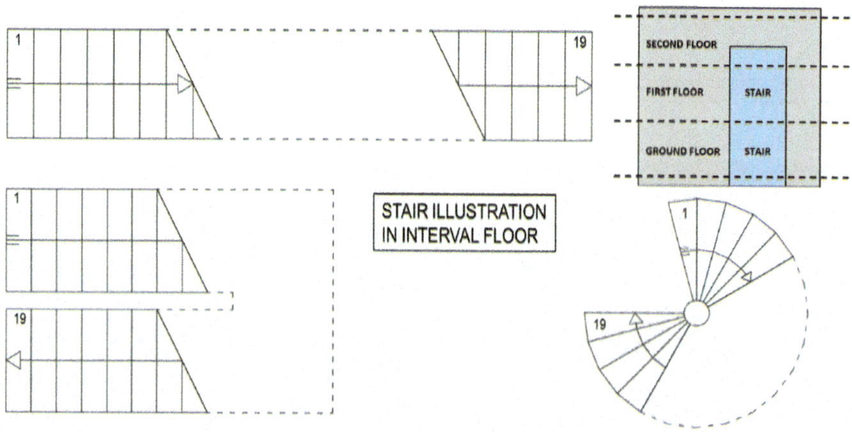

Fig. 161 Stair illustration at interval floor

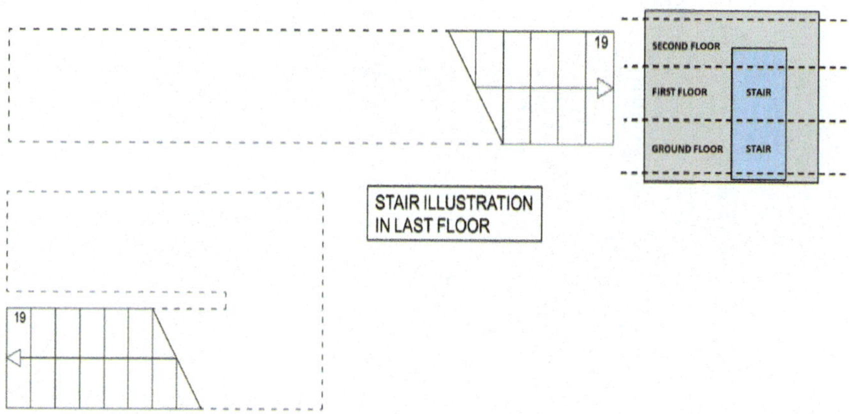

Fig. 162 Stair illustration on the last floor

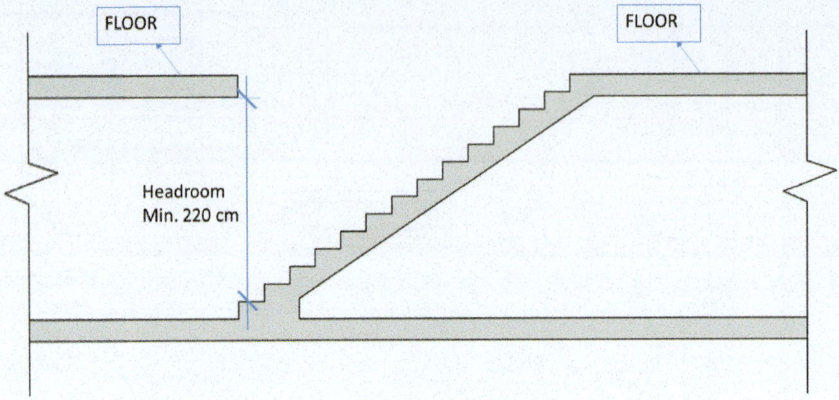

Fig. 163 Headroom in a stair. Min. headroom is 220 cm

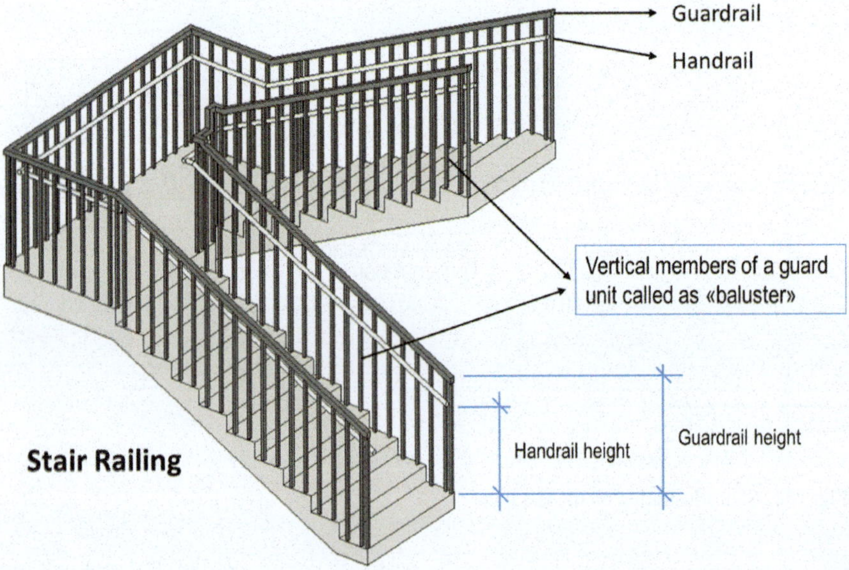

Fig. 164 Stair railing and its terminology

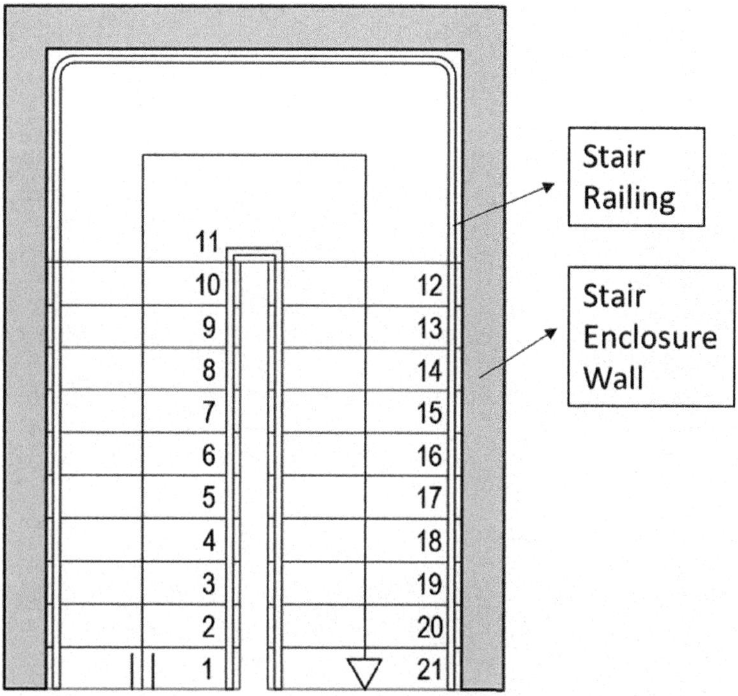

Fig. 165 Plan view of a stair with railing

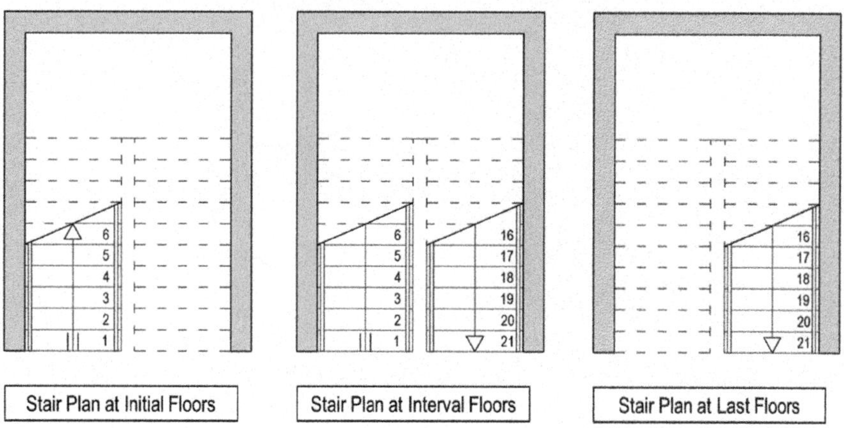

Fig. 166 Plan views of stairs at the starting, interval and last floor

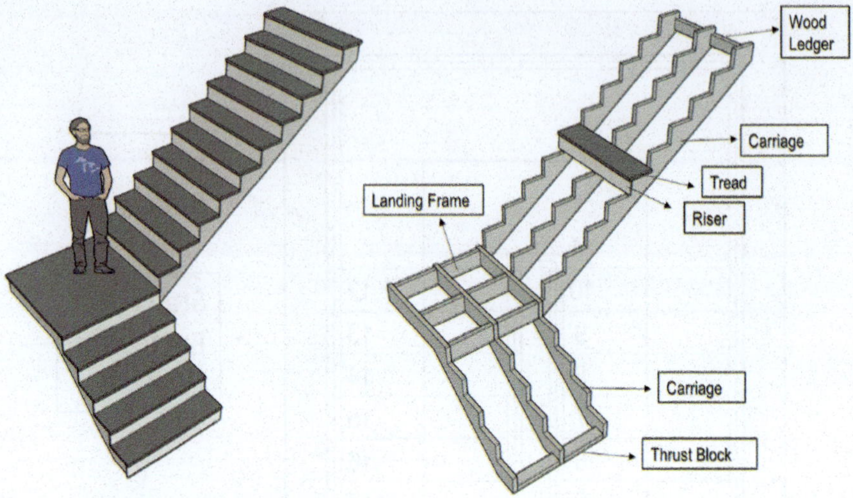

Fig. 167 Wood stair components

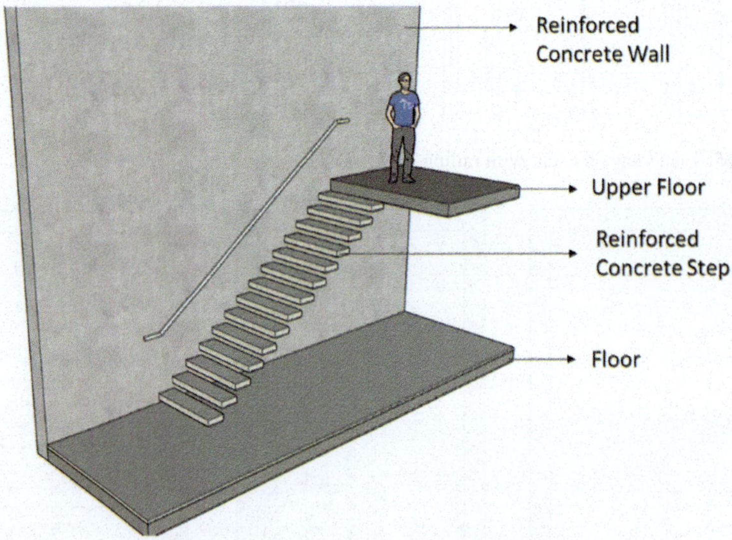

Fig. 168 Reinforced concrete stair attached to a structural wall

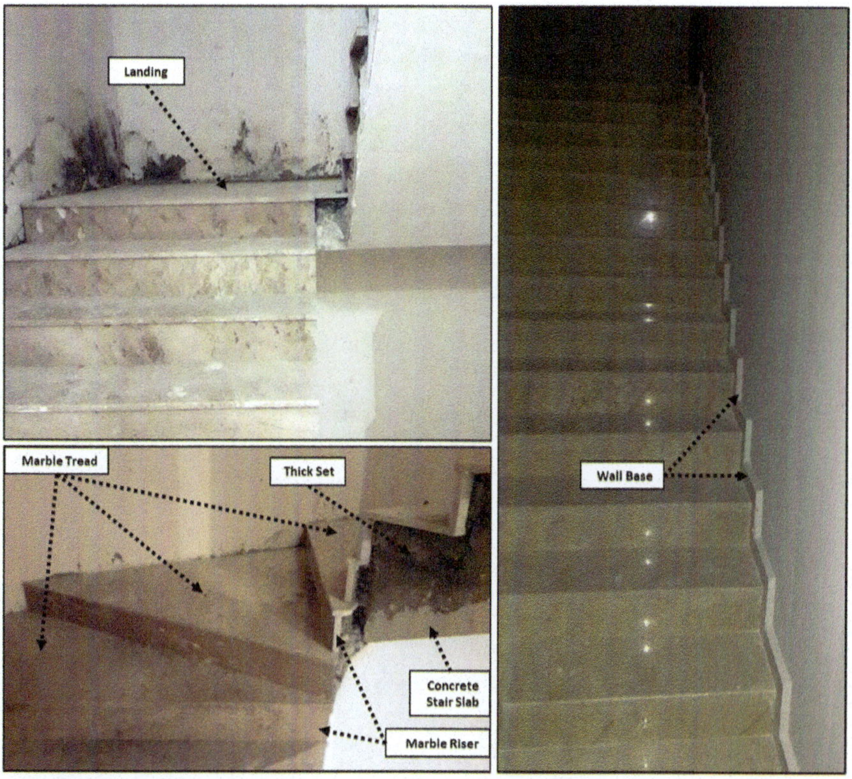

Fig. 169 Marble-covered reinforced concrete stair applications

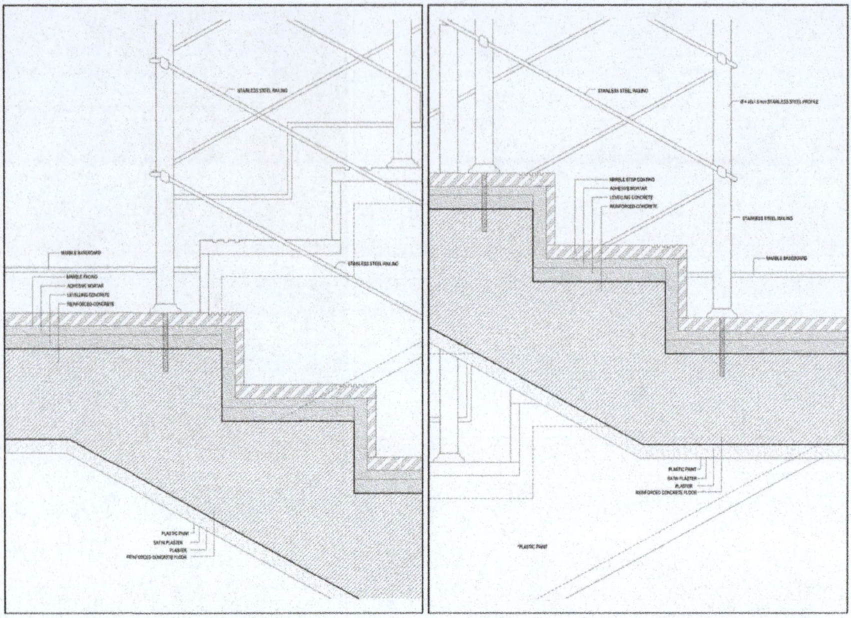

Fig. 170 Marble stair covering detail

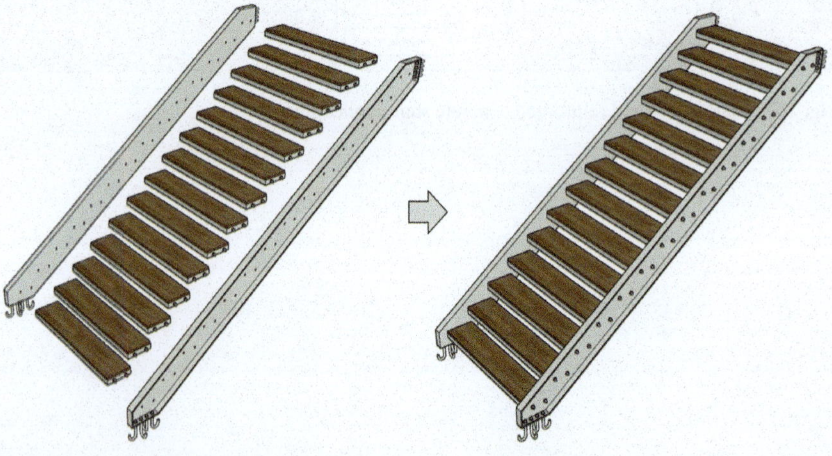

Fig. 171 Components of a steel stair

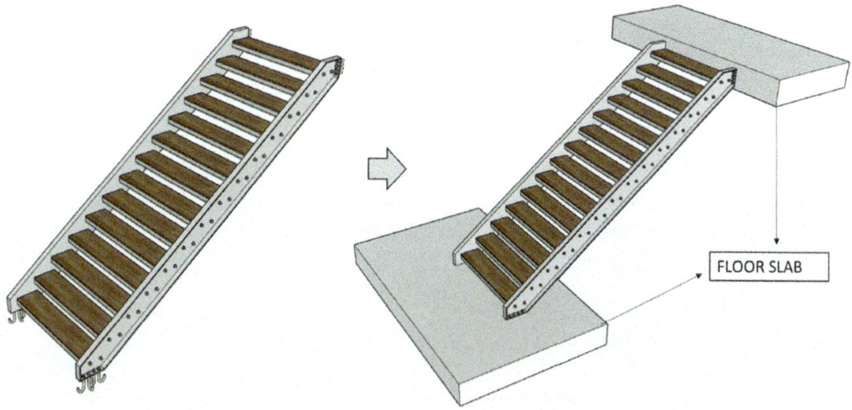

Fig. 172 Connection of steel stair to floor slabs

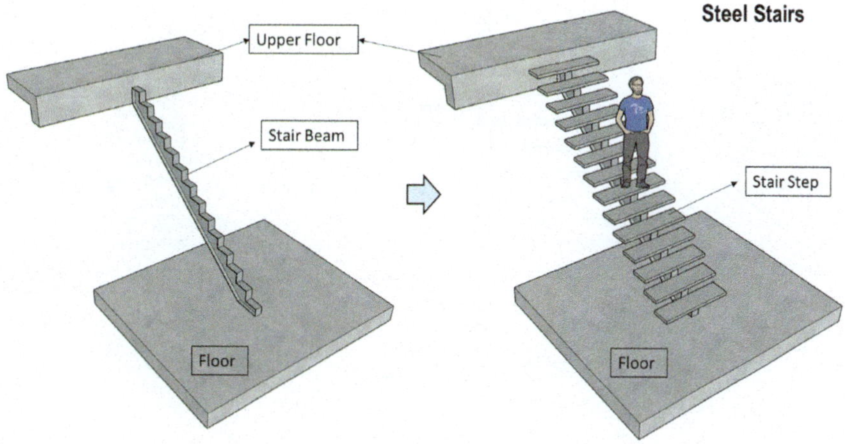

Fig. 173 Components and connection of steel stair to floor slab

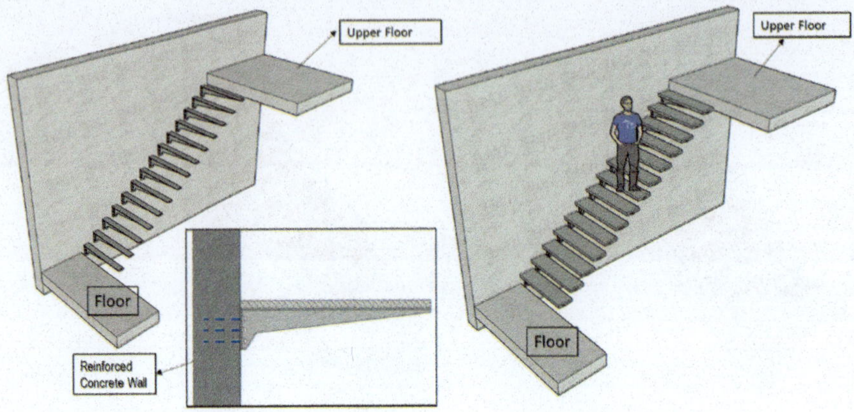

Fig. 174 Steel stair attached to the structural wall

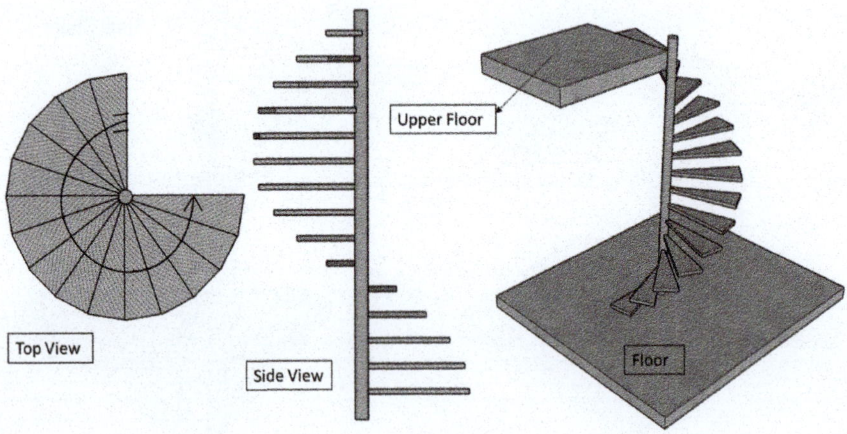

Fig. 175 Steel circular stair

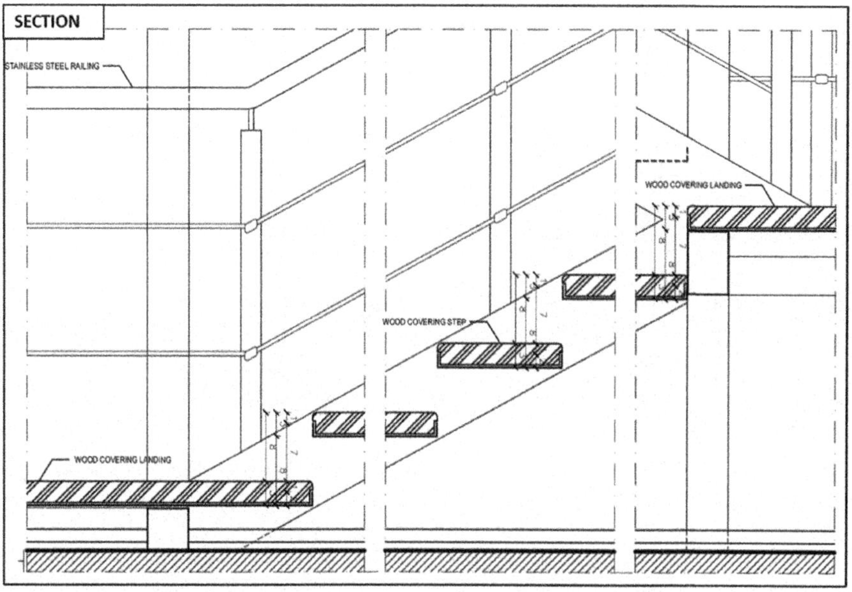

Fig. 176 A detailed drawing to illustrate steel structured stair with wood steps and landing

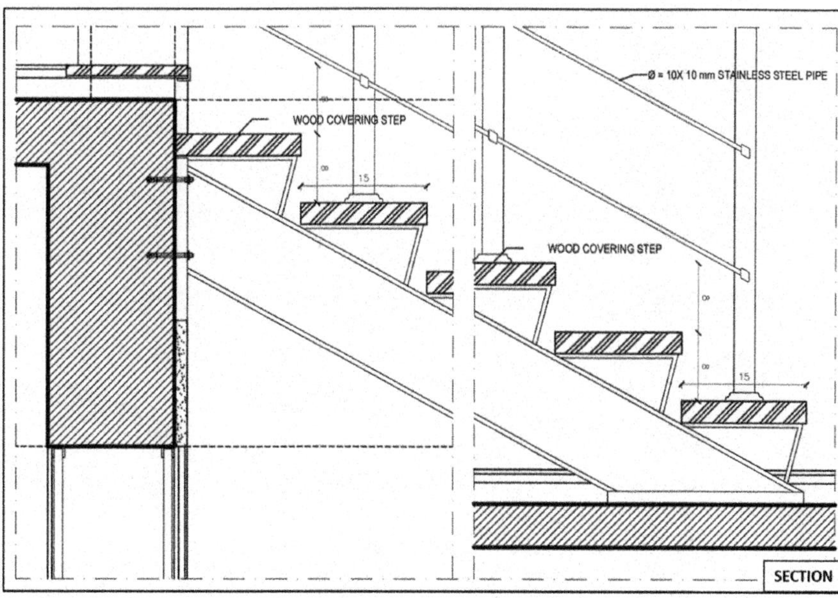

Fig. 177 A detailed drawing to represent a steel-structured stair with wood steps

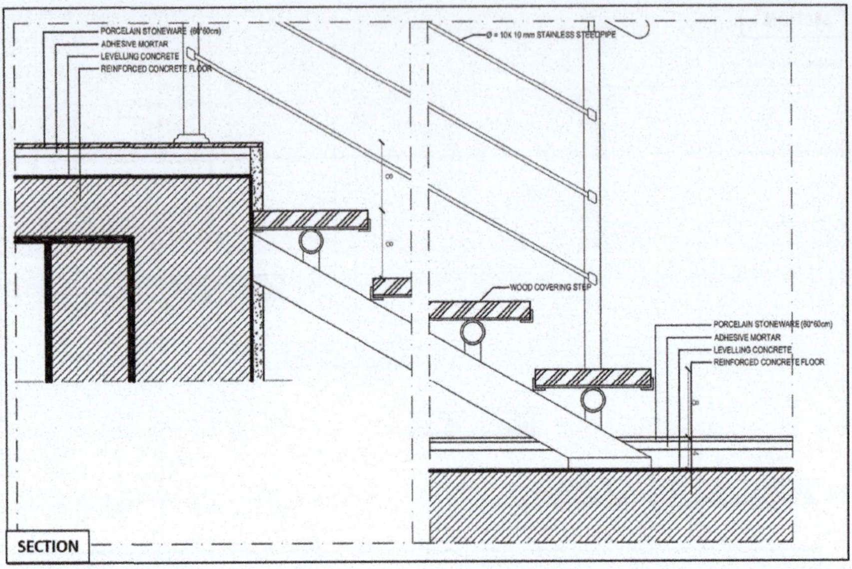

Fig. 178 A detailed drawing to demonstrate steel-structured stairs, wood steps, and ceramic floor coverings

Fig. 179 U-shaped steel structured stair application. The steps were covered with laminated wood. The railings were made of stainless steel profiles and attached to the steps

Fig. 180 A straight stair application

6 Ramps

6.1 Ramp Standards and Terminologies

See Figs. 189, 190, 191, 192 and 193.

6.2 Ramp Structures

See Figs. 194, 195, 196 and 197.

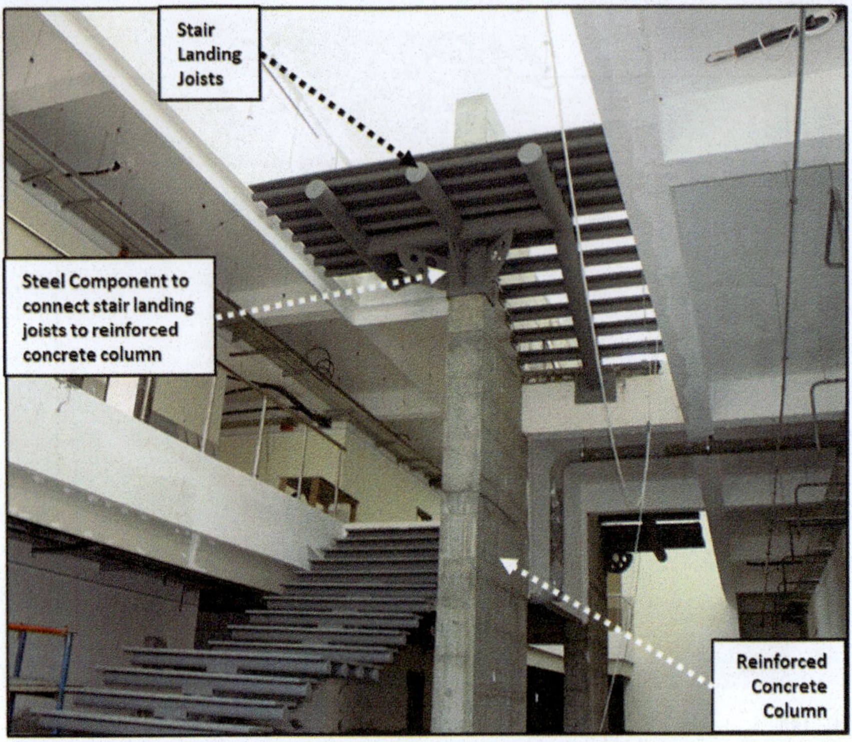

Fig. 181 A steel-structured stair was attached to a reinforced concrete column

7 Lift/Elevator

7.1 Lift/Elevator Terminology

Lift and elevator are synonyms. The lift system consists of three components: structural system, cabinet, and weight. Lift is designed regarding these three components. Lift is movable and operable in the vertical direction. Thus, it must have a rigid and stable structural system. A steel column-beam framing or reinforced concrete shear wall system is used. The minimum shear wall width for the reinforced concrete shear wall system is 30 cm. Space design must be provided for both cabinet and weight. There are no standard dimensions for the cabinet design. However, there are certain limitations regarding the countries building codes and regulations. In Turkey, for the unique lifts, the shortest side of the cabinet cannot be less than 1.20 m, the area of the cabinet cannot be less than 1.80 m^2, and the opening of this cabinet cannot be less than 0.90 m. In Turkey, the minimum landing width cannot be less than 1.20 m for the landings that lift doors open if the door is sliding. The minimum landing width

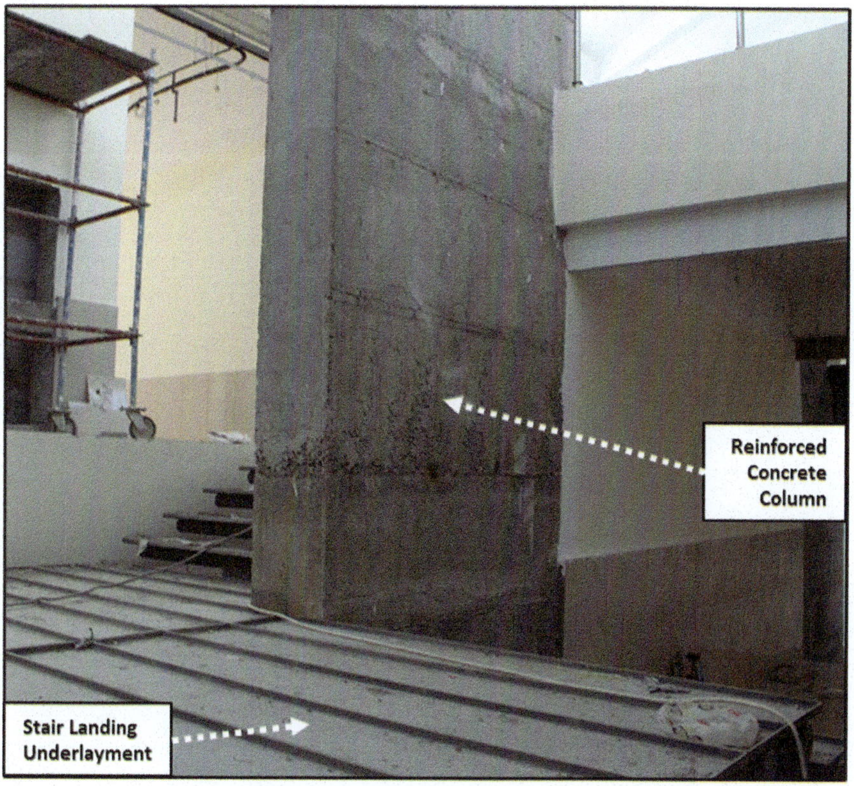

Fig. 182 The steel-structured stair was attached to a reinforced concrete column

cannot exceed 1.5 m if the doors open to the landing area. Manufacturers can provide the required space dimensions for the weight of the lift (Fig. 198).

7.2 Lift/Elevator Structures

See Figs. 199, 200, 201, 202.

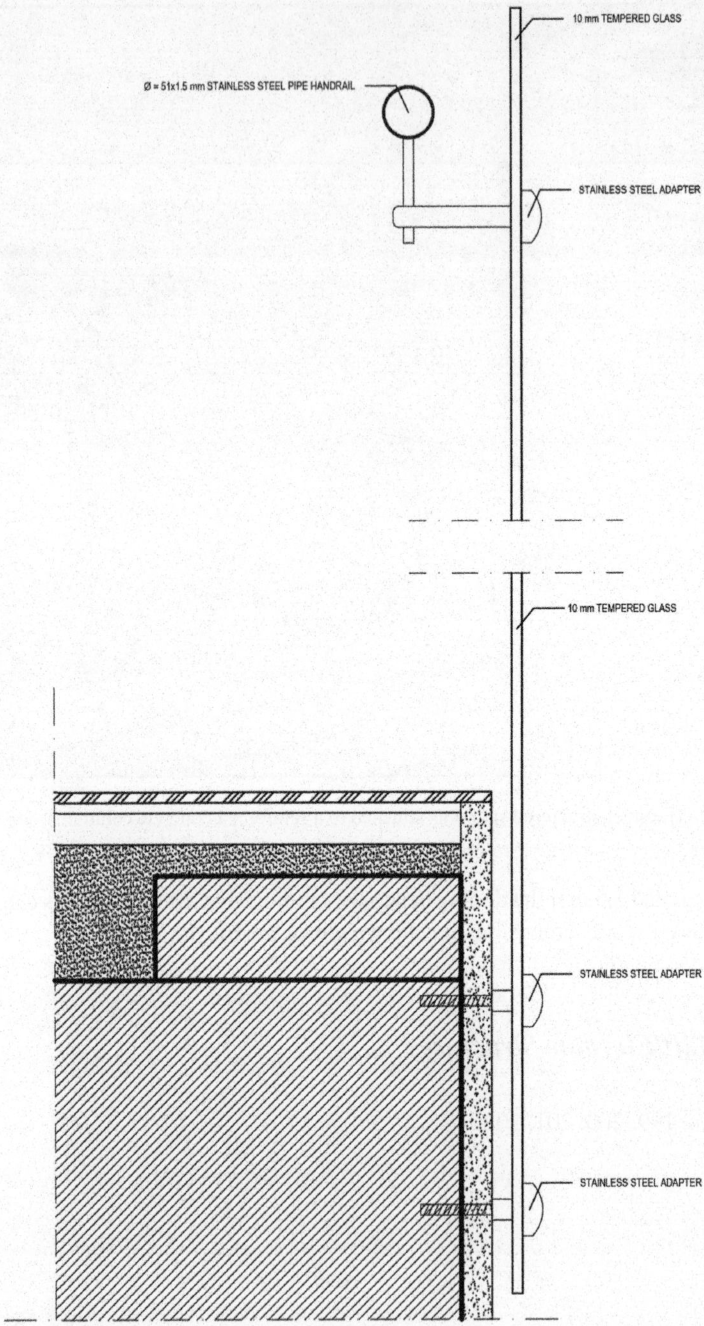

Fig. 183 A glass railing detail

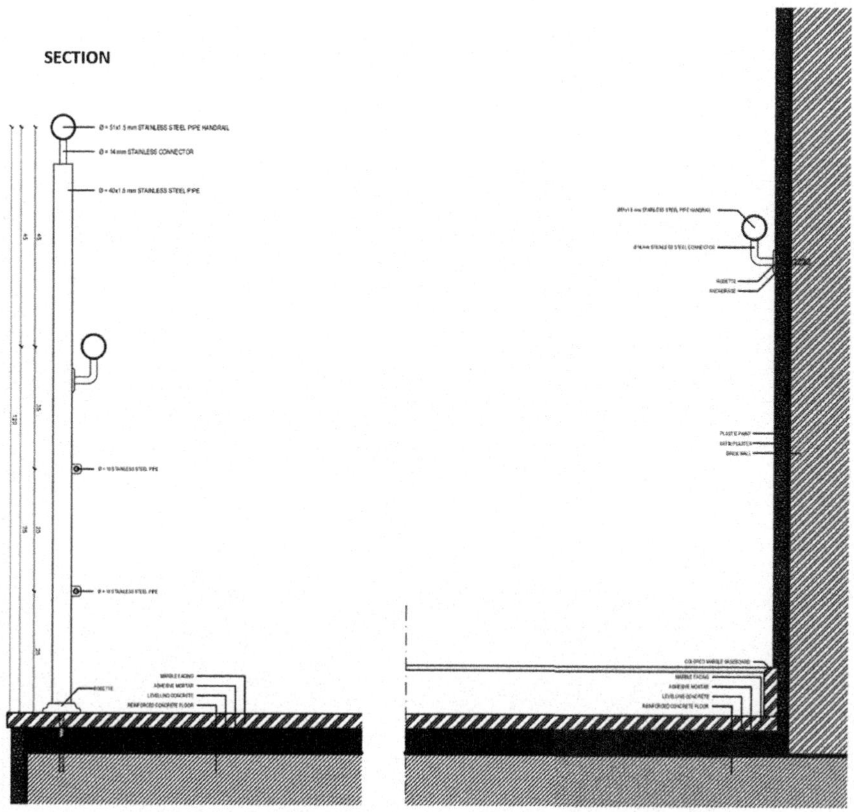

Fig. 184 A stainless-steel railing detail

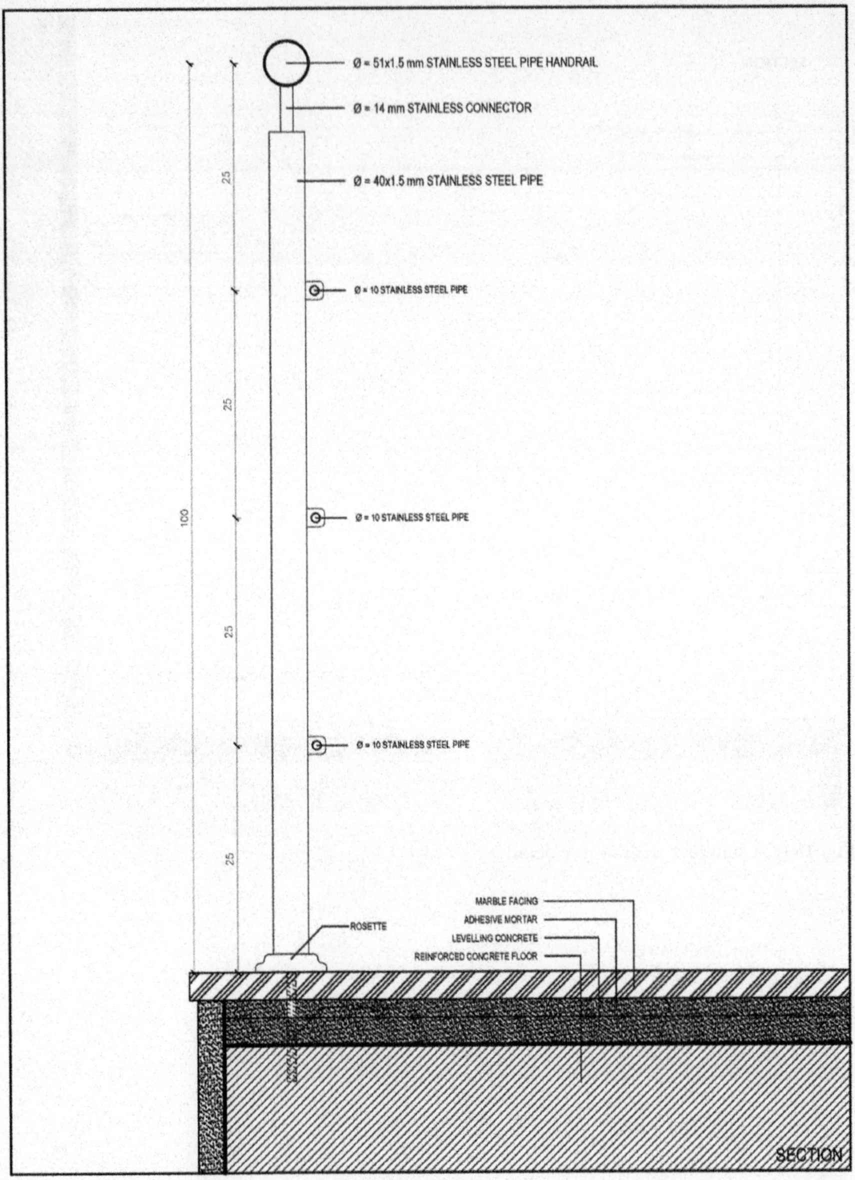

Fig. 185 A top-mounted stainless-steel railing detail

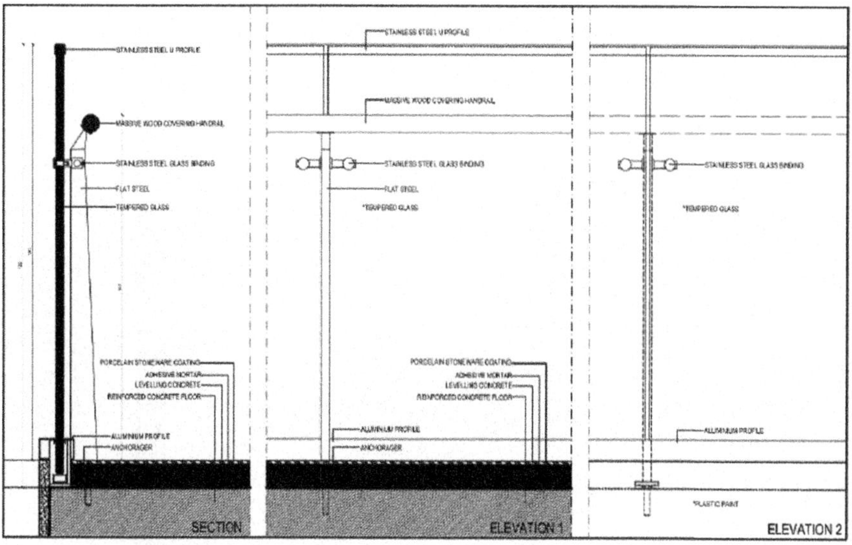

Fig. 186 A floor-mounted stainless-steel detail

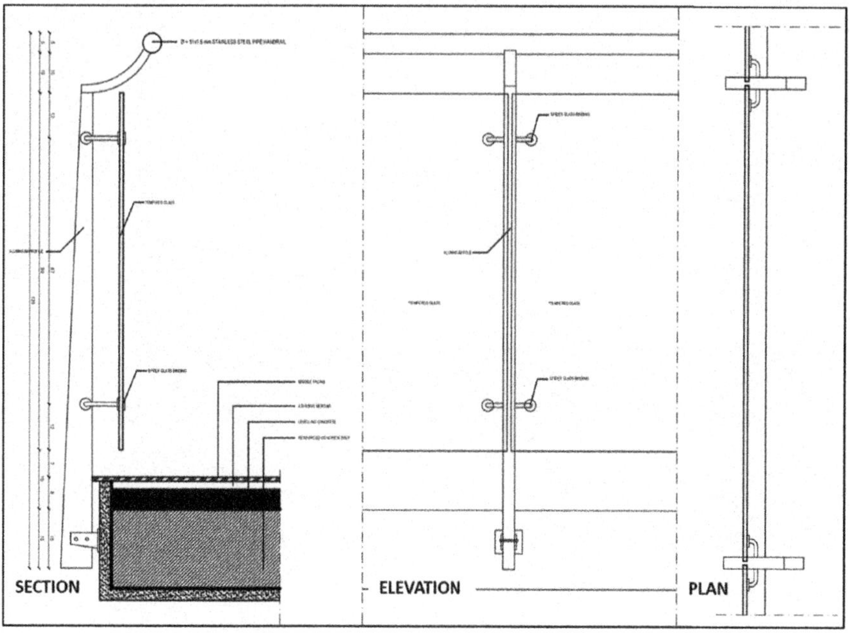

Fig. 187 A lama-baluster stainless-steel and glass railing detail

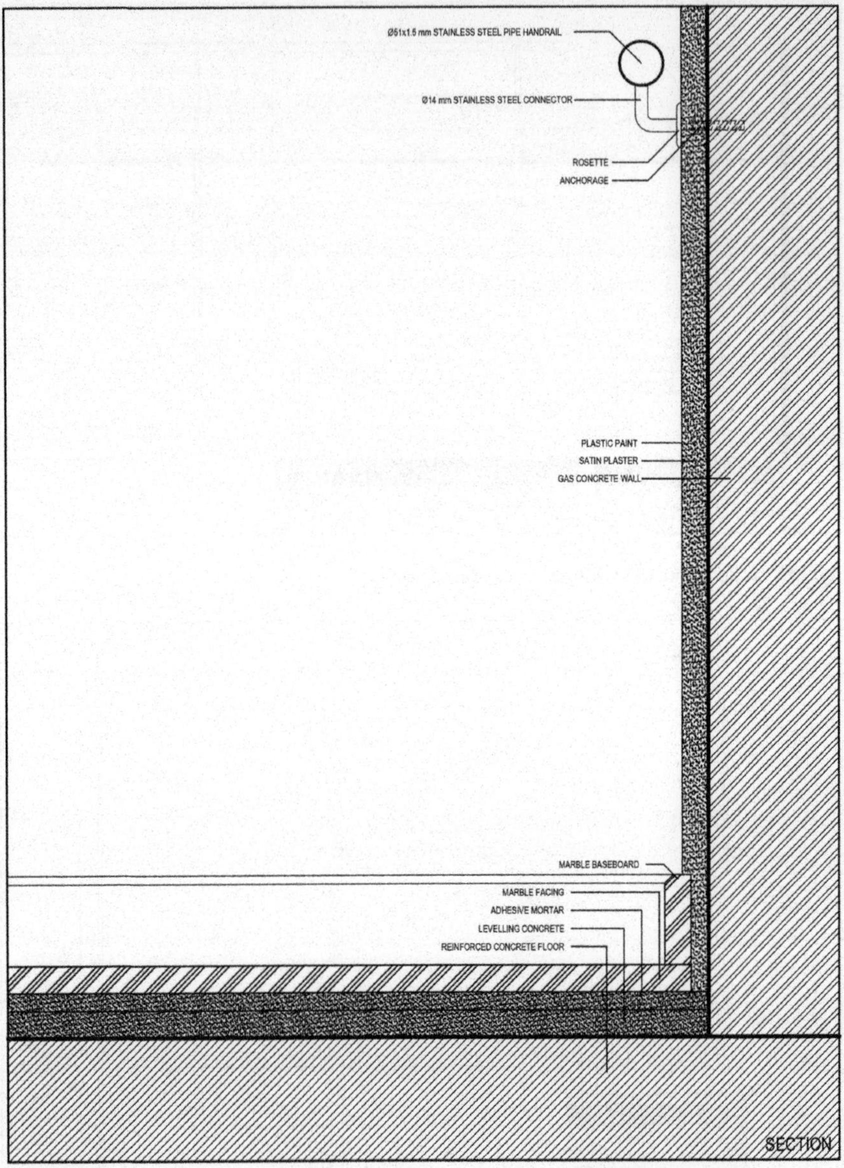

Fig. 188 A wall-mounted stainless-steel railing detail

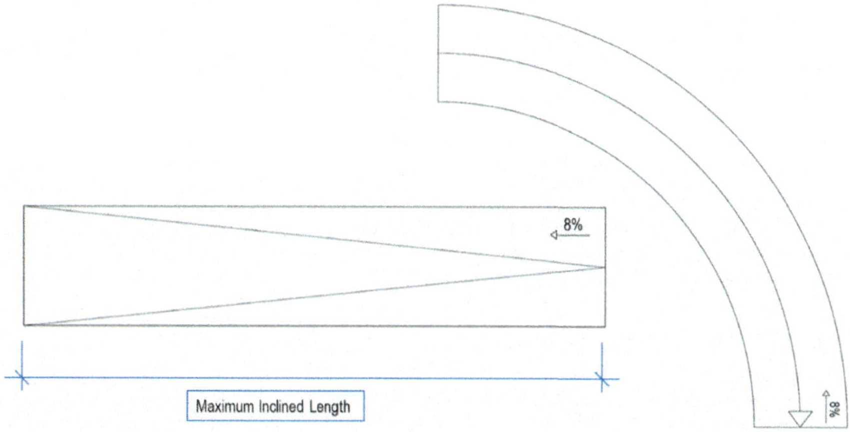

Fig. 189 Illustration of the straight and circular ramp

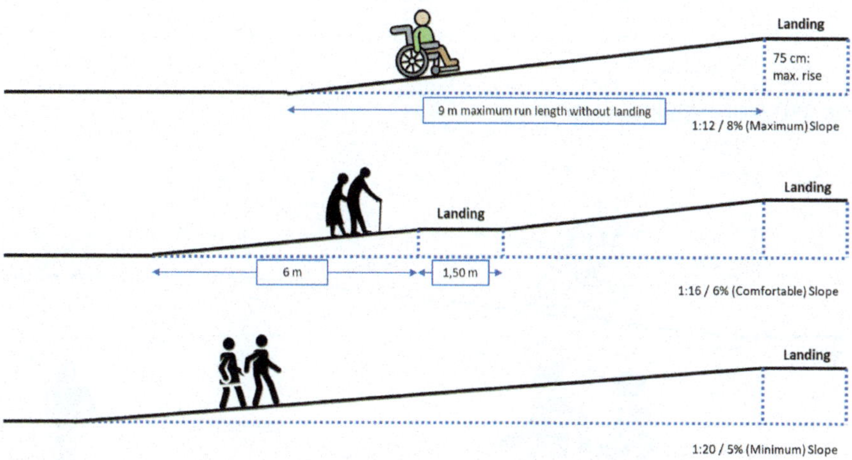

Fig. 190 Slope ratio configurations for ramps

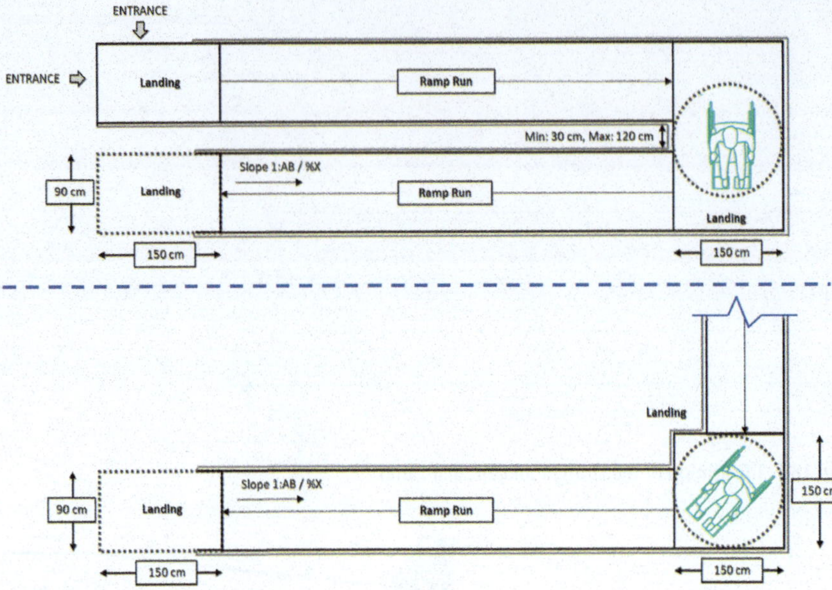

Fig. 191 Plan configurations for ramps

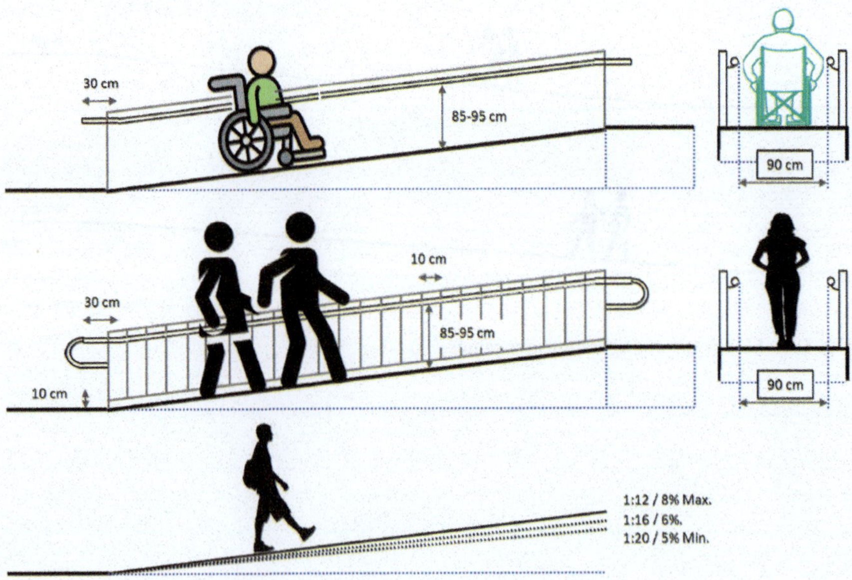

Fig. 192 Section configurations for ramps

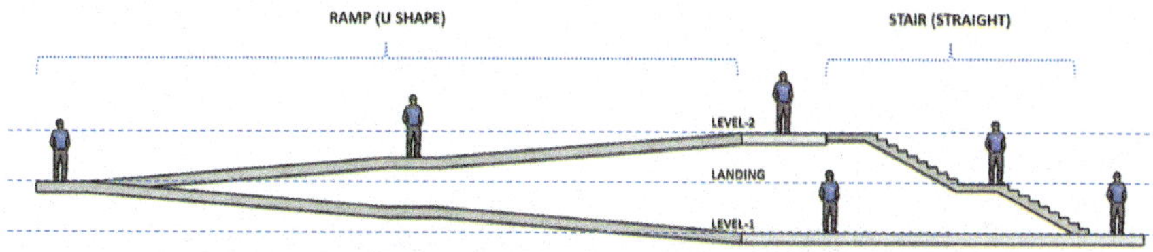

Fig. 193 Comparison of U-shape ramp versus straight stair

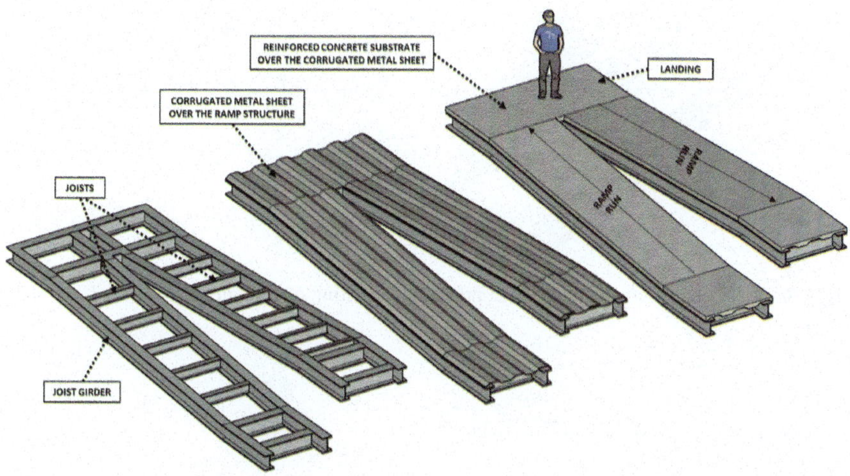

Fig. 194 A steel structure proposal consisting of double joist girders/beams at the edges of the ramp run

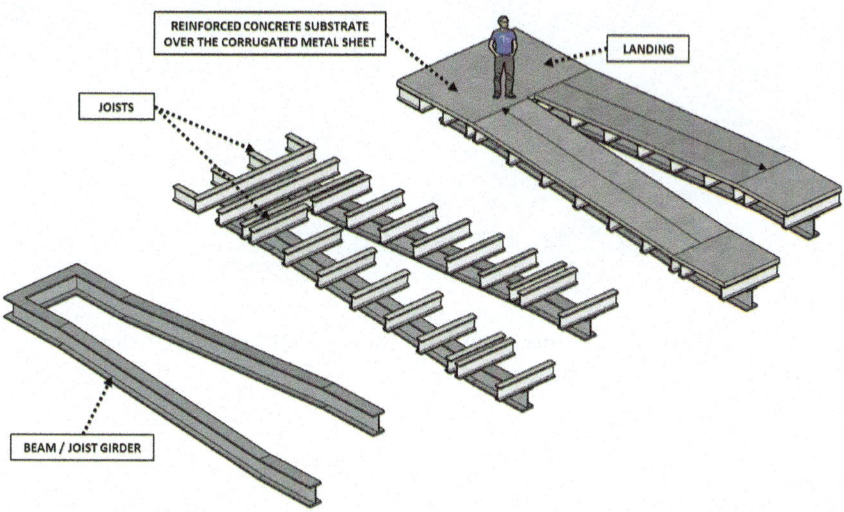

Fig. 195 A steel structure proposal consisting of single joist girders/beams at the center of the ramp run

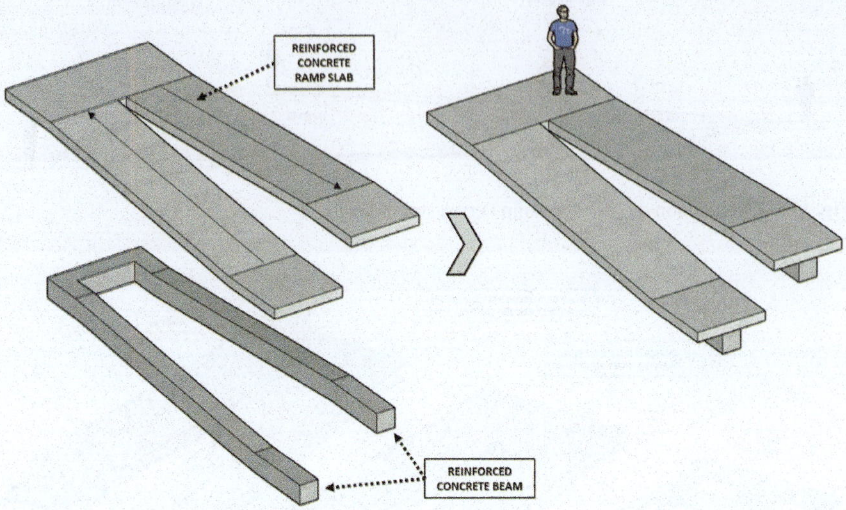

Fig. 196 A reinforced concrete structure proposal consisting of a single beam at the ramp slab's center

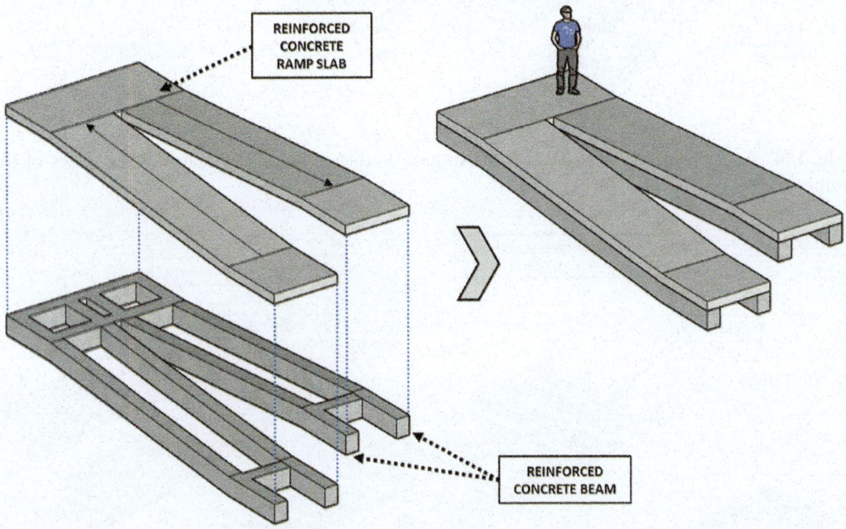

Fig. 197 A reinforced concrete structure proposal consisting of a double beam at the edges of the ramp slab

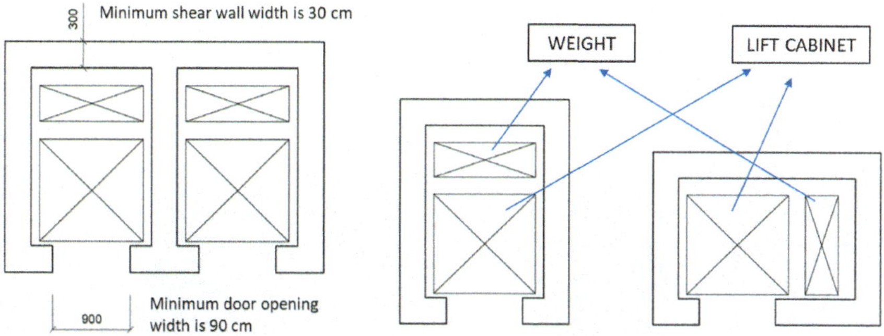

Fig. 198 Plan view of lift/elevator

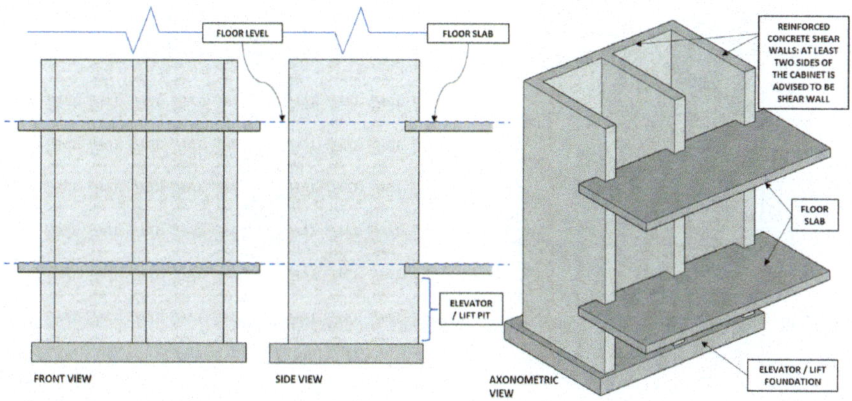

Fig. 199 Reinforced concrete lift/elevator structure sample

8 Mixed Super-Structures

In the aforementioned sections, material-based super-structures were introduced. In application and practice, designing and constructing mixed super-structures is possible for various reasons. This section introduces design considerations and application details of the mixed super-structures.

8.1 Design Considerations

The design considerations explain the necessity and requirements to establish a mixed super-structure. A mixed super-structure could be necessary for the following reasons: (i) structural necessities and (ii) decreasing the project construction costs. On the other hand, designing a mixed super-structure has specific requirements such

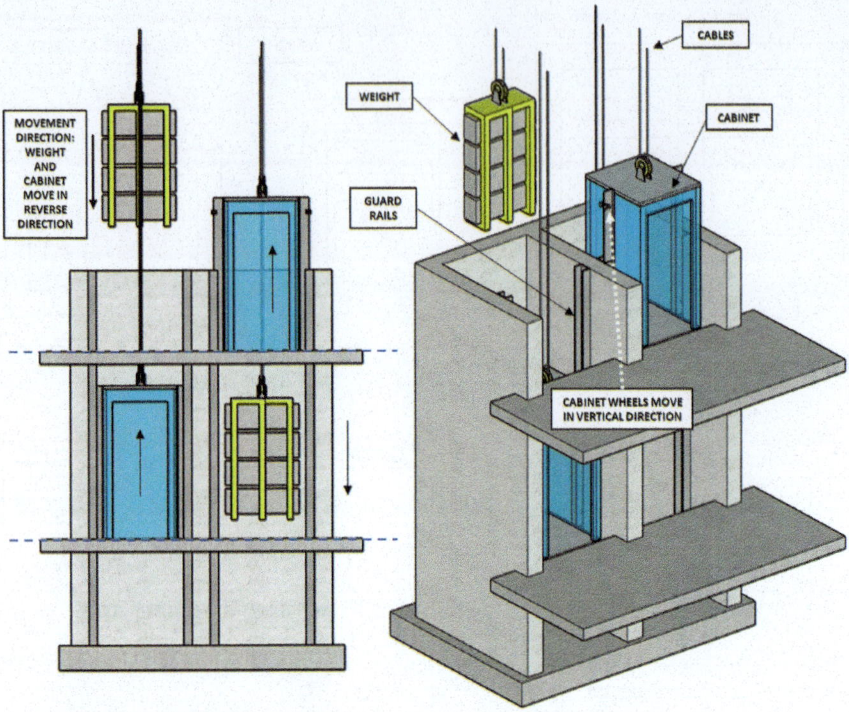

Fig. 200 Basics of lift/elevator system in reinforced concrete structure

as (i) design-based, (ii) construction-based, and (iii) in-use-based. The following subsections explain each necessity and requirement in detail.

8.1.1 Structural Necessities

As mentioned in the first chapter, when the building load is higher than the soil load-bearing capacity, it is necessary to decrease or increase the soil load-bearing capacity in foundation design. From this regard, a mixed super-structure design can decrease the building dead load. The design and application principles rely on establishing a hierarchy in the structure from the base to the roof regarding the load of each super-structure segment. In this case, the super-structure type with a higher dead load must be located close to the earth level, while the super-structure types with a lesser dead load must be located at higher building levels, as indicated in Fig. 203.

The proposed super-structure hierarchy decreases sheer and momentum forces applied to the building structure as illustrated in Fig. 204. This is because the momentum is correlated with distance and the amount of the weight. When the weight of the super-structure is heavy, and the distance of the weight is far, then both the momentum and sheer force of the super-structure to ground level become

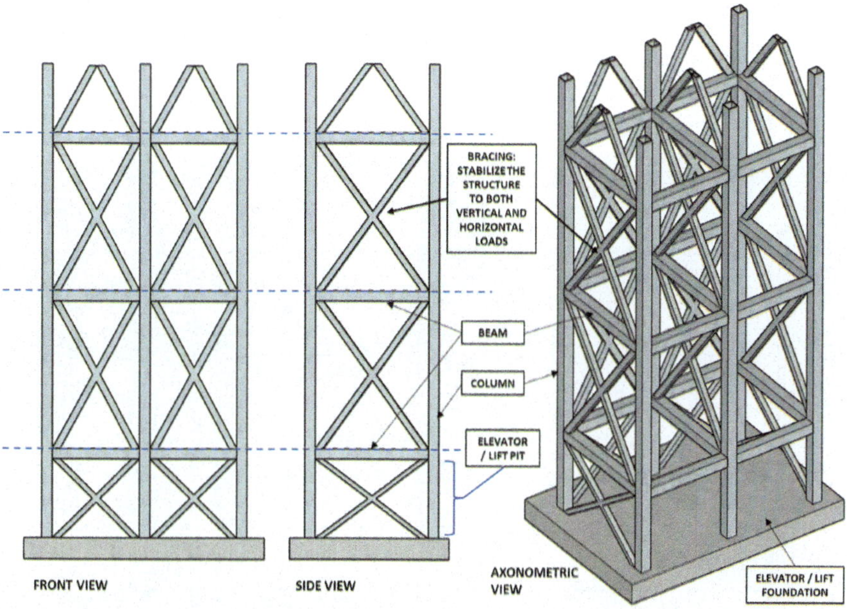

Fig. 201 Steel lift/elevator structure sample

high, requiring additional structural design and construction practices. On the other hand, when a concrete and steel mixed super-structure is applied, as indicated in Fig. 205, the proposal structure presents benefits on dead load weight, sheer force, momentum force, and cost for sub-structure while the proposal may present disadvantages in overall project cost due to use of steel since steel is expensive than concrete as a structural material. However, these outcomes could be valid for low to mid-rise buildings. The results could change when soil load-bearing capacity or building height is extremely low (Figs. 203, 204 and 205).

8.1.2 Decreasing the Project Construction Cost

The most commonly used structural material is reinforced concrete in the world. The reinforced concrete is mixed in a concrete plant, transformed into the construction site, and pumped into the formwork. The process of concrete mixing, pumping, and pouring have particular limitations. Reinforced concrete is cheaper when construction is limited to regular curing, pumping, and pouring. For example, the concrete plant has a distance limit to the construction site. When the concrete is loaded into the mixer truck, it must be mixed regularly; otherwise, either concrete could be poured inside the mixer truck, or the homogeneity of the concrete could be deformed due to excessive mixing time.

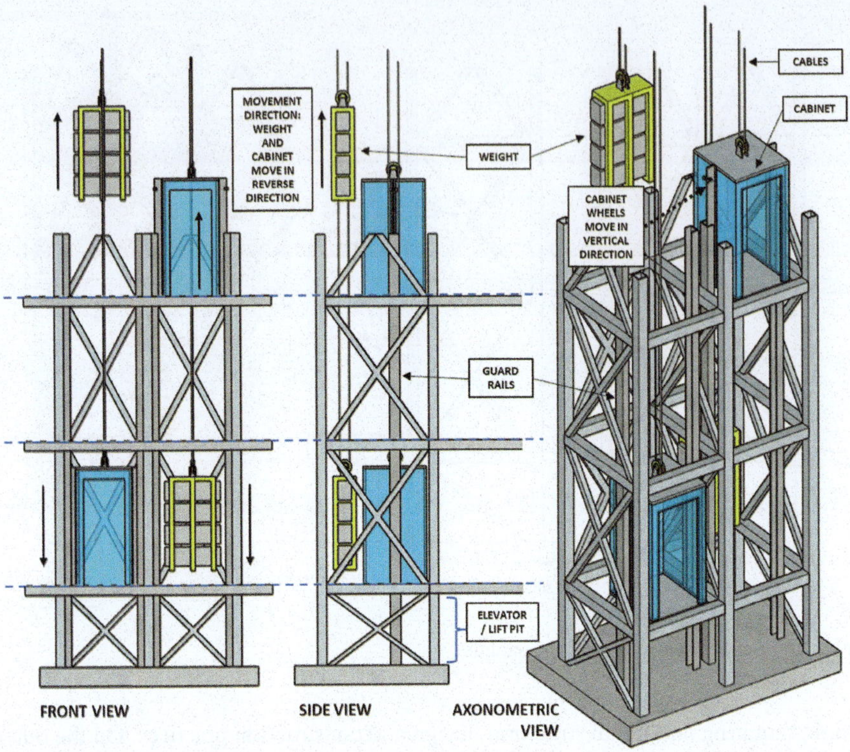

Fig. 202 Basics of lift/elevator system in steel structure

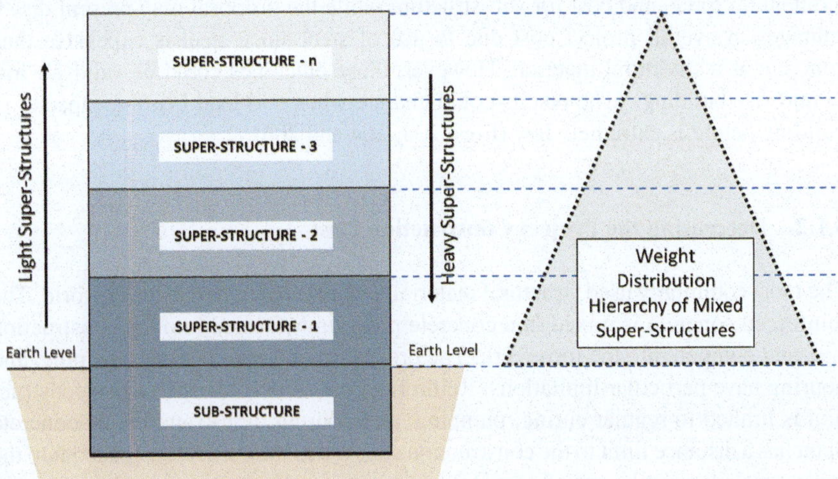

Fig. 203 Weight distribution hierarchy of a mixed super-structure

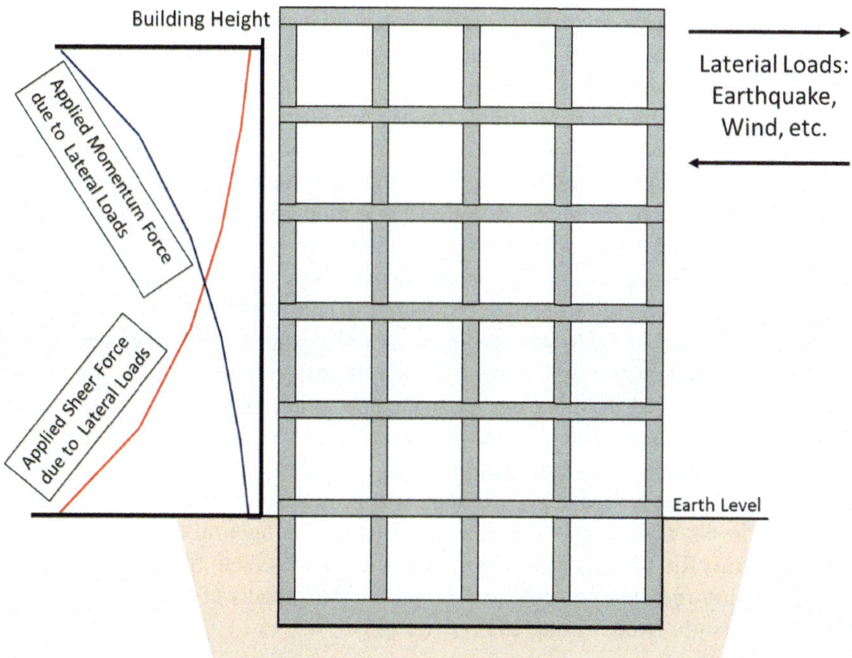

Fig. 204 Lateral loads force the building structure to deform by moment and shear force

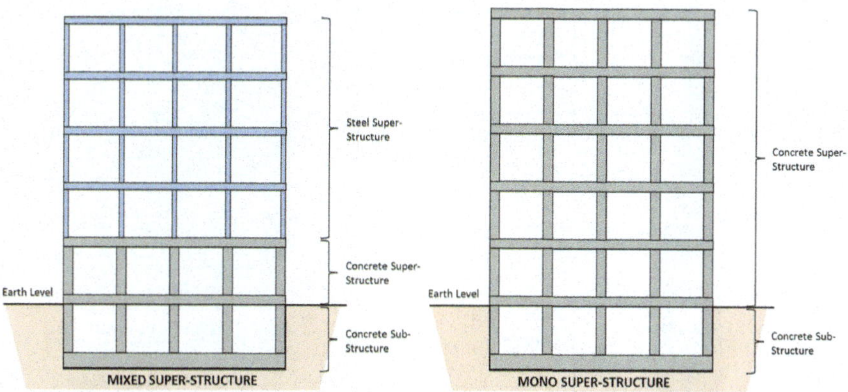

Parameters	Mixed Super-Structure	Mono Super-Structure
Dead Load Weight	Low	High
Sheer Force	Low	High
Momentum Force	Low	High
Cost for Super Structure	High	Low
Cost for Sub-Structure	Low	High

Fig. 205 A mixed super-structure design and comparison with mono-super-structure

Another example is that the formwork height must be within the height limit of the pumping truck. Otherwise, concrete is required to be pumped by use of other methods. Regular concrete prepared in concrete plants includes a workable ingredient after being pumped by a pumping truck. Other pumping options may require changing concrete ingredient levels to allow concrete workability after pouring. For example, in high-rise building construction, regular pumping trucks cannot reach the upper levels of the building. In this case, the concrete mixture is transported using other options. This may be the implementation of temporary pumping pipes allowing the transportation of concrete mixture to the upper floors of the building. The water content may be increased to allow both pumping and workability of concrete, and admixtures may be added to increase the workability and strength of concrete.

However, the construction cost could significantly increase when this regular process is overpassed. In this case, rather than insisting on mono-super structures, mixed superstructures could increase constructability and decrease the construction budget. For example, as indicated in Fig. 206, the Burj Khalifa building in Dubai is a concrete and steel mixed super-structure where the last floors were made of steel, and other levels were made of reinforced concrete. Using a mixed super-structure system in Burj Khalifa eliminates the constructability problem of the last floors due to the inability to pump concrete and decreases the impacts of lateral loads on the building structure (Abdelrazaq, 2012) (Fig. 206).

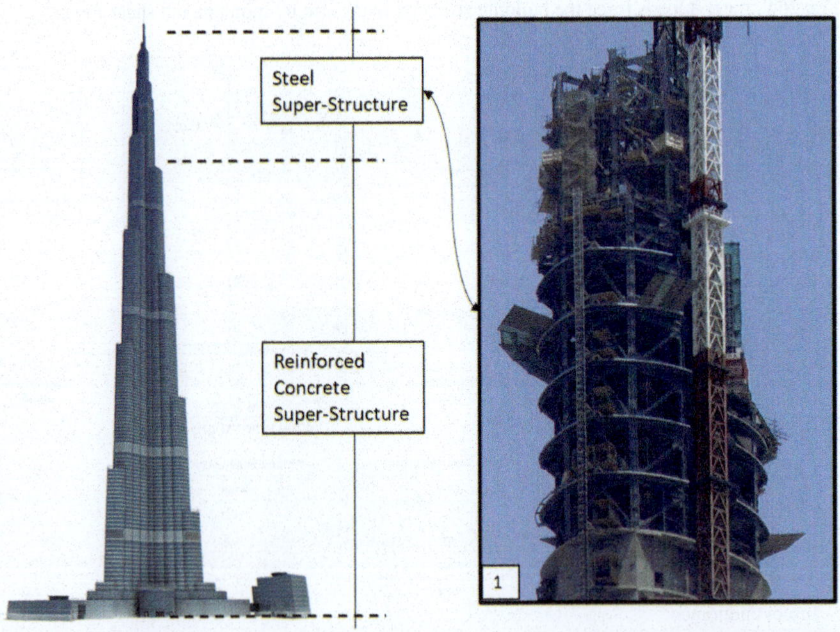

Fig. 206 Burj Khalifa is a mixed super-structure. *Source* (1)

8.1.3 Design and Construction Requirements

A mixed super-structure design needs to consider the collaboration and compatibility of different systems with each other. Collaboration and compatibility address the reflection of similar or parallel behavior against environmental impact. For example, the thermal expansion of steel due to seasonal changes is higher than concrete's. Similarly, steel vibration occurs due to lateral loads (such as wind), which is more sensible than concrete. The lack of consideration for collaboration and compatibility has caused particular problems in the endurance of the building components.

Material and system properties and attributes of masonry, concrete, wood, and steel construction methods are depicted in Table 2. The table stated that masonry and concrete construction systems represent many similar characteristics. The sharing of similar characteristics increases the compatibility and collaboration of the systems, allowing intertwined design and construction of the coupled methods. It is possible to see masonry and concrete intertwined structural designs worldwide. Although wood and steel share many common attributes, the fire resistance of the wood and steel intertwined structure would be extremely low. This is because the thermal conductivity of steel is high, and the fire endurance of steel and wood is extremely low, which may cause easy combustion of the structure.

The masonry concrete's diversity in material/system properties coupled with wood and steel does not prevent the design and construction of a mixed super-structure. Instead, handicaps and problems of the mixed super-structure design could

Table 2 Evaluation of material and system properties of masonry, concrete, wood, and steel construction methods

Material/system property	Masonry	Concrete	Wood	Steel
Thermal expansion	Low	Low	High	High
Thermal conductivity	Low	Medium	Low	High
Vibration	Low	Low	High	High
Durability against vibration	Low	Low	High	High
Weight	High	High	Low	Low
Tensile strength	Low	Low	High	High
Compressive strength	High	High	High	High
Maintenance requirement	Low	Low	High	High
Life endurance	High	Low	High	High
Fire endurance	High	High	Low	Low
Corrosion endurance	High	Medium	High	Low
Swelling endurance	High	High	Low	High
Insect endurance	High	High	Low	High
Material cost	Low	Low	Medium/high	Medium/high
Workmanship cost	Low	Low	Medium/high	Medium/high
Flexibility of use	Low	High	High	High

be overcome by implementing particular application details. In the following section, sample mixed super-structure designs were introduced, and application details were referenced regarding the evaluation parameters presented in Table 2.

8.2 Sample Mixed Super-Structure Designs and Application Details

Seven mixed super-structure samples were prepared and introduced in the following sub-sections. Each sample presents unique characteristics and attributes while preserving handicaps and problems. Proper application details were proposed and explained to overcome the handicaps and problems.

8.2.1 Masonry and Concrete Mixed Super-Structure

The sample mixed super-structure presented in Figs. 207 and 208 consists of a masonry structure in the basement and ground floor while a reinforced concrete frame structure on the first and second floors. Since the masonry structure is heavier than the reinforced concrete frame, the basement and ground floor are devoted to masonry; another option could consist of masonry in the basement and reinforced concrete frame in the ground, first, and second floors. The flooring of the whole structure was designed to be a flat slab. However, other flooring options include one-way solid, two-way solid slab, one-way joist, and two-way joist slabs. Immersive virtual models presented in Figs. 209 and 210 depicts a navigable interior view of the structure from different locations.

Table 3 illustrates and explains the material and system properties of masonry & and concrete mixed super-structure by comparing the mixed system with masonry and concrete mono-structures. The mixed system reflects low performance in thermal expansion, vibration, durability against vibration, tensile strength, maintenance requirement, material cost, and workmanship cost. Similarly, the mixed system represents high performance in weight, compressive strength, fire endurance, corrosion endurance, swelling endurance, and insect endurance. On the other hand, unlike masonry and concrete material system characteristics, the mixed system demonstrates intermediary performance at different levels. For instance, thermal conductivity, life endurance, corrosion endurance, and flexibility are the characteristics of the mixed super-structure showing performance between masonry and concrete.

8.2.2 Masonry and Steel Mixed Super-Structure

The sample mixed super-structure presented in Figs. 211 and 212 consists of a masonry structure in the basement and ground floor while a steel frame structure

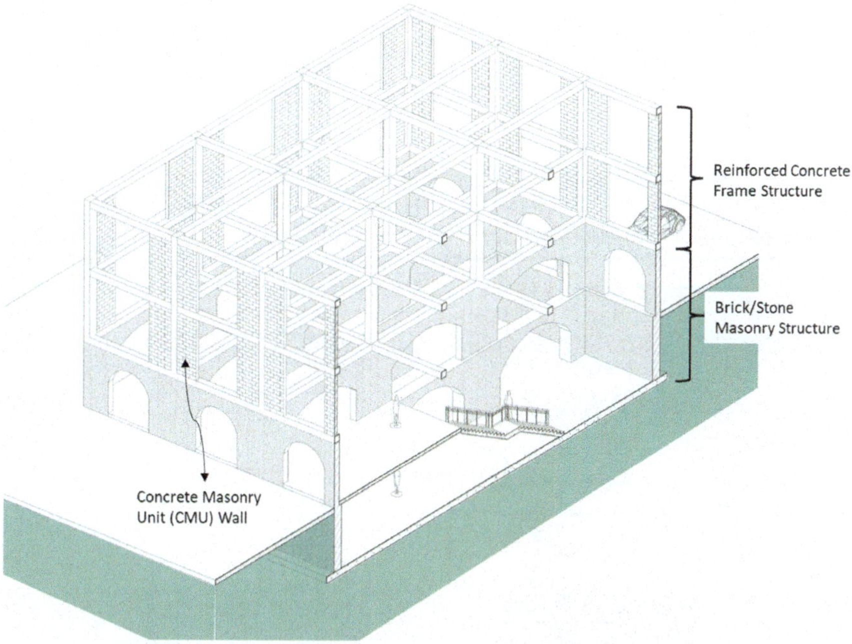

Fig. 207 A sample 3-D section of masonry and concrete mixed super-structure

in the first and second floors. Since masonry structure is heavier than steel frame, the basement and ground floor are devoted to masonry. Another option could include masonry in the basement and steel frame in the ground, first and second floors. Immersive virtual models were presented in Figs. 211, 212, and 213, depicting navigable interior views of the structure from different locations. The flooring in the masonry structure was designed to be a flat slab, while a one-way joist system was used in the steel structure. Immersive virtual models presented in Figs. 213, 214, and 215 depict navigable interior views of the structure from different locations.

Table 4 illustrates and explains the material and system properties of masonry and steel mixed super-structure by comparing the mixed system with masonry and steel mono-structures. The mixed system represents high performance in compressive strength, life endurance, swelling endurance, and insect endurance. On the other hand, different than masonry and steel mono-structural characteristics, the mixed system demonstrates intermediary performance in thermal expansion, thermal conductivity, vibration, durability against vibration, weight, tensile strength, maintenance requirement, fire endurance, corrosion endurance, material cost, workmanship cost and flexibility of use.

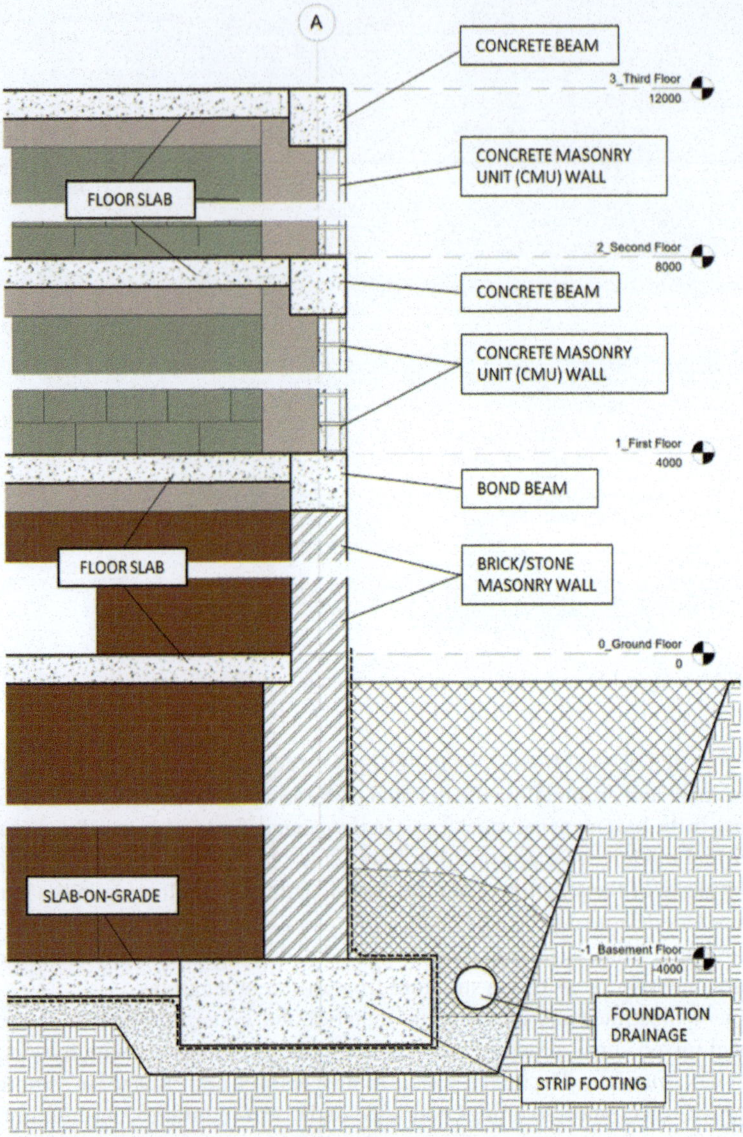

Fig. 208 The system detail of the masonry and concrete mixed super-structure

8.2.3 Masonry and Wood Mixed Super-Structure

The sample mixed super-structure presented in Figs. 216 and 217 consists of a masonry structure in the basement and ground floor while a wood frame structure in the first and second floors. Since masonry structure is heavier than wood frame, the basement and ground floor are devoted to masonry. Another option could include

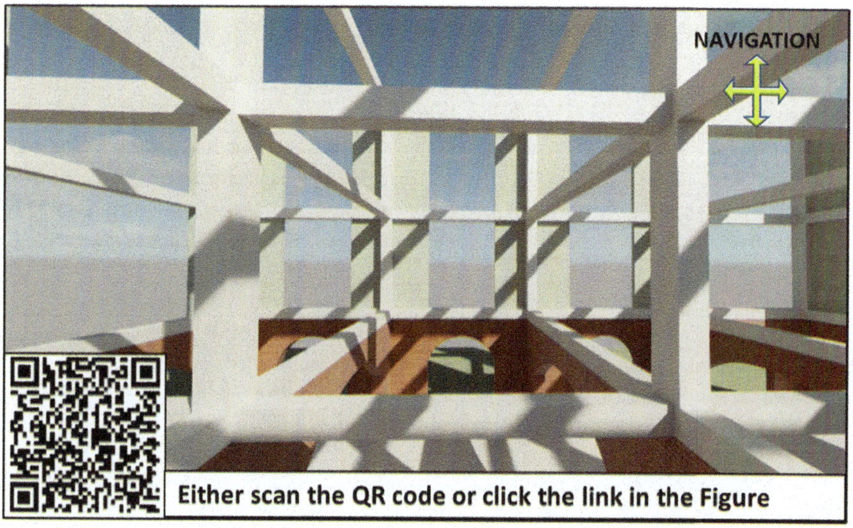

Fig. 209 The Immersive Virtual Model of the sample masonry and concrete mixed super-structure. Click the link: https://pano.autodesk.com/pano.html?url=jpgs/5d70a85c-82cf-4070-88a2-8efd0dadfed4&version=2

Fig. 210 The Immersive Virtual Model of the sample masonry and concrete mixed super-structure. Click the link: https://pano.autodesk.com/pano.html?url=jpgs/427d8f6b-96f0-43c6-a3ef-63f3fcc068e8&version=2

Table 3 Evaluation of material and system properties of masonry and and concrete mixed superstructure

Material/system property	Masonry	Concrete	Mixed system	Notes
Thermal expansion	Low	Low	Low	Both concrete and masonry structures have low thermal expansion
Thermal conductivity	Low	Medium	Low/ medium	The thermal conductivity of the mixed system is medium due to the use of concrete
Vibration	Low	Low	Low	The mixed system has strong resistance to vibration
Durability against Vibration	Low	Low	Low	The masonry structure is fragile against vibration in the long term. Furthermore, cracks and deformations occur due to vibration and require maintenance
Weight	High	High	High	The mixed system's weight is significantly high
Tensile strength	Low	Low	Low	The mixed system could gain tensile strength when reinforced with steel rebar
Compressive strength	High	High	High	The compressive strength of the mixed system is significantly high
Maintenance requirement	Low	Low	Low	The maintenance requirement of the mixed system is low
Life endurance	High	Low	Medium	The life endurance of the mixed system is at a medium level due to the use of concrete. Compared to the life of the masonry, the concrete structure needs to be renovated a couple of times
Fire endurance	High	High	High	The fire endurance of the mixed system is high
Corrosion endurance	High	Medium	Medium / High	The corrosion endurance of the proposed system is high
Swelling endurance	High	High	High	Both masonry and concrete have high swelling endurance
Insect endurance	High	High	High	Both masonry and concrete have high insect endurance
Material cost	Low	Low	Low	Both masonry and concrete material costs are particularly low in many countries
Workmanship cost	Low	Low	Low	Except for a few countries, both masonry and concrete workmanship are low in many countries
Flexibility of use	Low	High	Medium	The concrete has high flexibility of use when reinforced, while the masonry has low

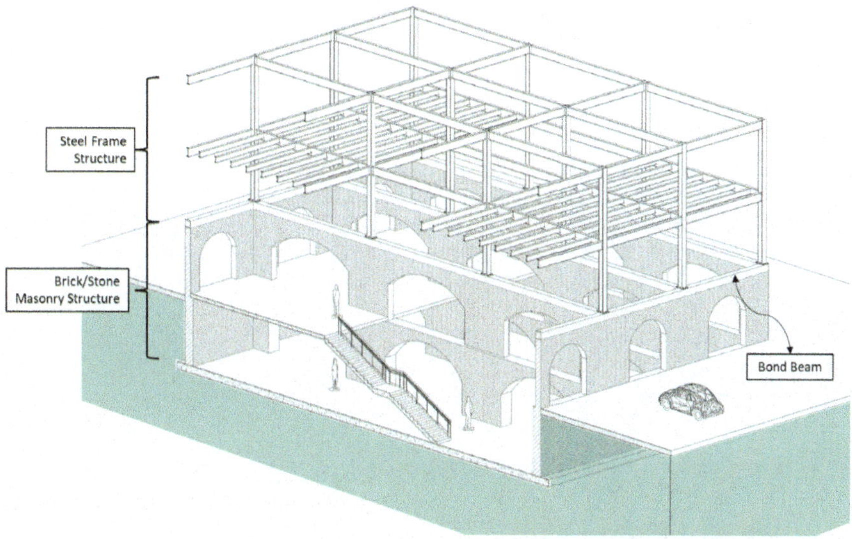

Fig. 211 A sample 3-D section of masonry and concrete mixed super-structure

masonry in the basement and wood frame in the ground, first and second floors. The flooring in the masonry structure was a flat plate in the sample model. On the other hand, the flooring system in the wood frame structure was a one-way joist system. Furthermore, the flooring option in the masonry structure could be one-way joist wood flooring as another option. Immersive virtual models were presented in Figs. 218, 219, and 220, depicting navigable interior views of the structure from different locations.

Table 5 illustrates and explains the material and system properties of masonry & and steel mixed super-structure by comparing the mixed system with masonry and wood mono-structures. The mixed system reflects low performance in thermal conductivity, whereas it represents high performance in compressive strength, life endurance, and corrosion endurance. On the other hand, different than masonry and wood mono-structural characteristics, the mixed system demonstrates intermediary performance in thermal expansion, vibration, durability against vibration, weight, tensile strength, maintenance requirement, fire endurance, swelling endurance, insect endurance, material cost, workmanship cost and flexibility of use.

8.2.4 Concrete and Steel Mixed Super-Structure

The sample mixed super-structure presented in Figs. 221 and 222 consists of concrete structures in the basement and ground floor while steel frame structures on the first and second floors. Since the concrete structure is heavier than the steel frame, the basement and ground floor are devoted to concrete. Another option could consist of a

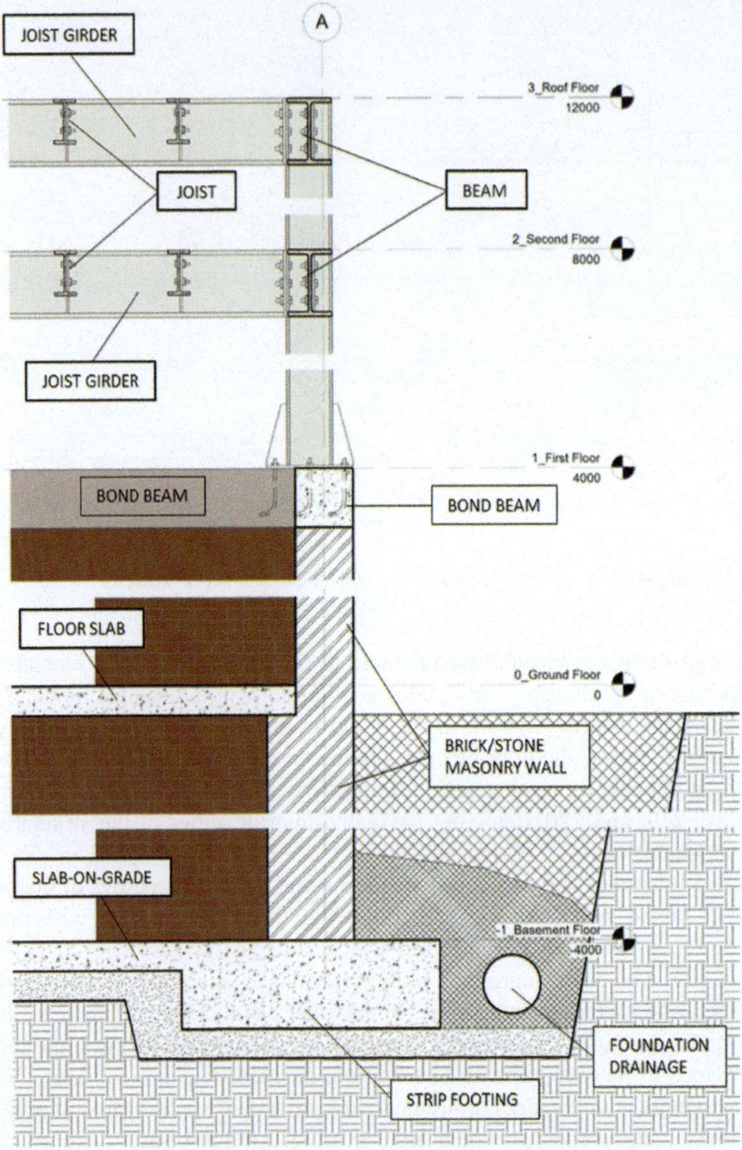

Fig. 212 The system detail of the masonry and steel mixed super-structure

concrete structure in the basement and a steel frame structure in the ground, first and second floors. The flooring in the concrete structure was a two-way solid slap in the sample model. Other concrete flooring systems could also be used as other options. On the other hand, the flooring system in the steel frame structure was a one-way joist

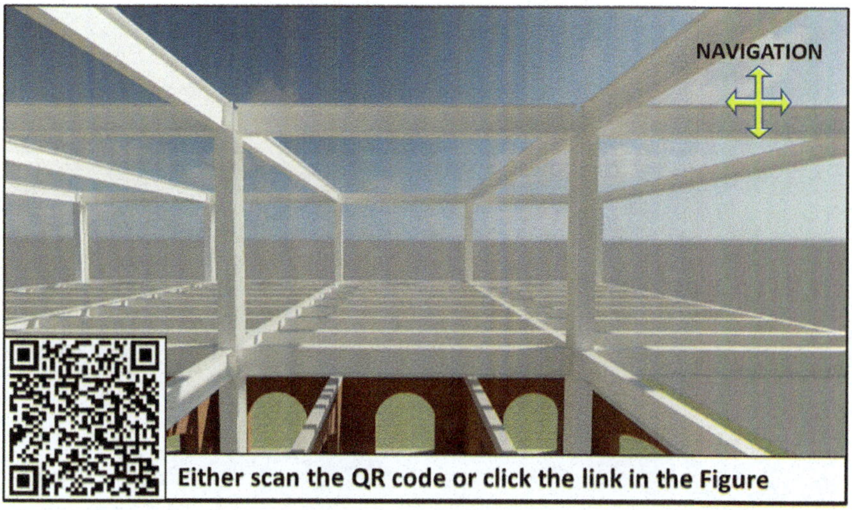

Fig. 213 The Immersive Virtual Model of the sample masonry and steel mixed super-structure. Click the link: https://pano.autodesk.com/pano.html?url=jpgs/de669864-75ca-4cb2-9fd0-4e50419f6a59&version=2

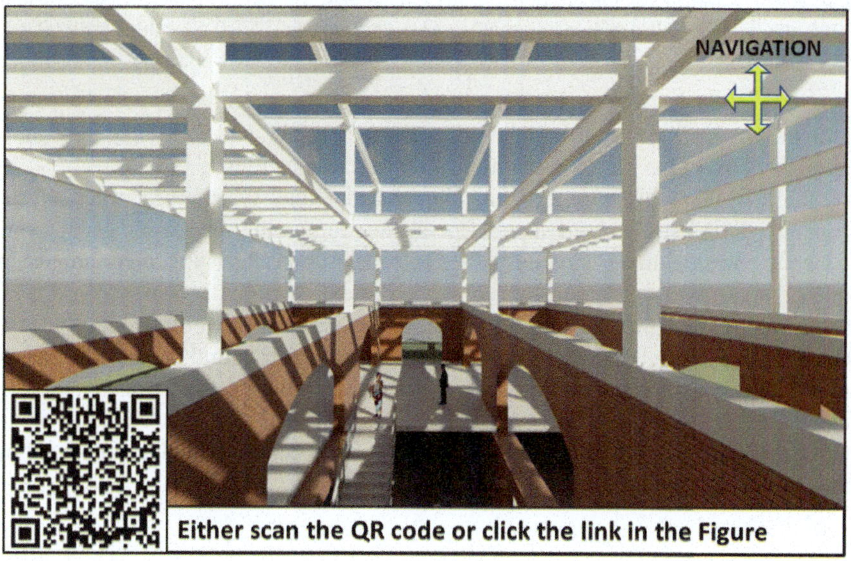

Fig. 214 The Immersive Virtual Model of the sample masonry and steel mixed super-structure. Click the link: https://pano.autodesk.com/pano.html?url=jpgs/0ff8d22a-efcf-4477-8709-c8af57007978&version=2

Fig. 215 The Immersive Virtual Model of the sample masonry and steel mixed super-structure. Click the link: https://pano.autodesk.com/pano.html?url=jpgs/90377fc0-60ac-49a2-ab49-009675a175da&version=2

system. The immersive virtual models in Figs. 223, 224, and 225 depict navigable interior views of the mixed structure from different locations.

Table 6 illustrates and explains the material and system properties of the concrete & and steel mixed super-structure by comparing the mixed system with concrete and steel mono-structures. The mixed system reflects low/medium performance in corrosion endurance, whereas it represents high performance in compressive strength, swelling endurance, insect endurance, and flexibility of use. On the other hand, different than the concrete and steel mono-structural characteristics, the mixed system demonstrates intermediary performance in thermal expansion, thermal conductivity, vibration, durability against vibration, weight, tensile strength, maintenance requirement, life endurance, fire endurance, corrosion endurance, swelling endurance, material cost, and workmanship cost .

8.2.5 Concrete and Wood Mixed Super-Structure

The sample mixed super-structure presented in Figs. 226 and 227 consists of concrete structures in the basement and ground floor while wood frame structures in the first and second floors. Since the concrete structure is heavier than the wood frame, the basement and ground floor are devoted to concrete. Another option could consist of a concrete structure in the basement and a wood frame structure in the ground, first

Table 4 Evaluation of material and system property of masonry and steel mixed super-structure

Material/system property	Masonry	Steel	Mixed system	Notes for application
Thermal expansion	Low	High	Medium	The thermal expansion performance of the mixed system is at a medium level
Thermal conductivity	Low	High	Medium	The thermal conductivity of the mixed system is at medium level
Vibration	Low	High	Medium	The mixed system performs medium-level shaking due to vibration
Durability against vibration	Low	High	Medium	The masonry structure is fragile against vibration in the long term. Furthermore, cracks and deformations occur due to vibration requiring maintenance
Weight	High	Low	Medium	The mixed system's weight is significantly at medium level when compared with the masonry & concrete mixed super structure
Tensile strength	Low	High	Medium	The mixed system tensile strength is at a medium level. It would be better when the entire tensile force in the structure is carried by steel for durability
Compressive strength	High	High	High	The compressive strength of the mixed system is significantly high
Maintenance requirement	Low	High	Medium	The maintenance requirement of the mixed system is at a medium level due to the use of steel since the steel periodically requires corrosion protection practices
Life endurance	High	High	High	The life endurance of the mixed system is high when corrosion protection practices are applied
Fire endurance	High	Low	Medium	The fire endurance of the mixed system is at a medium level since the steel's fire endurance is weaker than the masonry
Corrosion endurance	High	Low	Medium	The corrosion endurance of the proposed system is at a medium level. This is because steel requires corrosion protection practices to be periodically applied
Swelling endurance	High	High	High	Both masonry and steel have high swelling endurance

(continued)

Table 4 (continued)

Material/system property	Masonry	Steel	Mixed system	Notes for application
Insect endurance	High	High	High	Both masonry and steel have high insect endurance
Material cost	Low	Medium/ high	Medium	The material cost of steel is at medium and high levels in many countries
Workmanship cost	Low	Medium/ high	Medium	Except for a few countries, the steel's workmanship cost is high in many countries
Flexibility of use	Low	High	Medium	The mixed system has a medium level of flexibility of use

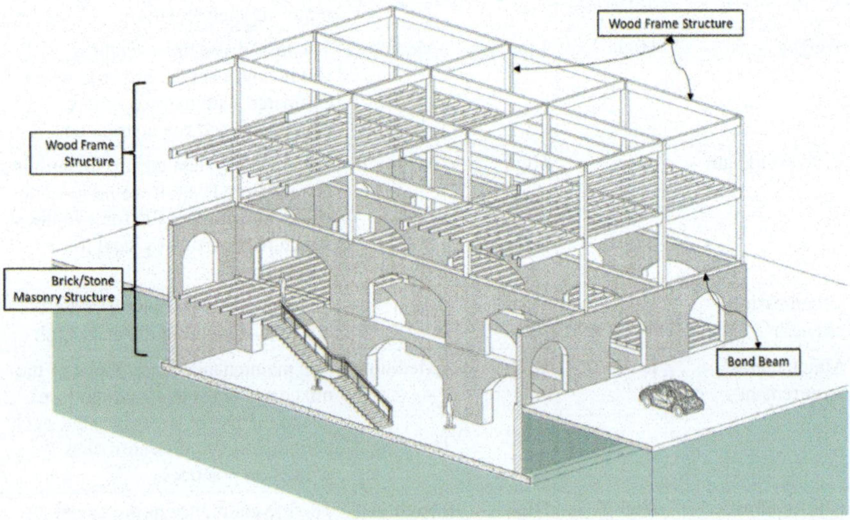

Fig. 216 A sample masonry and wood mixed super-structure

and second floors. The flooring in the concrete structure was a two-way solid slap in the sample model. Other concrete flooring systems could also be used as other options. On the other hand, the flooring system in the wood frame structure was a one-way joist system. The immersive virtual models in Figs. 228, 229, and 230 depict navigable interior views of the mixed structure from different locations.

Table 7 illustrates and explains the material and system properties of the concrete & steel mixed super-structure by comparing the mixed system with concrete and steel mono-structures. The mixed system reflects low/medium performance in thermal conductivity while representing high performance in compressive strength and flexibility of use. On the other hand, different than the concrete and wood mono-structural characteristics, the mixed system demonstrates intermediary performance

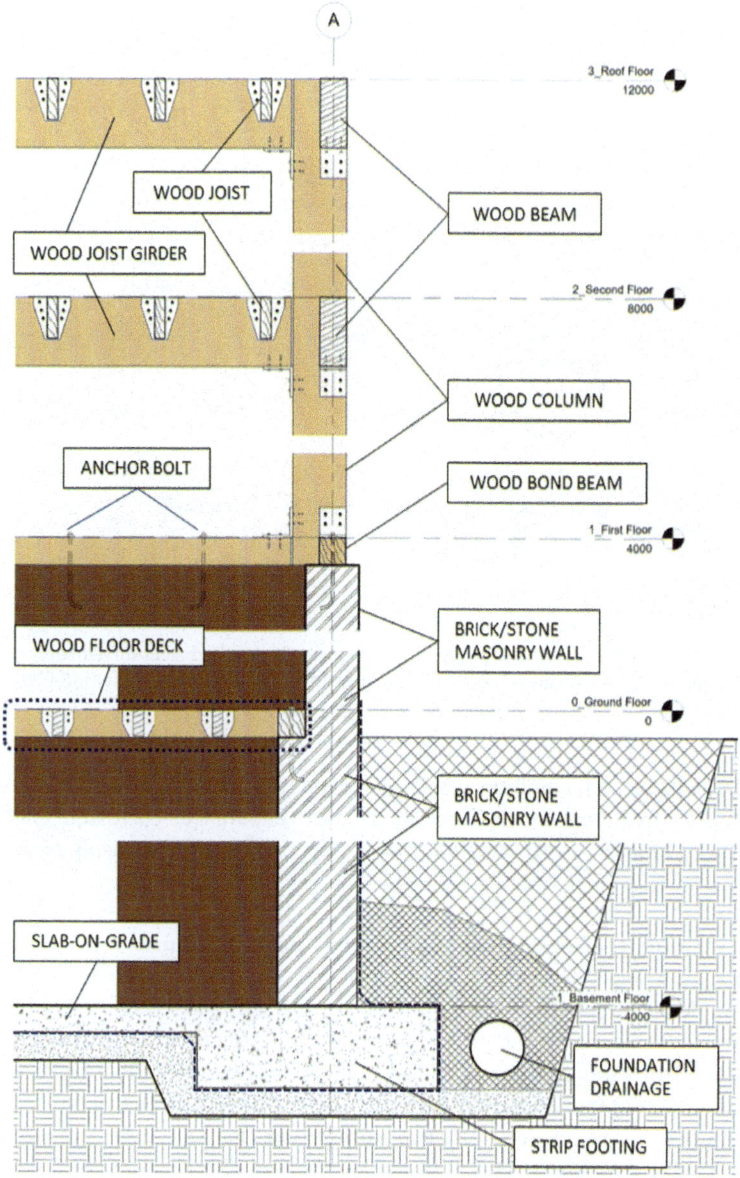

Fig. 217 The system detail of the masonry and wood mixed super-structure

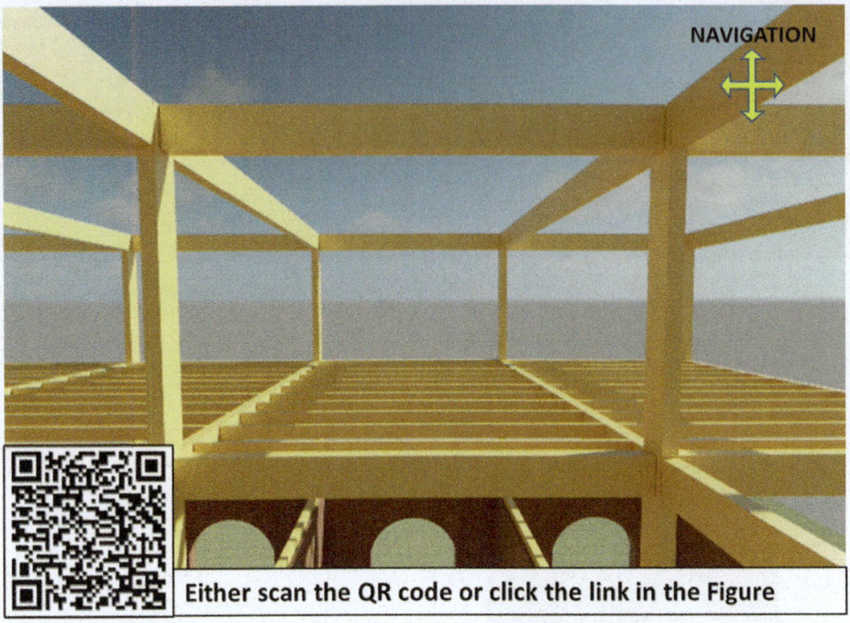

Fig. 218 The Immersive Virtual Model of the sample concrete and steel mixed super-structure. Click the link: https://pano.autodesk.com/pano.html?url=jpgs/2cc7e519-328d-43d6-9b96-b5e7b1b953ed&version=2

in thermal expansion, vibration, durability against vibration, weight, tensile strength, maintenance requirement, life endurance, fire endurance, corrosion endurance, swelling endurance, insect endurance, material cost, and workmanship cost.

8.2.6 Masonry—Concrete—Wood Mixed Super-Structure

The sample mixed super-structure presented in Figs. 231 and 232 consists of masonry structure in the basement and ground floor, concrete structure on the first and second floors, and wood frame structure on the third and fourth floors. Since masonry is the heaviest structure in the mixed system, the basement and ground floor are devoted to masonry, the concrete structure is devoted to the first and second, and the wood frame structure is devoted to the third and fourth floors. Another option could include a masonry structure in the basement, a concrete structure on the ground floors, and a wood frame structure on the remaining floors. The flooring in the masonry structure was flat plate, whereas the two-way solid slab in the concrete structure was in the sample model. Other concrete flooring systems could also be used as options in masonry and concrete structures in the mixed system.

On the other hand, the flooring system in the wood frame structure was a one-way joist system. Furthermore, the entire flooring system in the mixed structure could be

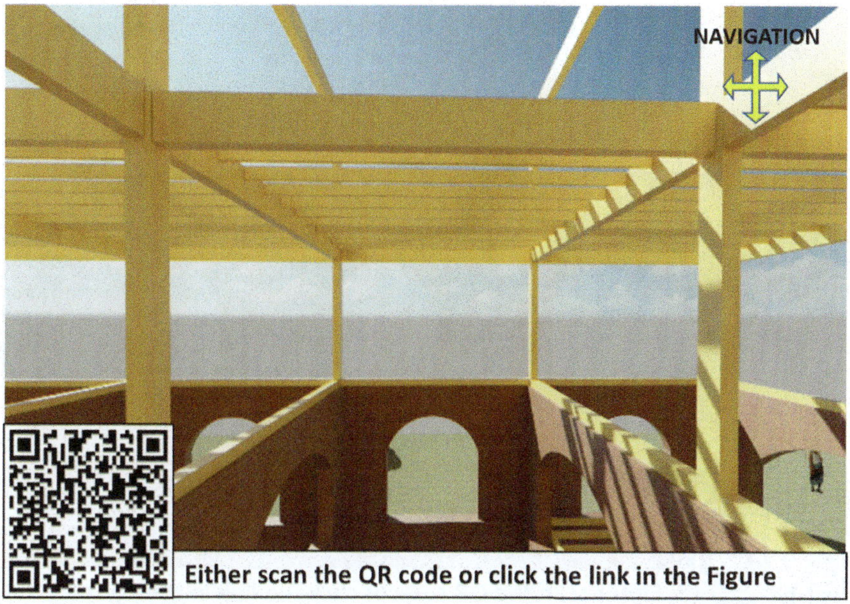

Fig. 219 The Immersive Virtual Model of the sample concrete and steel mixed super-structure. Click the link: https://pano.autodesk.com/pano.html?url=jpgs/34d69420-17b1-4c1d-ac18-40bb8fddc1f3&version=2

one-way joist wood frame flooring as an alternative option. The immersive virtual models in Figs. 233, 234, 235, and 236 depict navigable interior views of the mixed structure from different locations.

Table 8 illustrates and explains the material and system properties of the masonry-concrete-wood mixed super-structure by comparing the mixed system with masonry, concrete, and wood mono-structures. The mixed system reflects low thermal conductivity performance while representing high weight, compressive strength, and corrosion endurance performance. On the other hand, different than each mono-structural characteristics, the mixed system demonstrates intermediary performance in thermal expansion, vibration, durability against vibration, tensile strength, maintenance requirement, life endurance, fire endurance, swelling endurance, insect endurance, material cost, workmanship cost and flexibility of use.

8.2.7 Masonry—Concrete—Steel Mixed Super-Structure

The sample mixed super-structure presented in Figs. 237 and 238 consists of masonry structure in the basement and ground floor, concrete structure on the first and second floors, and steel frame structure on the third and fourth floors. Since masonry is the heaviest structure in the mixed system, the basement and ground floor are devoted to masonry, the concrete structure is devoted to the first and second, and the steel

Fig. 220 The Immersive Virtual Model of the sample concrete and steel mixed super-structure. Click the link: https://pano.autodesk.com/pano.html?url=jpgs/b317cd32-ba77-4eac-9f9a-56aefbad2061&version=2

frame structure is for the third and fourth floors. Another option could consist of a masonry structure in the basement, a concrete structure on the ground floor, and a steel frame structure on the remaining floors. The flooring in the masonry structure was flat plate, whereas the two-way solid slab in the concrete structure was in the sample model. Other concrete flooring systems could also be used as options in masonry and concrete structures in the mixed system.

On the other hand, the flooring system in the steel frame structure was a one-way joist system. Furthermore, the entire flooring system in the mixed structure could be one-way joist steel frame flooring as an alternative option. The immersive virtual models in Figs. 239, 240, 241, and 242 depict navigable interior views of the mixed structure from different locations.

Table 9 illustrates and explains the material and system properties of the masonry-concrete-steel mixed super-structure by comparing the mixed system with masonry, concrete, and steel mono-structures. The mixed system reflects low/medium performance in material cost while representing high performance in weight, compressive strength, swelling endurance, and insect endurance. On the other hand, different than each mono-structural characteristic, the mixed system demonstrates intermediary performance in thermal expansion, thermal conductivity, vibration, durability against vibration, tensile strength, maintenance requirement, life endurance, fire endurance, corrosion endurance, workmanship cost, and flexibility of use.

Table 5 Evaluation of material and system property of masonry and wood mixed super-structure

Material/system property	Masonry	Wood	Mixed system	Notes for application
Thermal expansion	Low	High	Medium	The mixed system has medium-level thermal expansion
Thermal conductivity	Low	Low	Low	The thermal conductivity of the mixed system is at a low level
Vibration	Low	High	Medium	The mixed system reflects medium-level shaking against vibration
Durability against vibration	Low	High	Medium	The masonry structure is fragile against vibration in the long term. Furthermore, cracks and deformations occur due to vibration requiring maintenance, while wood has high structural durability against vibration
Weight	High	Low	Medium	The mixed system's weight is at a medium level due to the use of the wood being particularly light
Tensile strength	Low	High	Medium	The mixed system tensile strength is at a medium level. It would be better when entire tensile forces in the structure are carried by wood for durability
Compressive strength	High	High	High	The compressive strength of the mixed system is significantly high
Maintenance requirement	Low	High	Medium	The maintenance requirement of the mixed system is at a medium level since the wood structures require protection from insects, water, and fire, which will be periodically applied
Life endurance	High	High	High	The life endurance of the mixed system is at a high level when the maintenance requirement of the wood structure is applied
Fire endurance	High	Low	Medium	The fire endurance of the mixed system is at a medium level. The wood structure requires specific fire protection methods
Corrosion endurance	High	High	High	The corrosion endurance of the proposed system is high when the maintenance requirements of the wood structure are applied

(continued)

Table 5 (continued)

Material/system property	Masonry	Wood	Mixed system	Notes for application
Swelling endurance	High	Low	Medium	The swelling endurance of the wood structure is low, requiring periodical maintenance. Thus, the swelling endurance of the mixed system is at a medium level
Insect endurance	High	Low	Medium	The insect endurance of the wood is particularly low, requiring periodical maintenance. Thus, the swelling endurance of the mixed system is at a medium level
Material cost	Low	Medium/ high	Medium	The material cost of wood is at a medium level in many countries
Workmanship cost	Low	Medium/ high	Medium	The workmanship cost of the wood is medium level in many countries
Flexibility of use	Low	Medium	Medium	The masonry has low while wood has medium level flexibility of use

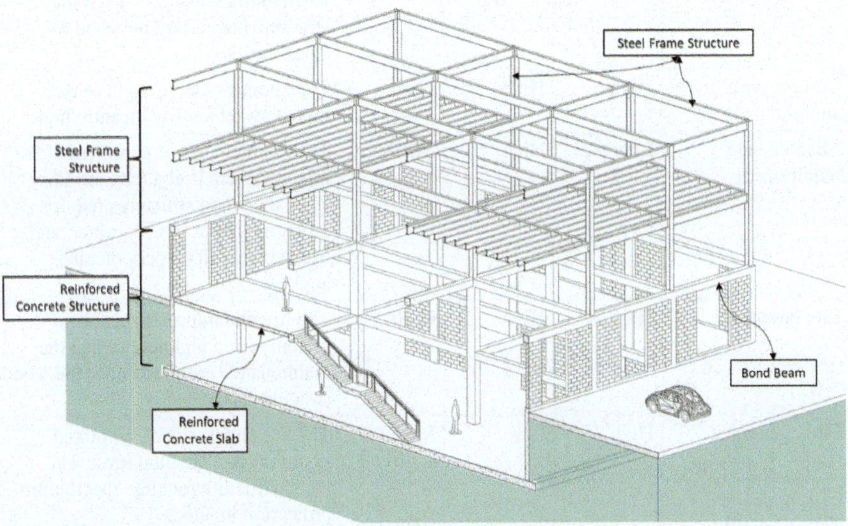

Fig. 221 A sample concrete and steel mixed super-structure

Reference

https://commons.wikimedia.org/wiki/File:Burj_Dubai_Under_Construction_on_16_May_2008.jpg

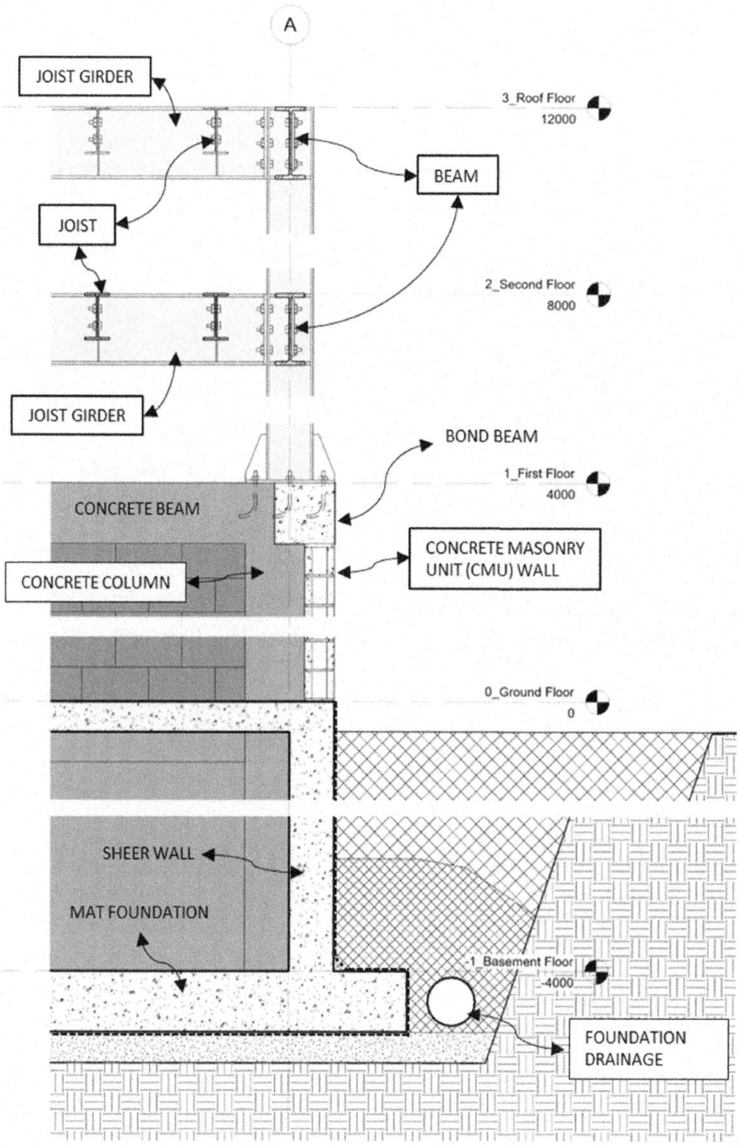

Fig. 222 The system detail of the concrete and steel mixed structure

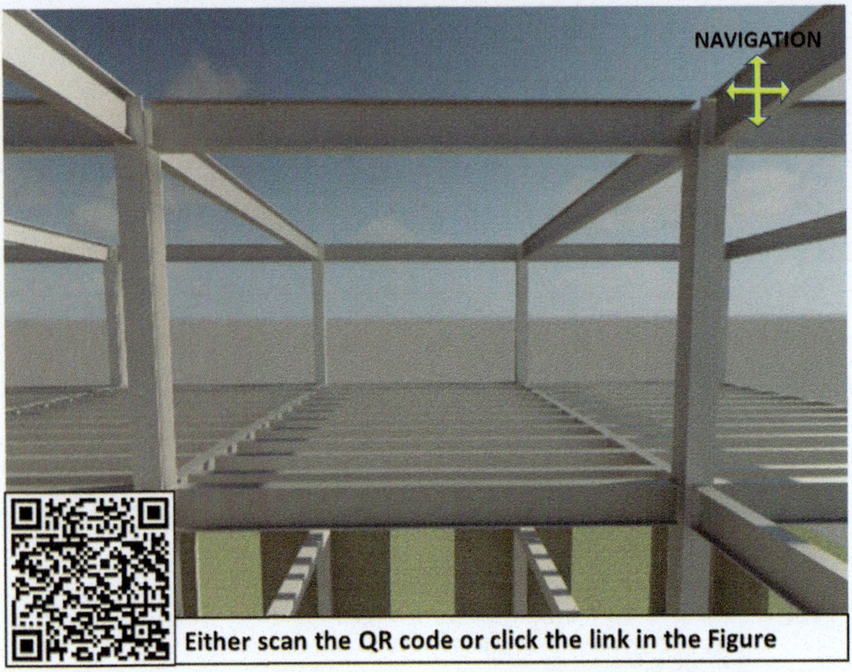

Fig. 223 The Immersive Virtual Model of the sample concrete and steel mixed super-structure. Click the link: https://pano.autodesk.com/pano.html?url=jpgs/51972af6-9e77-42e9-b8dd-effe92961798&version=2

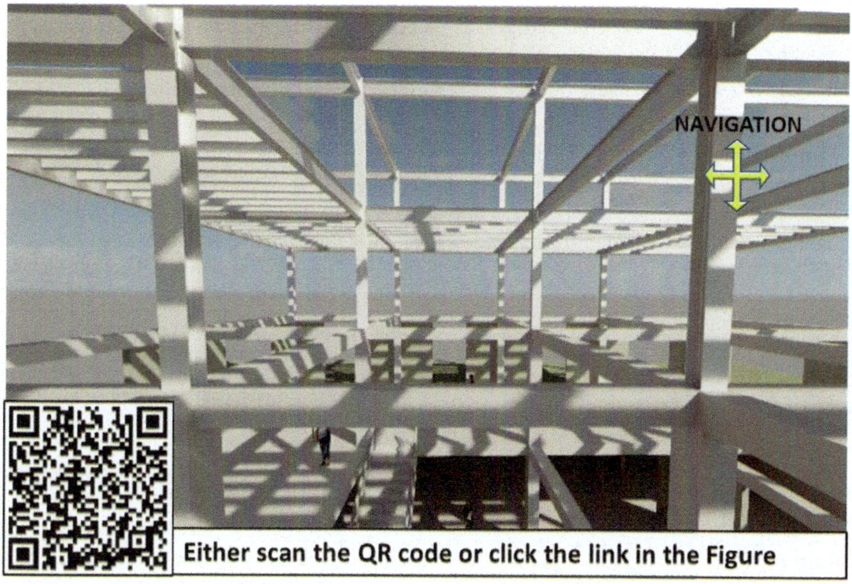

Fig. 224 The Immersive Virtual Model of the sample concrete and steel mixed super-structure. Click the link: https://pano.autodesk.com/pano.html?url=jpgs/8371e9ed-e7c5-47cb-9e4e-d67edffa3938&version=2

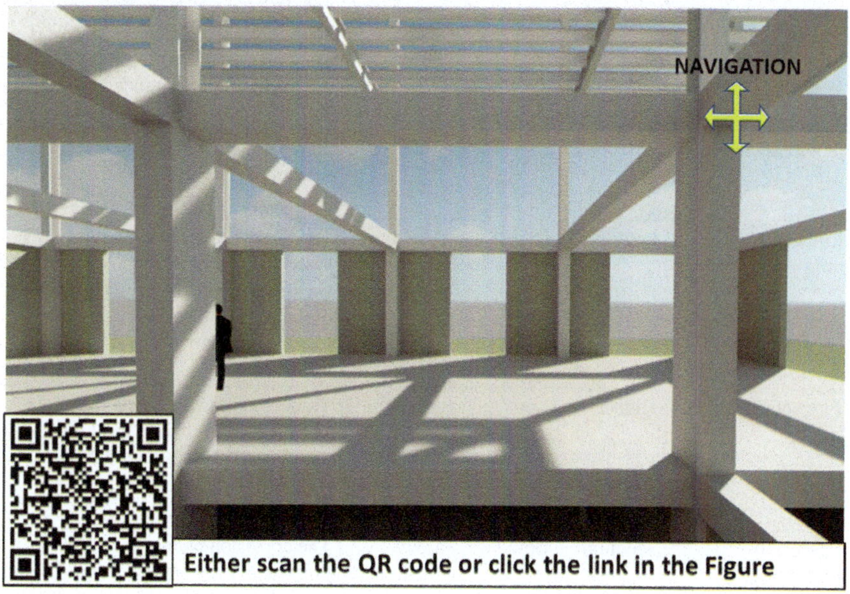

Fig. 225 The Immersive Virtual Model of the sample concrete and steel mixed super-structure. Click the link: https://pano.autodesk.com/pano.html?url=jpgs/57ea7901-d8bd-421c-9050-3fb3b980c13b&version=2

Table 6 Evaluation of material and system properties of the concrete and steel mixed superstructure

Material/system property	Concrete	Steel	Mixed system	Notes for application
Thermal expansion	Low	High	Medium	The thermal expansion performance of the mixed system is at a medium level
Thermal conductivity	Medium	High	Medium/ high	The thermal conductivity of the mixed system is medium/high due to the use of concrete
Vibration	Low	High	Medium	The mixed system performs medium-level shaking due to vibration
Durability against vibration	Low	High	Medium	The concrete structure is fragile against vibration in the long term. Furthermore, cracks and deformations occur due to vibration requiring maintenance
Weight	High	Low	Medium	The mixed system's weight is at medium level
Tensile strength	Low	High	Medium	The mixed system tensile strength is at a medium level. It would be better when the entire tensile force in the structure is carried by steel for durability
Compressive strength	High	High	High	The compressive strength of the mixed system is significantly high
Maintenance requirement	Low	High	Medium	The maintenance requirement of the mixed system is medium since the steel requires corrosion protection methods to be periodically applied
Life endurance	Low	High	Medium	The life endurance of the mixed system is at a medium level due to the use of concrete. Compared to the life of the steel, the concrete structure needs to be renovated a couple of times
Fire endurance	High	Low	Medium	The fire endurance of the mixed system is at a medium level due to the use of steel. This is due to the fact that steel fire resistance is significantly low
Corrosion endurance	Medium	Low	Low/ medium	The corrosion endurance of the proposed system is at a low/medium level. The steel requires corrosion protection methods to be periodically applied

(continued)

Table 6 (continued)

Material/system property	Concrete	Steel	Mixed system	Notes for application
Swelling endurance	High	High	High	Both steel and concrete have high swelling endurance
Insect endurance	High	High	High	Both steel and concrete have high insect endurance
Material cost	Low	Medium/ high	Medium	The steel's material cost is high in many countries while the concrete is low. Thus, the mixed system' material cost is at medium level
Workmanship cost	Low	High	Medium	The workmanship cost of steel is high in many countries while the concrete is at a low level, making the mixed system's material cost medium level
Flexibility of use	High	High	High	Both concrete and steel have a high level of flexibility of use

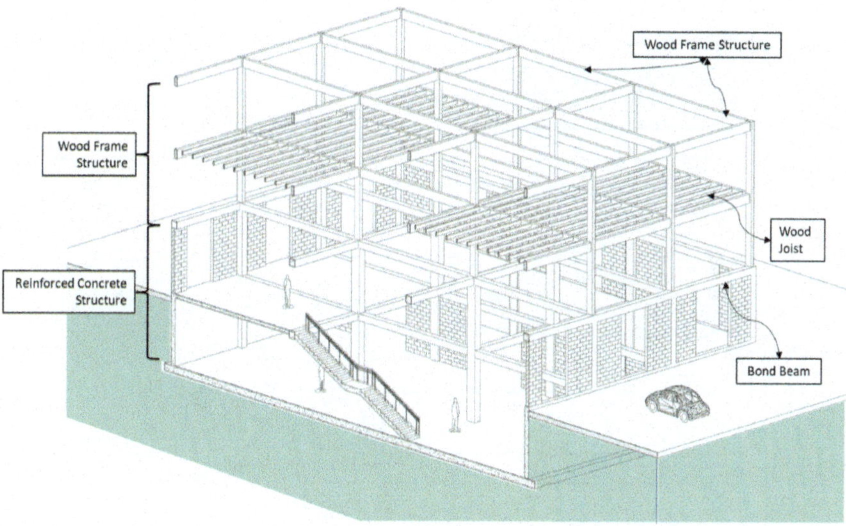

Fig. 226 A sample 3-D section of the concrete and wood mixed super-structure

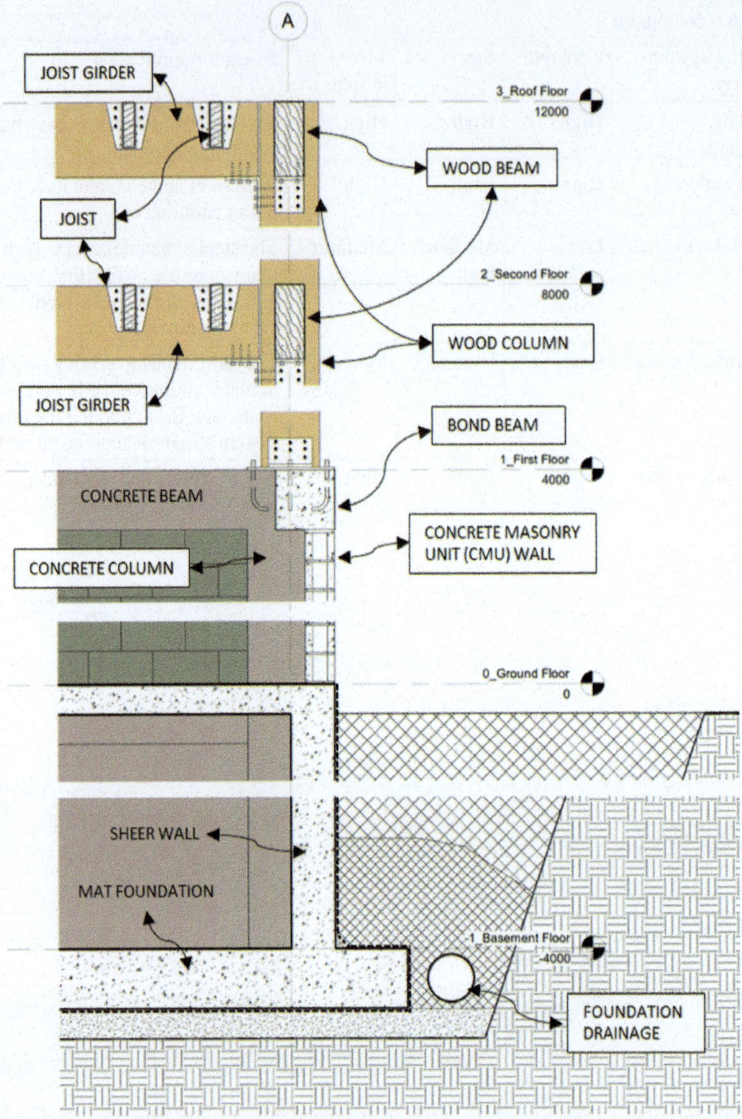

Fig. 227 The system detail of the concrete and wood mixed super-structure

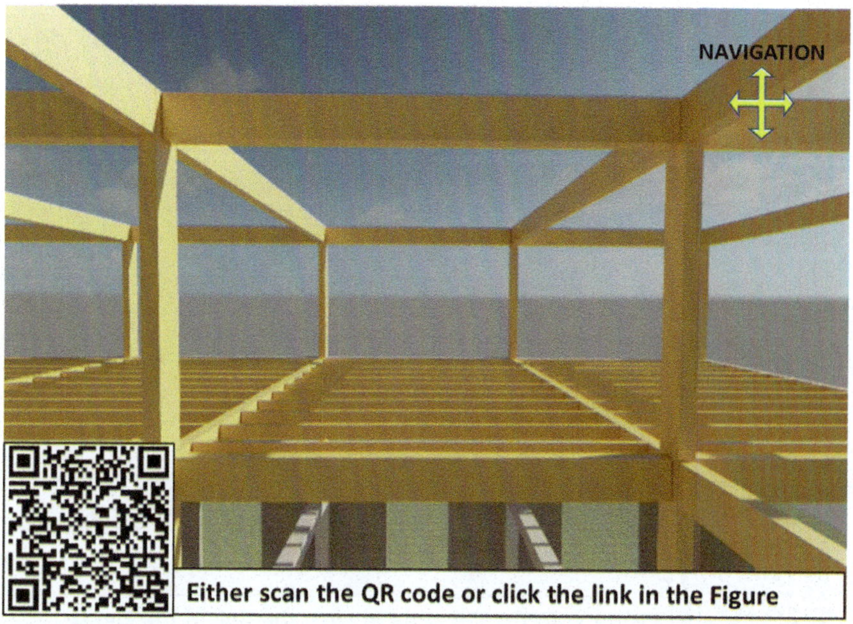

Fig. 228 The Immersive Virtual Model of the sample concrete and steel mixed super-structure. Click the link: https://pano.autodesk.com/pano.html?url=jpgs/35e680e4-076d-4162-a454-78edac4fb646&version=2

Fig. 229 The Immersive Virtual Model of the sample concrete and steel mixed super-structure. Click the link: https://pano.autodesk.com/pano.html?url=jpgs/842135c2-b36c-4a5e-bcd4-7dd7b0b7fb50&version=2

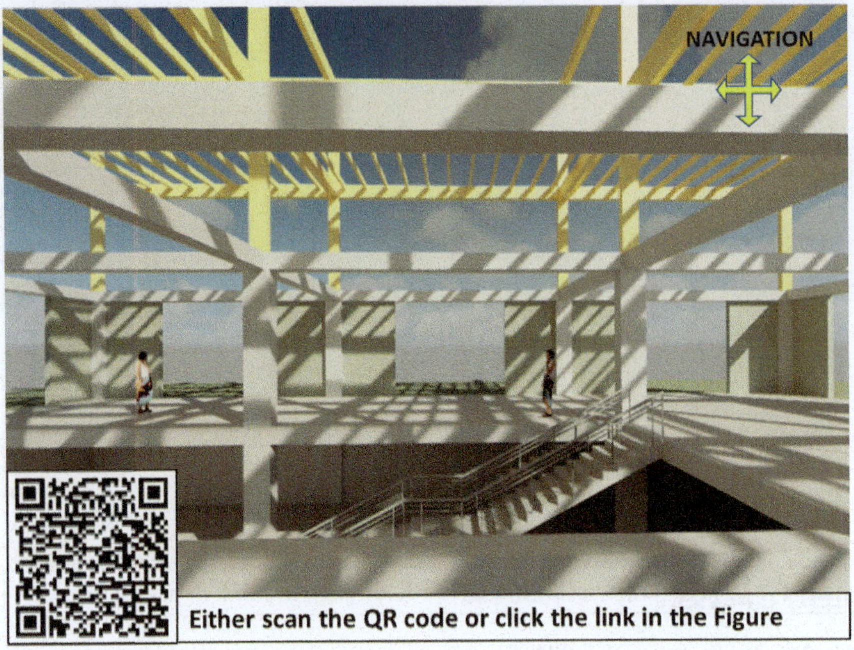

Fig. 230 The Immersive Virtual Model of the sample concrete and steel mixed super-structure. Click the link: https://pano.autodesk.com/pano.html?url=jpgs/bd9ad4ab-5927-433d-801a-935414406f3c&version=2

Table 7 Evaluation of material and system property of concrete and wood mixed super-structure

Material/system property	Concrete	Wood	Mixed system	Notes for application
Thermal expansion	Low	High	Medium	The mixed system has medium-level thermal expansion
Thermal conductivity	Medium	Low	Medium/low	The thermal conductivity of the mixed system is medium/low level
Vibration	Low	High	Medium	The mixed system reflects medium-level shaking against vibration
Durability against vibration	Low	High	Medium	The concrete structure is fragile against vibration in the long term. Furthermore, cracks and deformations occur due to vibration requiring maintenance, while wood has high structural durability against vibration
Weight	High	Low	Medium	The mixed system's weight is medium due to the use of particularly light wood
Tensile strength	Low	High	Medium	The mixed system tensile strength is at a medium level. It would be better when entire tensile forces in the structure are carried by wood for durability
Compressive strength	High	High	High	The compressive strength of the mixed system is significantly high
Maintenance requirement	Low	High	Medium	The maintenance requirement of the mixed system is at a medium level since the wood structures require protection from insects, water, and fire, which will be periodically applied
Life endurance	Low	High	Medium	The life endurance of the mixed system is at a medium level due to the use of concrete. Compared to the life of the wood, the concrete structure needs to be renovated a couple of times
Fire endurance	High	Low	Medium	The fire endurance of the mixed system is at a medium level. The wood structure requires specific fire protection methods
Corrosion endurance	Medium	High	Medium / High	The corrosion endurance of the proposed system is medium/high when the maintenance requirements of the wood structure are applied

(continued)

Table 7 (continued)

Material/system property	Concrete	Wood	Mixed system	Notes for application
Swelling endurance	High	Low	Medium	The swelling endurance of the wood structure is low, requiring periodical maintenance. Thus, the swelling endurance of the mixed system is at a medium level
Insect endurance	High	Low	Medium	The insect endurance of the wood is particularly low, requiring periodical maintenance. Thus, the swelling endurance of the mixed system is at a medium level
Material cost	Low	Medium/ high	Medium	The material cost of wood is at medium/high level in many countries
Workmanship cost	Low	Medium/ high	Medium	The workmanship cost of the wood is medium/high level in many countries
Flexibility of use	High	High	High	Both concrete and steel have a high level of flexibility of use

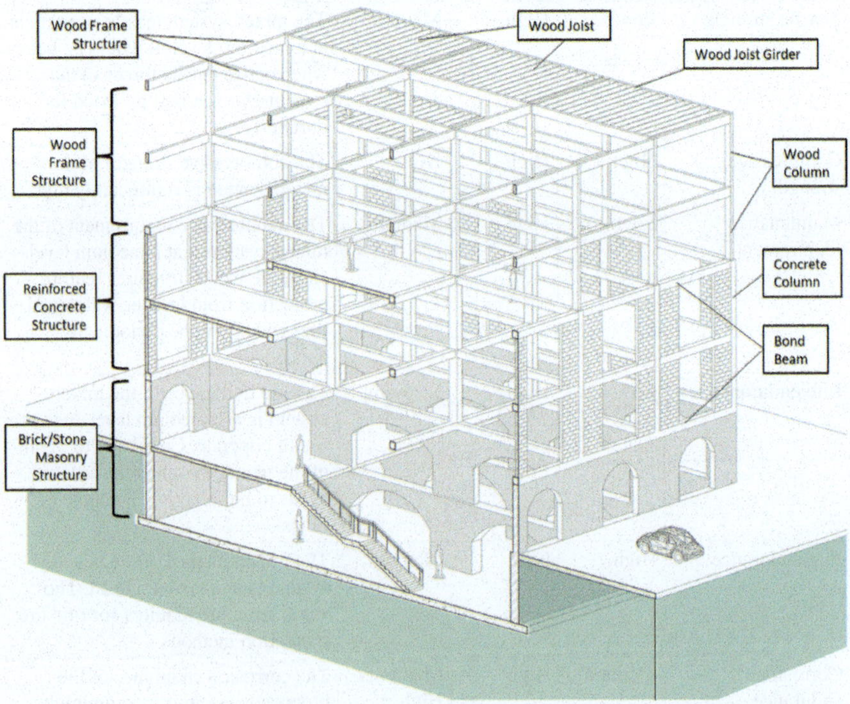

Fig. 231 The 3-D section of a sample masonry-concrete-wood mixed super-structure

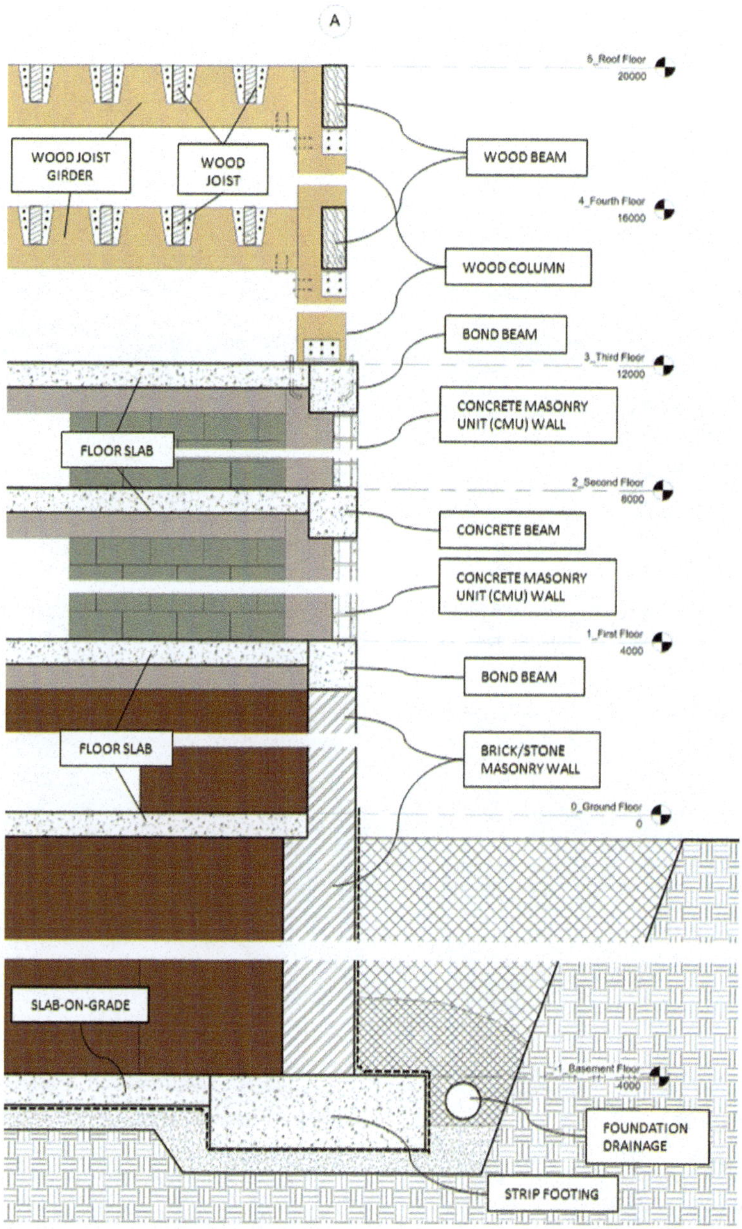

Fig. 232 The system detail of the masonry-concrete-wood mixed super-structure

Fig. 233 The Immersive Virtual Model of the sample concrete and steel mixed super-structure. Click the link: https://pano.autodesk.com/pano.html?url=jpgs/300a6ce8-af68-4ea1-8dc9-d4e503b99a86&version=2

Fig. 234 The Immersive Virtual Model of the sample concrete and steel mixed super-structure. Click the link: https://pano.autodesk.com/pano.html?url=jpgs/fc8e2af8-b346-4369-9237-bde8bae1a40e&version=2

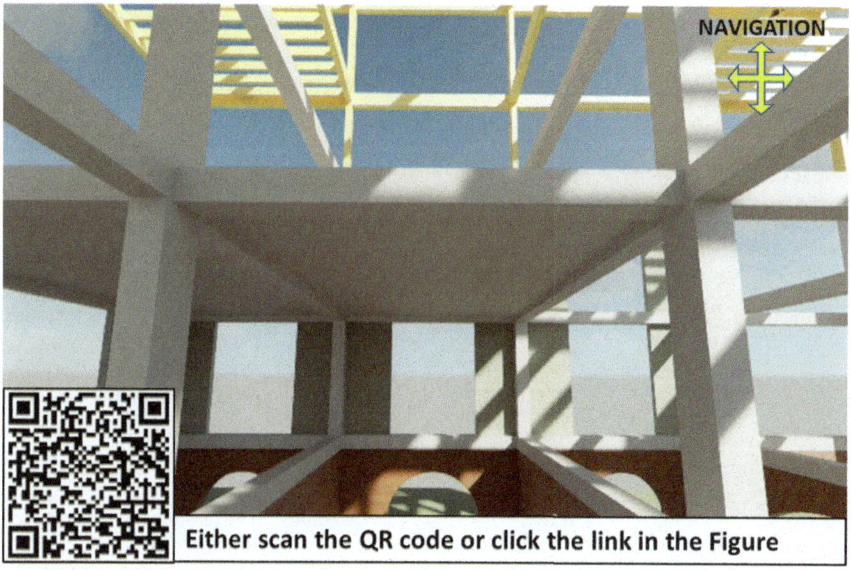

Fig. 235 The Immersive Virtual Model of the sample concrete and steel mixed super-structure. Click the link: https://pano.autodesk.com/pano.html?url=jpgs/364ecb1a-6895-4a12-9844-aefd5c317069&version=2

Fig. 236 The Immersive Virtual Model of the sample concrete and steel mixed super-structure. Click the link: https://pano.autodesk.com/pano.html?url=jpgs/c281efc7-d57e-49a4-b6bb-2446d2e57446&version=2

Table 8 Evaluation of material and system property of masonry-concrete-wood mixed superstructure

Material/system property	Masonry	Concrete	Wood	Mixed system	Notes for application
Thermal expansion	Low	Low	High	Medium	The masonry and concrete have low, while wood has high thermal expansion properties, making the mixed system medium level
Thermal conductivity	Low	Medium	Low	Low	The thermal conductivity of the mixed system is low since the masonry and the wood structures have low thermal conductivity characteristics
Vibration	Low	Low	High	Medium	The mixed system reflects medium-level shaking against vibration
Durability against vibration	Low	Low	High	Medium	The masonry and the concrete structures are fragile against vibration in the long term. Furthermore, cracks and deformations occur due to vibration, requiring maintenance
Weight	High	High	Low	High	The mixed system's weight is significantly high
tensile strength	Low	Low	High	Medium	It would be better when the entire tensile force in the structure is carried by wood for durability
Compressive Strength	High	High	High	High	The compressive strength of the mixed system is significantly high
Maintenance requirement	Low	Low	High	Medium	The maintenance requirement of the mixed system is medium
Life endurance	High	Low	High	Medium	The life endurance of the mixed system is at a medium level due to the use of concrete. Compared to the masonry and wood life, the concrete structure needs to be renovated several times
Fire endurance	High	High	Low	Medium	The fire endurance of the mixed system is medium due to the use of wood

(continued)

Table 8 (continued)

Material/system property	Masonry	Concrete	Wood	Mixed system	Notes for application
Corrosion endurance	High	Medium	High	High	The corrosion endurance of the proposed system is high in general
Swelling endurance	High	High	Low	Medium	Both masonry and concrete have high swelling endurance, while wood has low swelling endurance, requiring the application of specific protection methods
Insect endurance	High	High	Low	Medium	Both masonry and concrete have high insect endurance, while wood has particularly low, causing additional insect protection methods for extending the endurance of the mixed system
Material cost	Low	Low	Medium/ high	Medium	Both masonry and concrete material costs are particularly low, while wood is medium/ high in many countries
Workmanship cost	Low	Low	Medium/ high	Medium	Except in a few countries, masonry and concrete workmanship costs are low, while wood is medium/high in many countries
Flexibility of use	Low	High	Medium	Medium	The flexibility of use of each structure in the mixed system varies; thus, the overall characteristics of the system could be accepted as medium

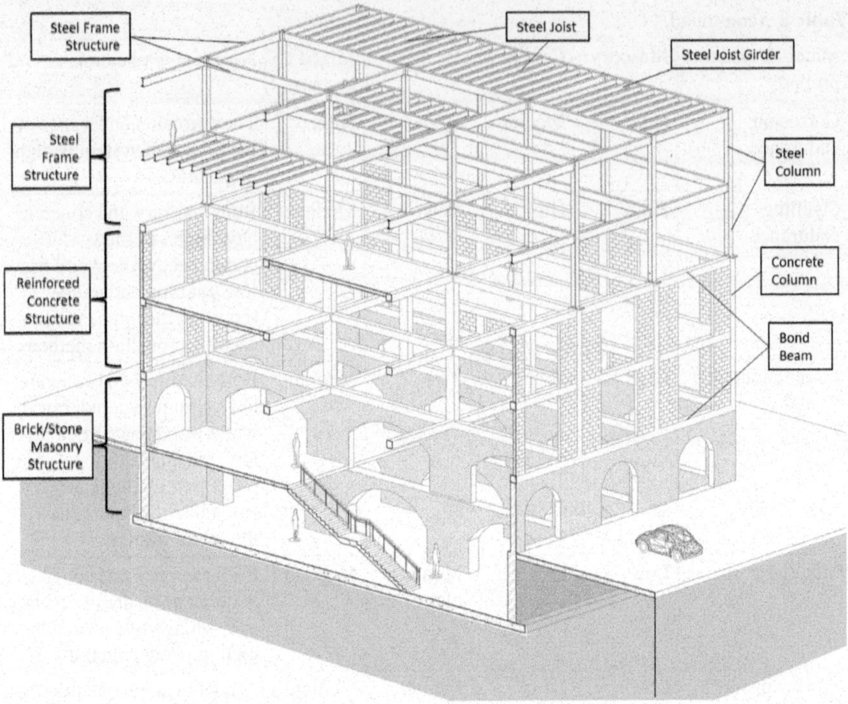

Fig. 237 The 3-D section of a sample masonry-concrete-steel mixed super-structure

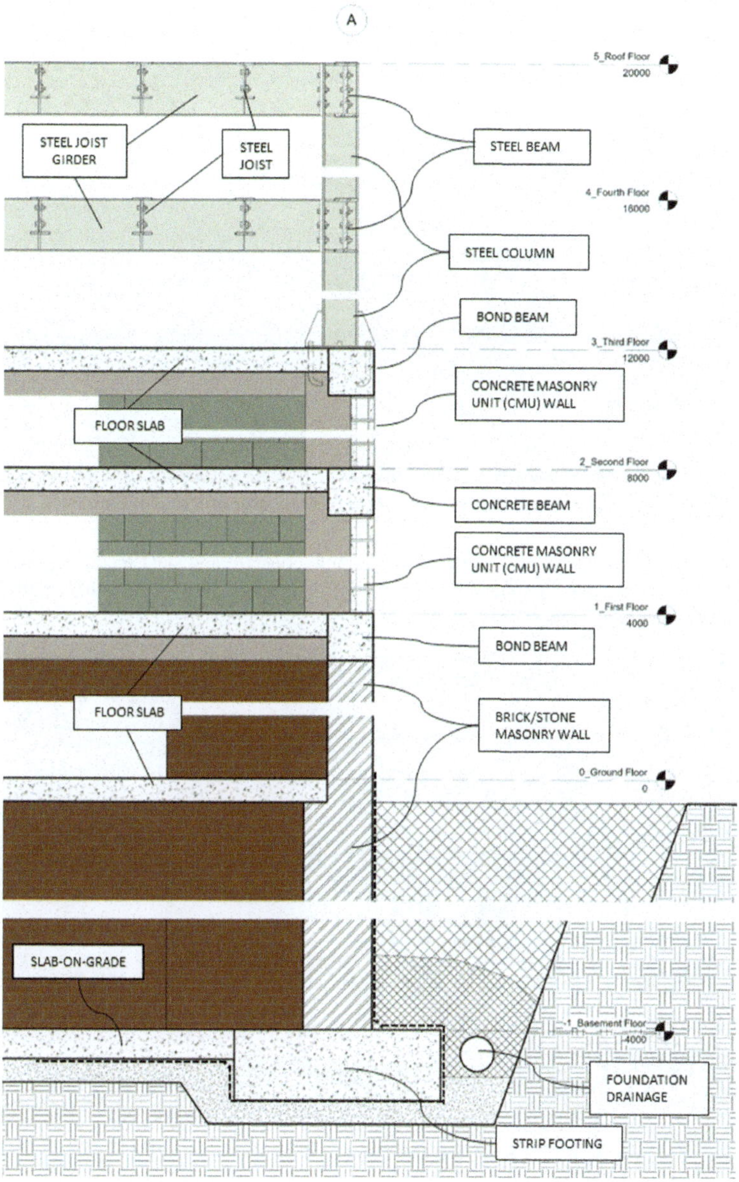

Fig. 238 The system detail of the masonry-concrete-steel mixed super-structure

Fig. 239 The Immersive Virtual Model of the sample concrete and steel mixed super-structure. Click the link: https://pano.autodesk.com/pano.html?url=jpgs/3039d1c9-e601-4071-b8f0-17b509715acd&version=2

Fig. 240 The Immersive Virtual Model of the sample concrete and steel mixed super-structure. Click the link: https://pano.autodesk.com/pano.html?url=jpgs/a158d5d3-b700-4a94-a463-bbee2c189289&version=2

Fig. 241 The Immersive Virtual Model of the sample concrete and steel mixed super-structure. Click the link: https://pano.autodesk.com/pano.html?url=jpgs/fc8fc1c9-eb4f-49a3-a89a-a9dff67c6647&version=2

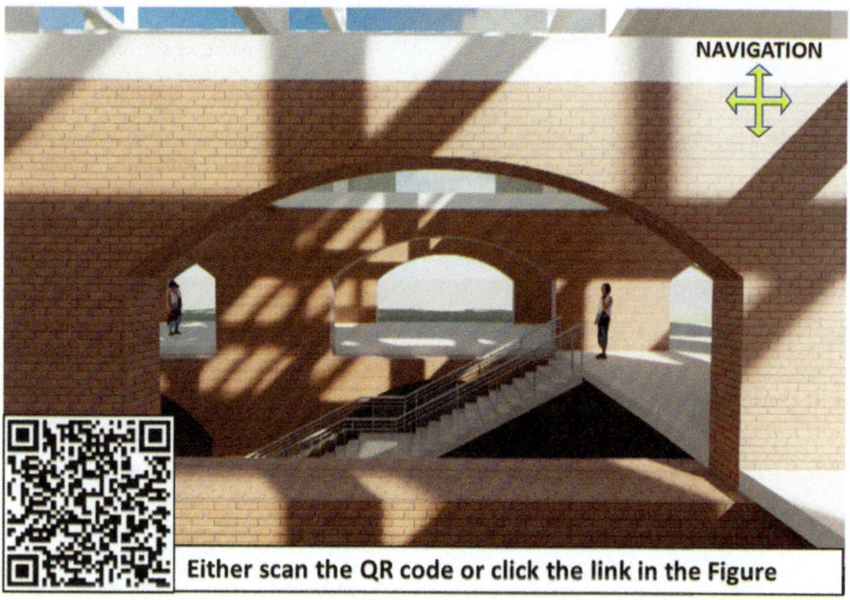

Fig. 242 The Immersive Virtual Model of the sample concrete and steel mixed super-structure. Click the link: https://pano.autodesk.com/pano.html?url=jpgs/6d791a64-3b4b-442c-bc69-b9284d5b9acf&version=2

Table 9 Evaluation of material and system property of masonry-concrete-steel mixed superstructure

Material/system property	Masonry	Concrete	Steel	Mixed system	Notes for application
Thermal expansion	Low	Low	High	Medium	The masonry and concrete have low, while steel has high thermal expansion properties, making the mixed system medium level
Thermal conductivity	Low	Medium	High	Medium	The thermal conductivity of the mixed system is medium since each structure's thermal conductivity varies
Vibration	Low	Low	High	Medium	The mixed system reflects low/medium level shaking against vibration
Durability against vibration	Low	Low	High	Medium	The masonry and the concrete structures are fragile against vibration in the long term. Furthermore, cracks and deformations occur due to vibration, requiring maintenance
Weight	High	High	Low	High	The mixed system's weight is significantly high
tensile strength	Low	Low	High	Medium	It would be better when the entire tensile force in the structure is carried by steel for durability
Compressive strength	High	High	High	High	The compressive strength of the mixed system is significantly high
Maintenance requirement	Low	Low	High	Medium	The maintenance requirement of the mixed system is medium. This is because the steel requires periodic corrosion protection methods to be applied
Life Endurance	High	Low	High	Medium/ high	The life endurance of the mixed system is at a medium level due to the use of concrete. Compared to the masonry and steel life, the concrete structure needs to be renovated several times
Fire endurance	High	High	Low	Medium	The fire endurance of the mixed system is medium due to the use of steel

(continued)

Table 9 (continued)

Material/system property	Masonry	Concrete	Steel	Mixed system	Notes for application
Corrosion endurance	High	Medium	Low	Medium	The corrosion endurance of the proposed system is medium in general. The steel is vulnerable to corrosion, requiring the application of corrosion protection methods
Swelling endurance	High	High	High	High	The swelling endurance of the mixed system is high
Insect endurance	High	High	High	High	The insect endurance of the mixed system is high
Material cost	Low	Low	Medium/ high	Low/ medium	Both masonry and concrete material costs are particularly low, while steel is medium/high in many countries
Workmanship cost	Low	Low/ High	Medium/ High	Medium	Except in a few countries, masonry and concrete workmanship costs are low, while steel is medium/high in many countries
Flexibility of use	Low	High	High	Medium	The masonry structure limits the flexibility of use, particularly in the mixed system; thus, overall characteristics could be accepted as medium

Chapter 4
Roofing Systems

Abstract One of the important parts of the building envelope is the roof layer, a horizontal and angular layer separating the building from the atmospheric conditions. This chapter presents the roof system with its layers and application details.

1 Roofing

1.1 Low-Sloped Roof

It is a water-resisting roof with a continuous roof membrane over a relatively flat surface. It has a slope less than 25% and greater than 2%. The fundamental components of low-sloped roofs are roof membrane & covering, insulation, roof deck, flashing, and vapor retarder.

1.1.1 Roof Membrane and Covering

It is the water-proofing layer. Due to being top layers, it is constantly subjected to the weathering effects of sun, rain, snow, hail, and wind. Roof membranes are divided into three general categories:

- **Built-up Roof Membrane** consists of three to five plies (layers) of felt with intervening mopping of bitumen (asphalt or coal tar). Both asphalt and coal tar are adversely affected by ultraviolet radiation, and an aggregate cover generally protects a built-up roof membrane. The aggregate cover also increases the roof's fire resistance.
- **A Modified Bitumen Roof Membrane** is similar to a built-up roof membrane and comprises two or three plies of modified bitumen sheets with intervening mopping of bitumen. A protective aggregate cover or a mineral granule-surfaced cap sheet is required.

R. Sarı and E. B. Çalışkan, *Building Construction Methods and Systems*,
https://doi.org/10.1007/978-3-031-50043-5_4

- **Single-Ply Roof Membrane** consists of only one sheet of a synthetic polymer (plastic) and does not require any protective cover. Although slightly variable, the relative market shares of built-up, modified bitumen, and single-ply roof membranes are approximately the same-nearly one-third each.

1.1.2 Insulation

Different than wall insulation, roof insulation requires extra attention. The loads and movements are parallel in wall layers, while all these efforts are perpendicular to the roof. Thus, special details are needed to be applied in flashing. The cracks and leakage at the roof surface may cause various leaks at the ceiling, making it hard to detect the exact leakage point. When such a problem is confronted, the entire roof water isolation system must be rehabilitated to remove water. Thermal insulation layers are the most vulnerable layer among roof layers against water leakage. The vapor retarder layer is generally placed at the thermal insulation's top and bottom surfaces.

Furthermore, these vapor retarder layers are connected at flashings without relying on chemical bonding. This is because, timely, chemical bonding is prone to be distorted due to seasonal temperature changes, causing not working of chemical conjunction. Mechanical connections are preferred besides chemical conjunction methods to avoid being impacted (Fig. 1).

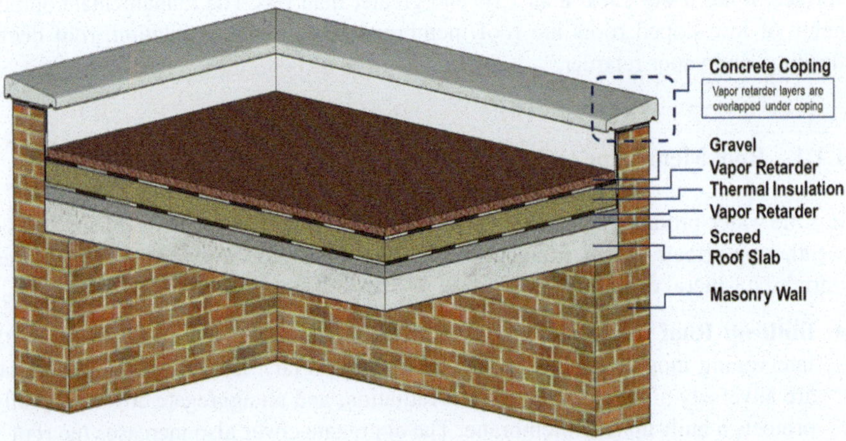

Fig. 1 Two layered vapor retarders at the roof deck

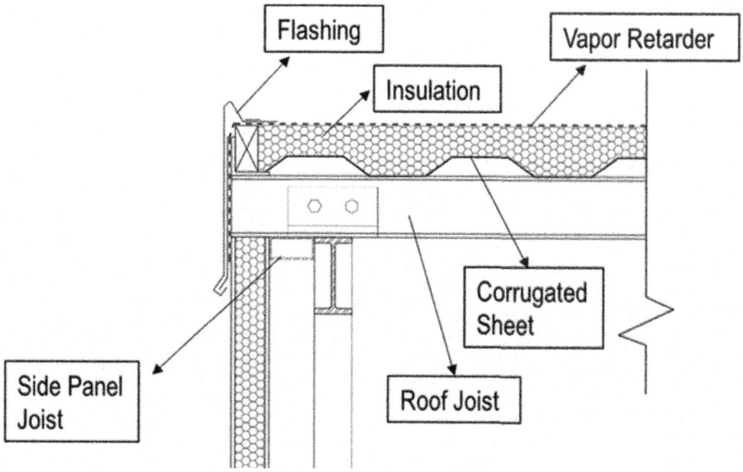

Fig. 2 Flashing detail at the steel roof deck

1.1.3 Roof Deck

The roof deck includes structural flooring layers placed for the roof level. Various flooring system is available besides long-span and steeped opportunities. No more figure is illustrated in this section regarding the aforementioned roof flooring systems.

1.1.4 Flashing

Flashing is used at the edges of the roof where the roof ends or there are expansion joints. Special flashing details are applied for each case. Some sample flashing details are illustrated in the Figs. 2, 3, 4, 5, 6, 7 and 8.

1.1.5 Vapor Retarder

A vapor retarder prevents vapor penetration and impacts the thermal insulation layer. It is advised by professionals to locate the thermal insulation layer in between two vapor retarder layers to take the best and most durable performance (Figs. 9, 10 and 11).

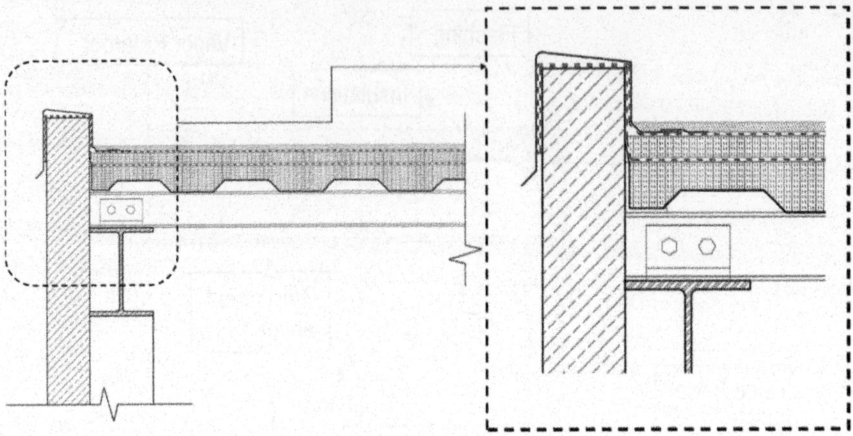

Fig. 3 Flashing detail for steel roof deck and CMU or brick parapet wall connection

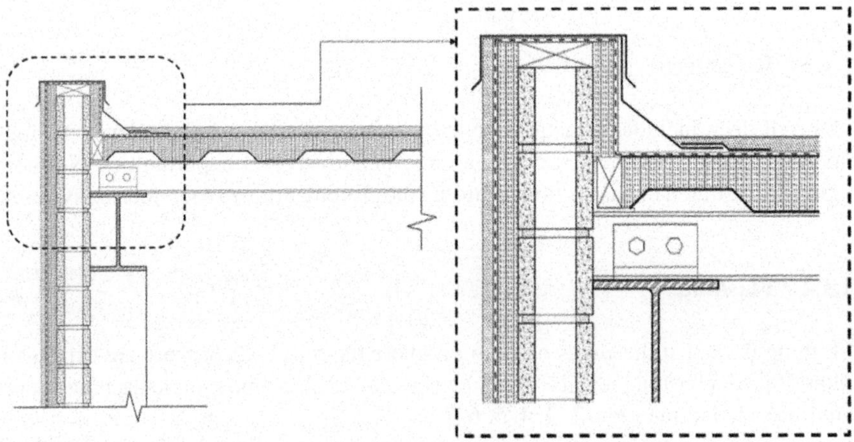

Fig. 4 Flashing detail for steel roof deck and CMU parapet wall connection

1.2 Steep Roof

It is also called a water-shedding roof and typically consists of small individual roofing units (shingles) overlapping. The fundamental components of a steep roof are roof shingles, roof underlayment, ice dam protection membrane when needed, roof deck, and flashing.

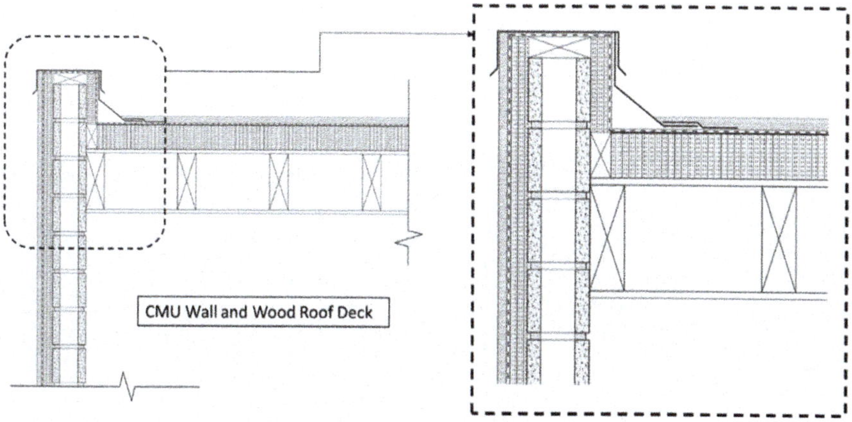

Fig. 5 Flashing detail for wood roof deck and CMU parapet wall connection

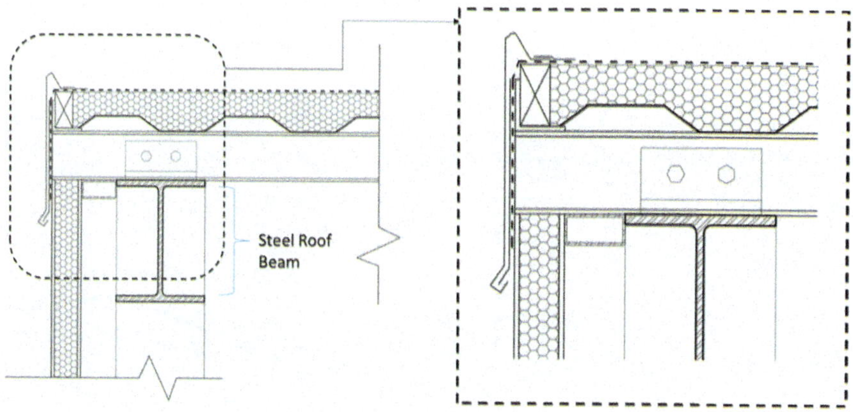

Fig. 6 Flashing detail for steel roof deck and insulated wall panel

1.2.1 Roof Shingles

Roofing shingles are flat or curved tiles that interlock or overlap in a way that channels water off of a pitched roof. They are made from materials that vary in cost, weight, durability, color, and style. Roofing shingles are generally comprised of a locally available material. Commonly used shingles are asphalt, slate, concrete, clay tiles, wood, and metal shingles.

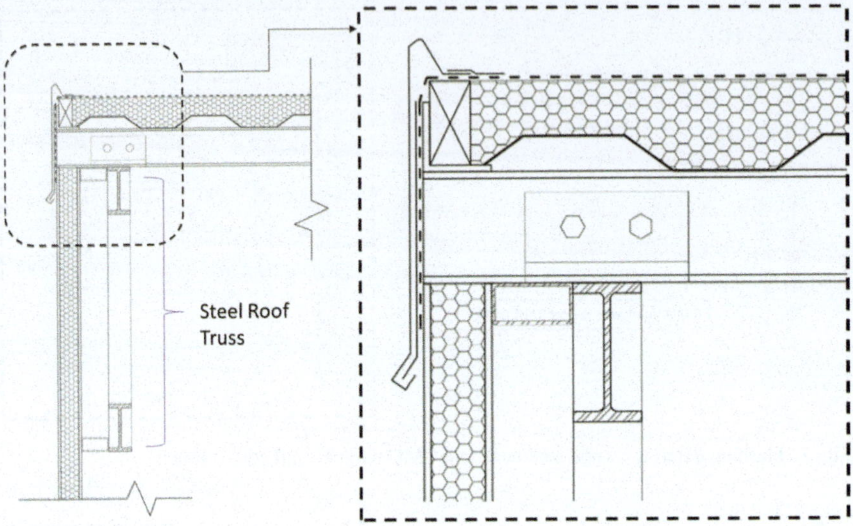

Fig. 7 Flashing detail for steel truss roof deck and insulated wall panel

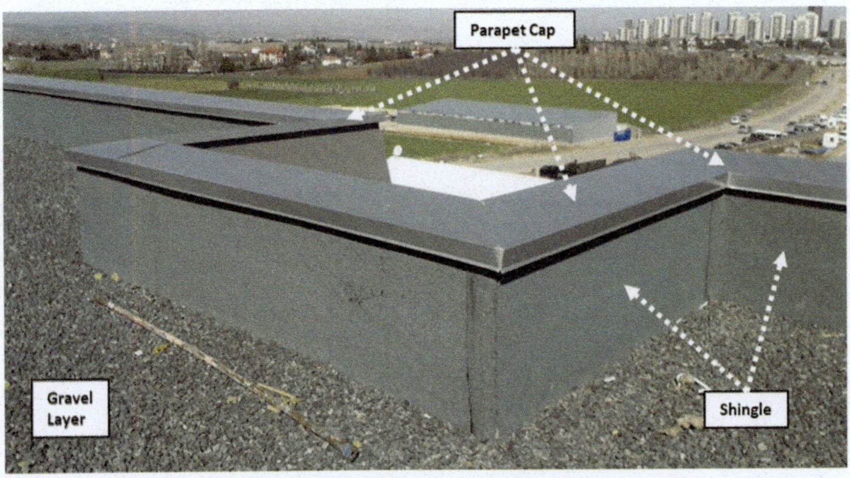

Fig. 8 Parapet cap (as flashing), shingle, and gravel layer application in a flat roof

1.2.2 Roof Underlayment

Roof underlayment is a water-resistant layer under the shingles as a second line of protection against water leakage (Fig. 12).

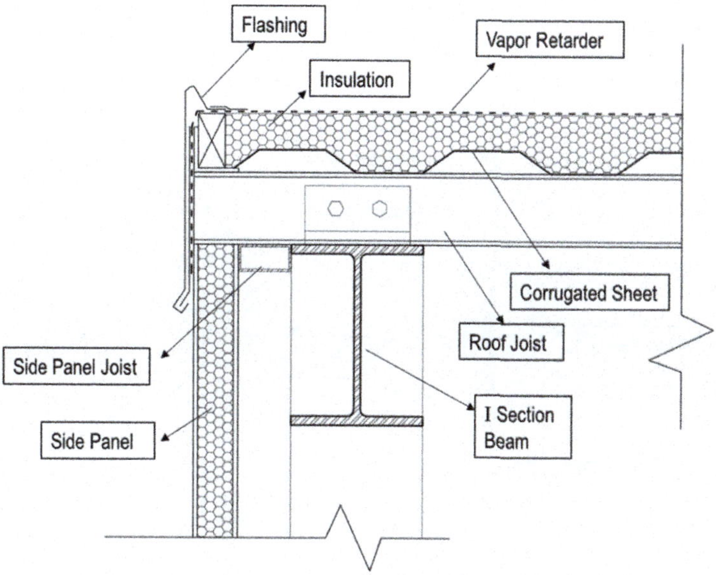

Fig. 9 Vapor retarder at steel roof deck and insulated wall panel connection

1.2.3 Roof Flashing

As in low-slope roofs, flashings are additional waterproofing components used to create a watertight seal at terminations and transitions in the roof. Critical areas for flashings are valleys, eaves, rakes, chimneys, and other penetrations (Fig. 13).

1.2.4 Roof Deck

The deck is a structural component of the roof. The deck generally consists of plywood or oriented strand board (OSB) panels to cover the structural components (Figs. 14, 15, 16, 17 and 18).

1.2.5 Some Application Details

See Figs. 19, 20, 21, 22, 23, 24, 25, 26, 27, 28, 29, 30 and 31.

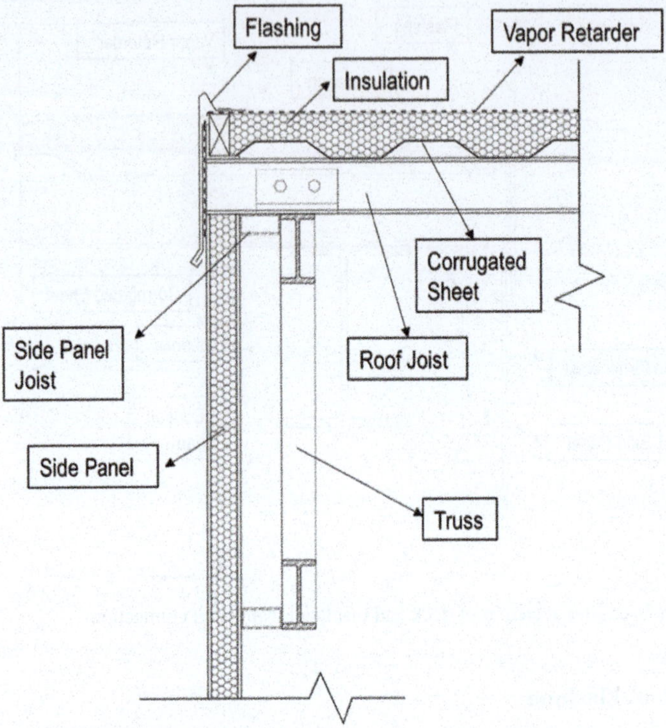

Fig. 10 Vapor retarder at steel roof deck and insulated wall panel connection

Fig. 11 Asphalt rolling is used as a vapor retarder in foundation and roof structures

Fig. 12 The membrane layer applied over the OSB panel is a roof underlayment

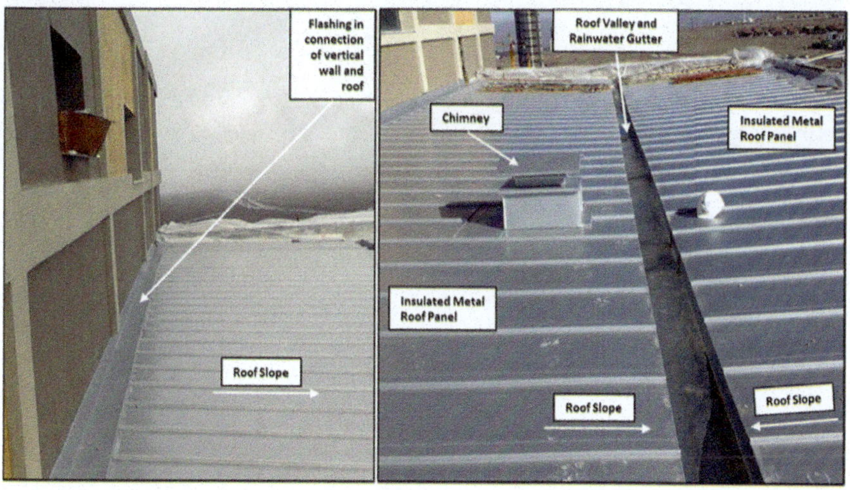

Fig. 13 Roof flashing and valley application

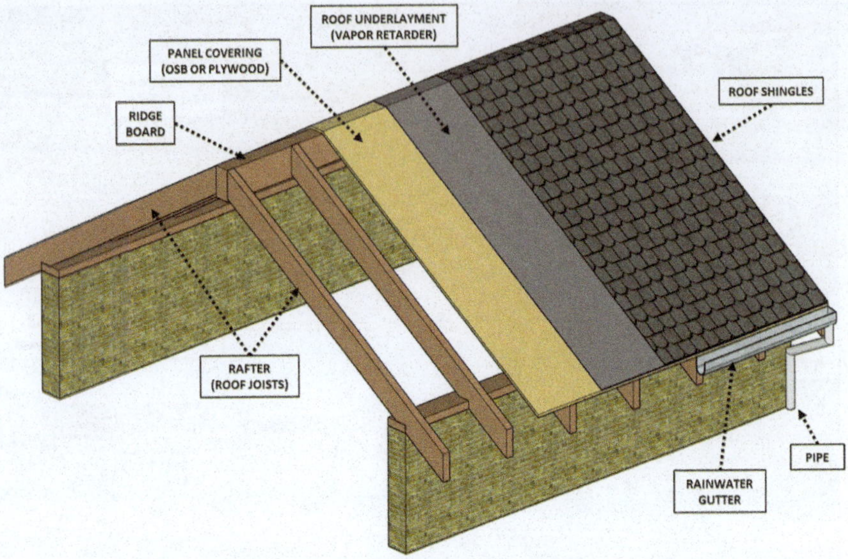

Fig. 14 Layers of a roof deck

Fig. 15 Wood strips are used at the roof deck

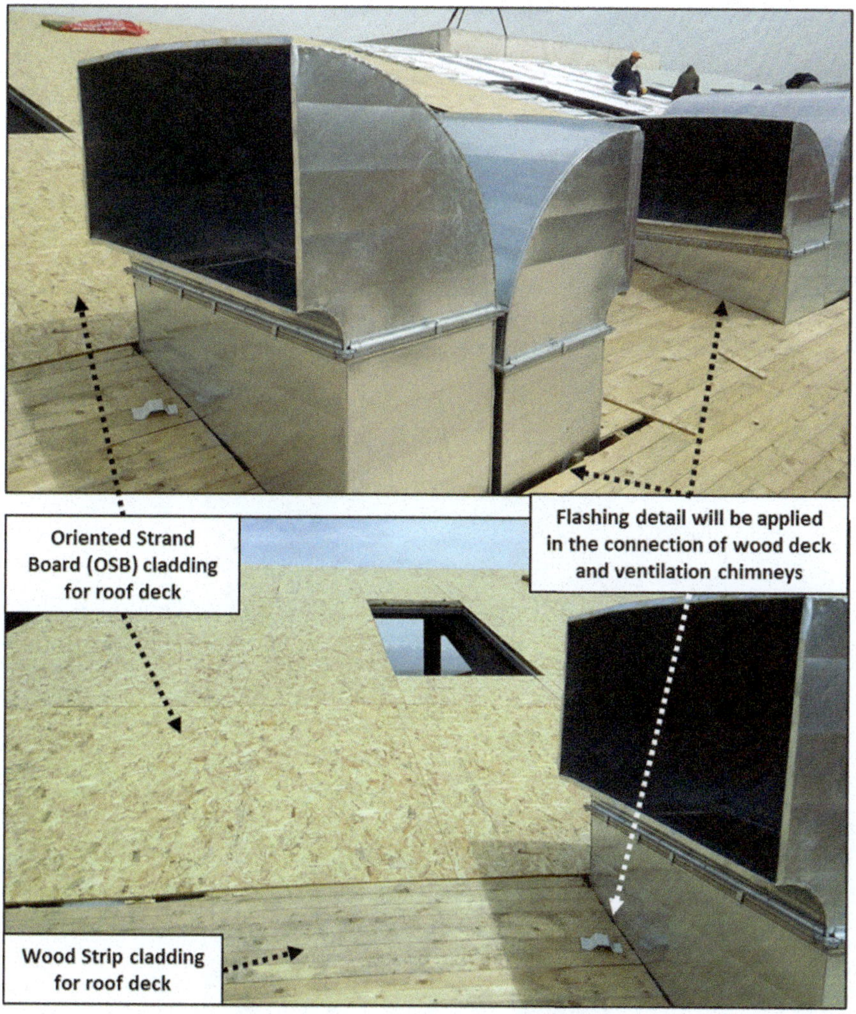

Fig. 16 OSB and wood strip roof deck

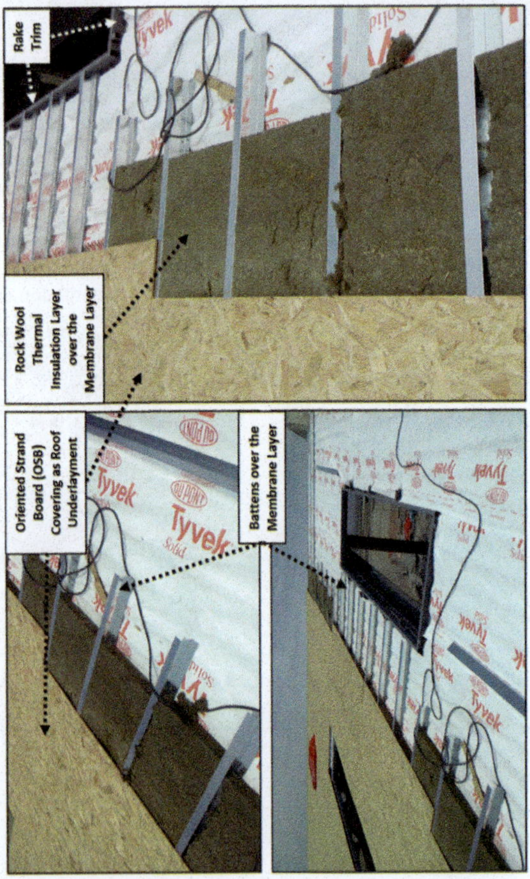

Fig. 17 Batten, membrane, rock wool thermal insulation, and OSB cladding application in roof construction

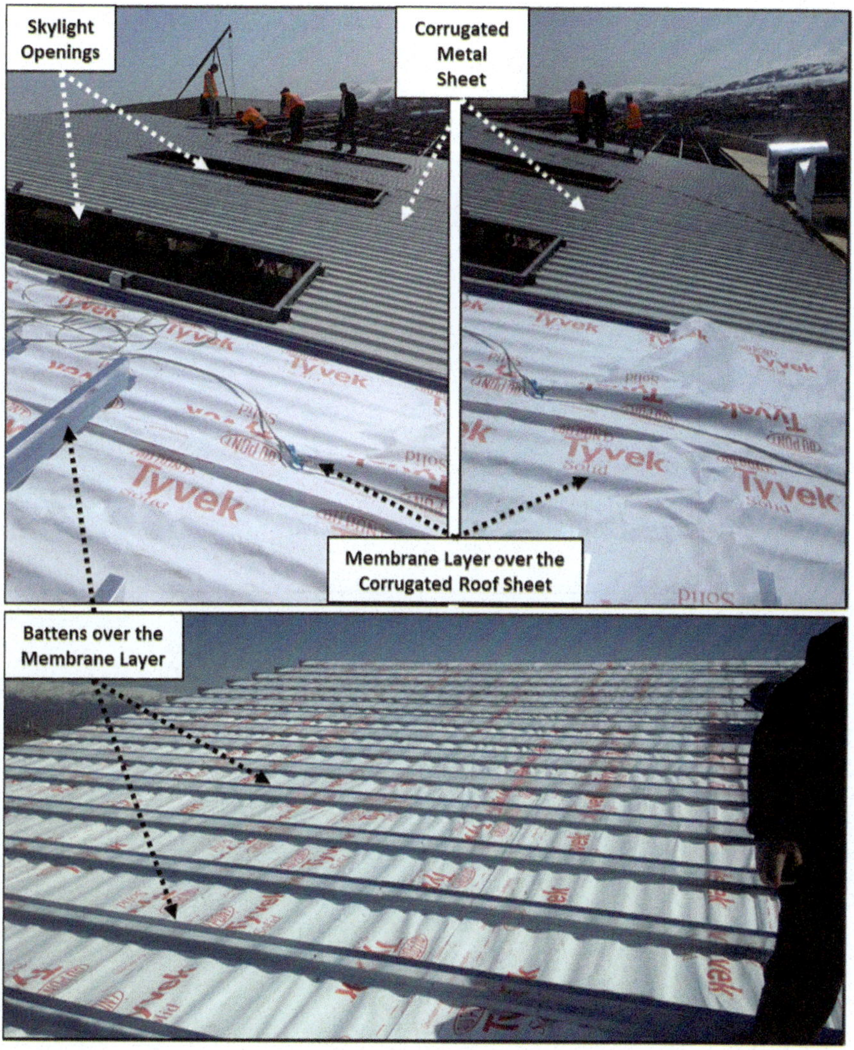

Fig. 18 Batten and membrane cladding in a steel roof deck

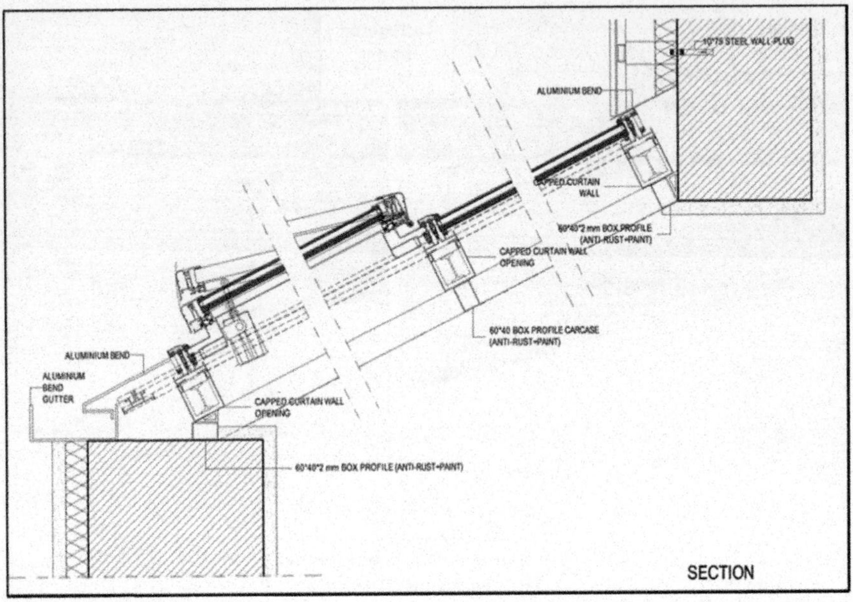

Fig. 19 A skylight detail in a roof

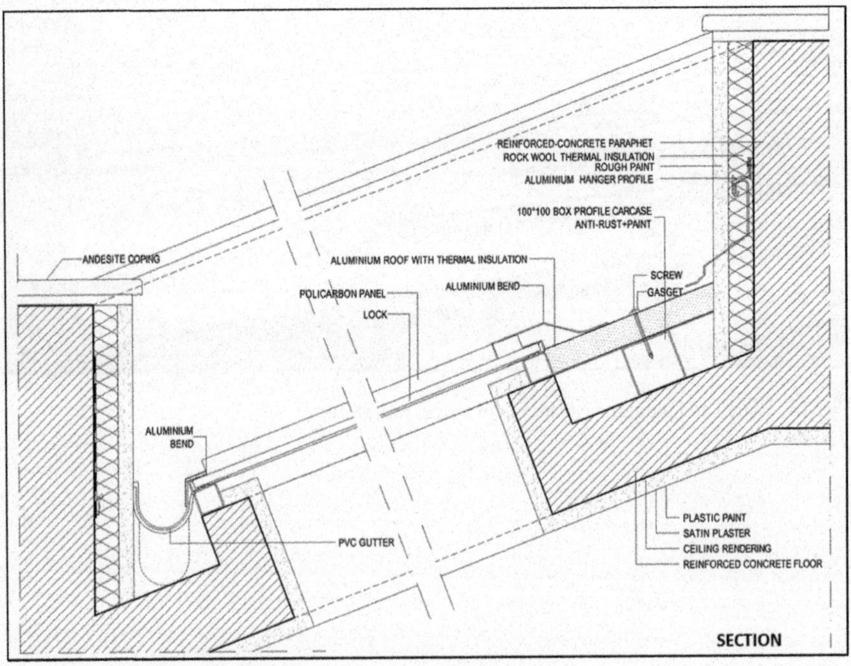

Fig. 20 A skylight, parapet, rainwater gutter, and flashing detail

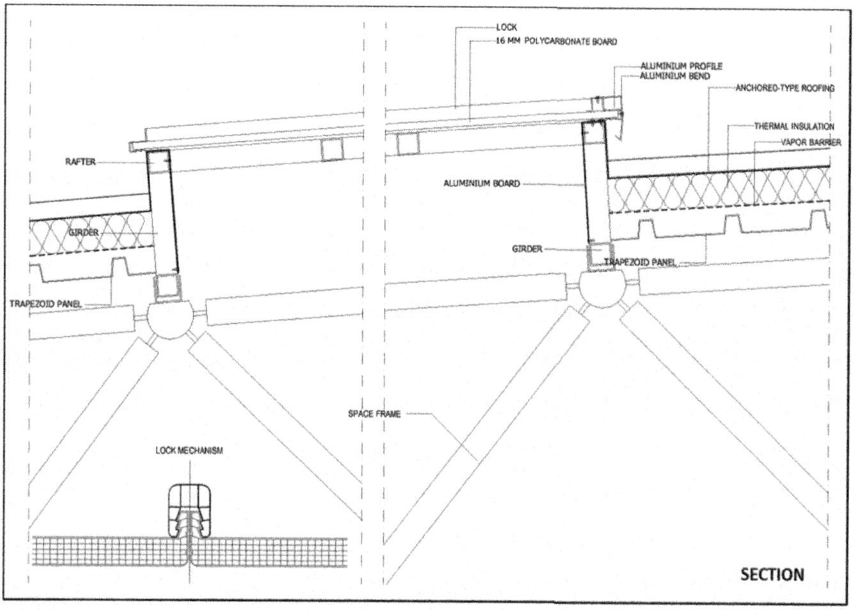

Fig. 21 A skylight detail is allocated on a framed steel roof

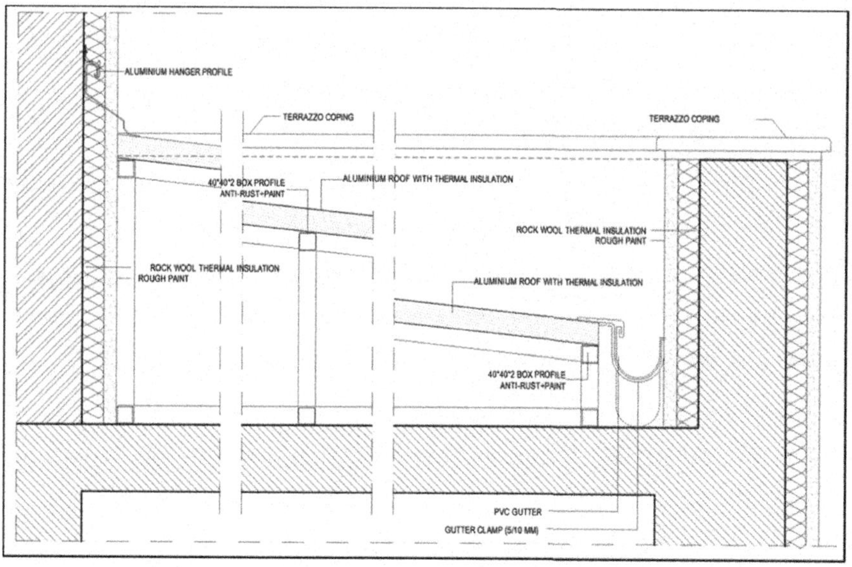

Fig. 22 A parapet, steep roof, rainwater gutter, and flashing detail

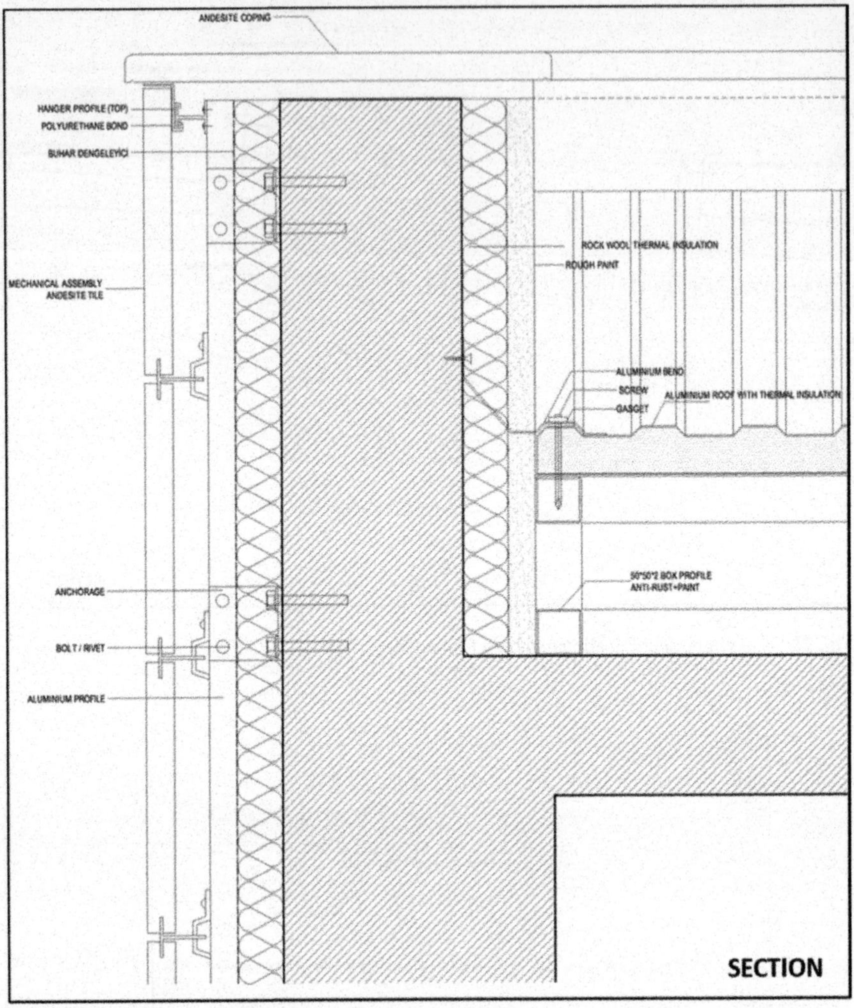

Fig. 23 A parapet, sandwich panel roof covering, and mechanical exterior cladding detail

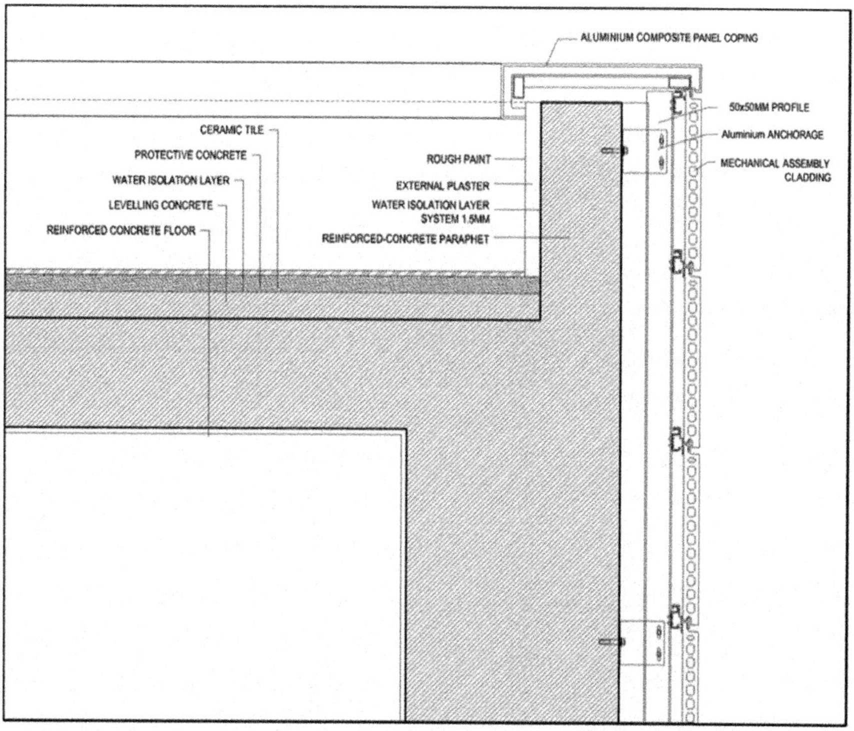

Fig. 24 A ceramic tile finished roof detail

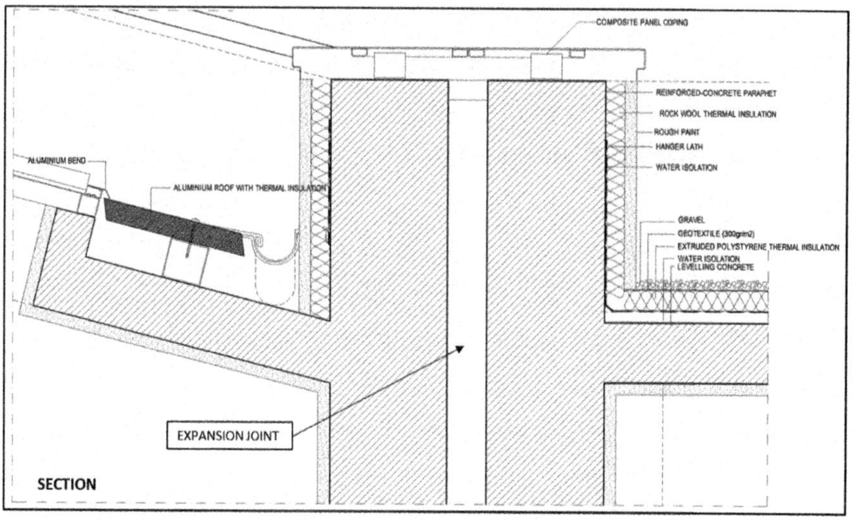

Fig. 25 An expansion joint detail in the roof

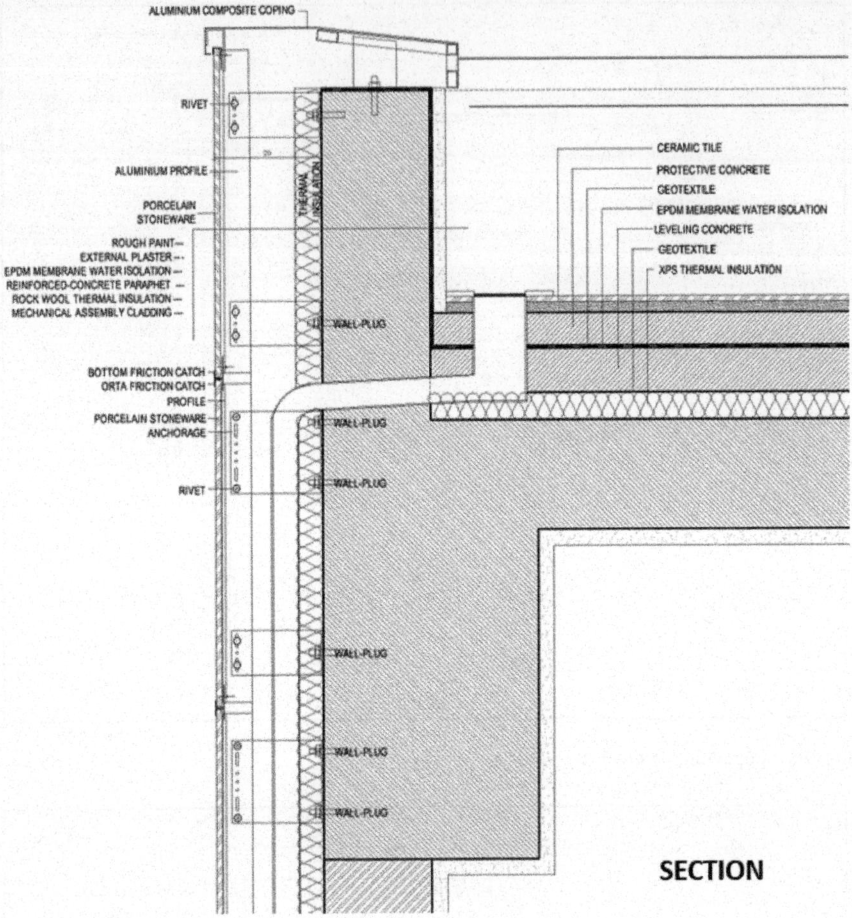

Fig. 26 Flat roof with parapet and rainwater gutter detail

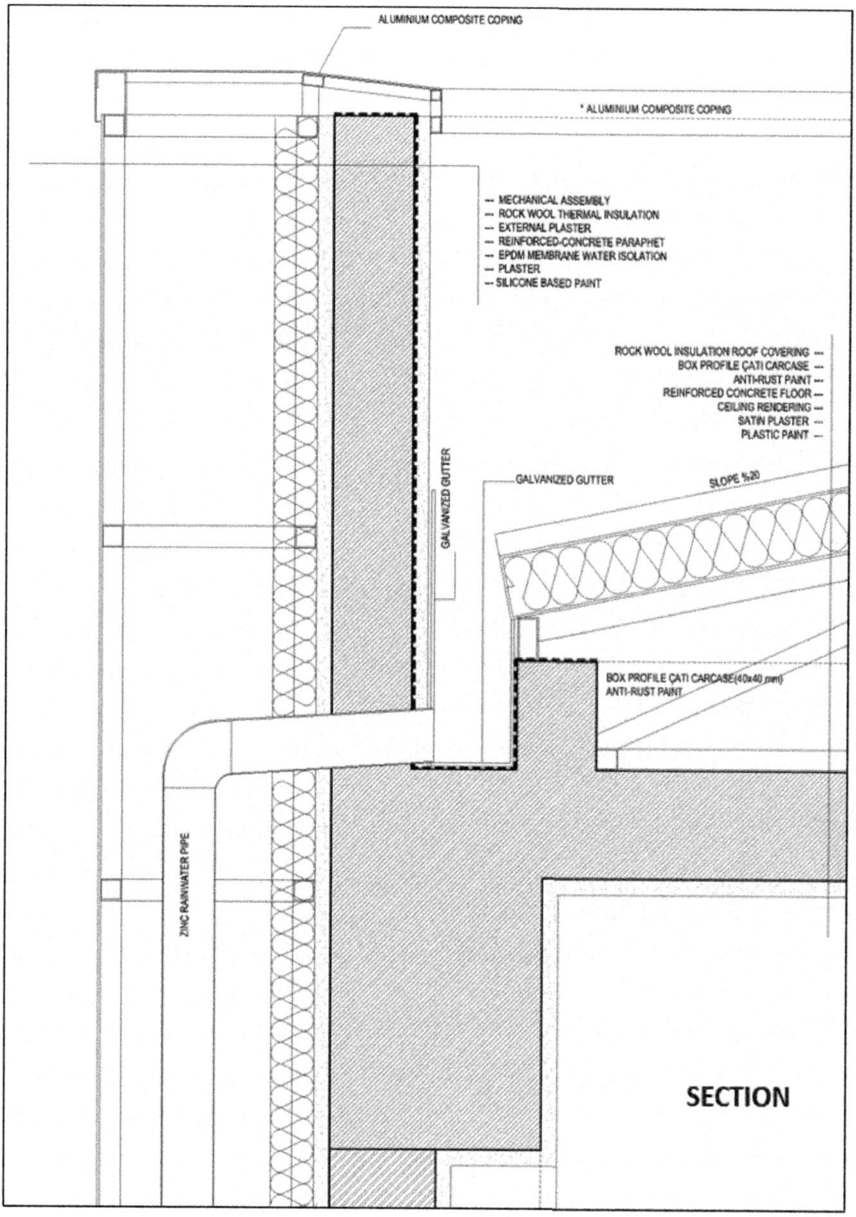

Fig. 27 A steep roof, parapet, and rainwater gutter detail

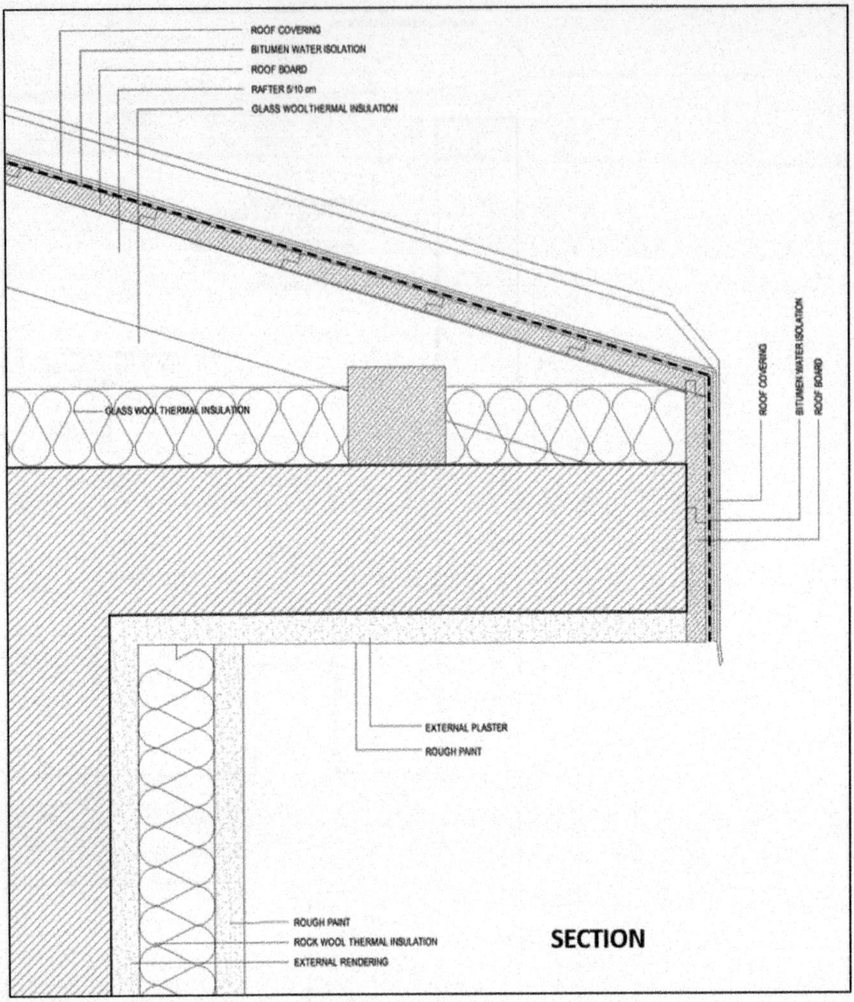

Fig. 28 A steep roof and eave detail

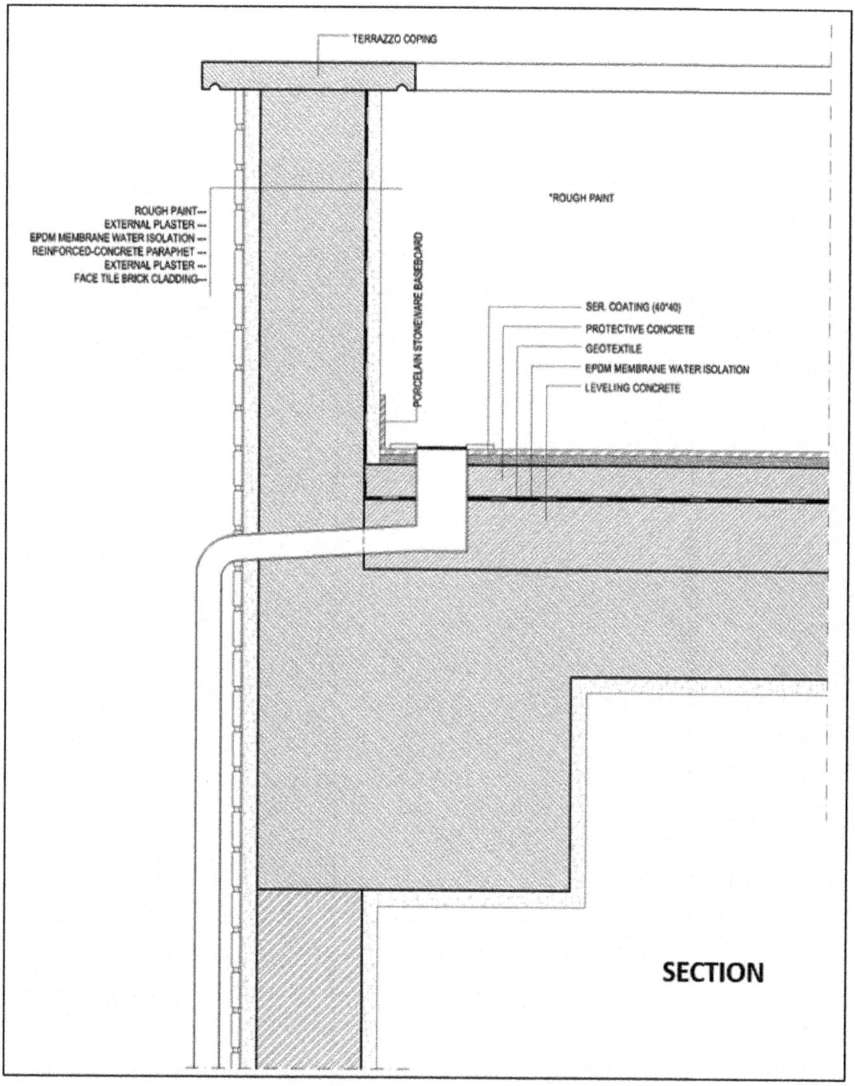

Fig. 29 A flat roof with ceramic tile finishing, parapet, rainwater gutter, and pipe detail

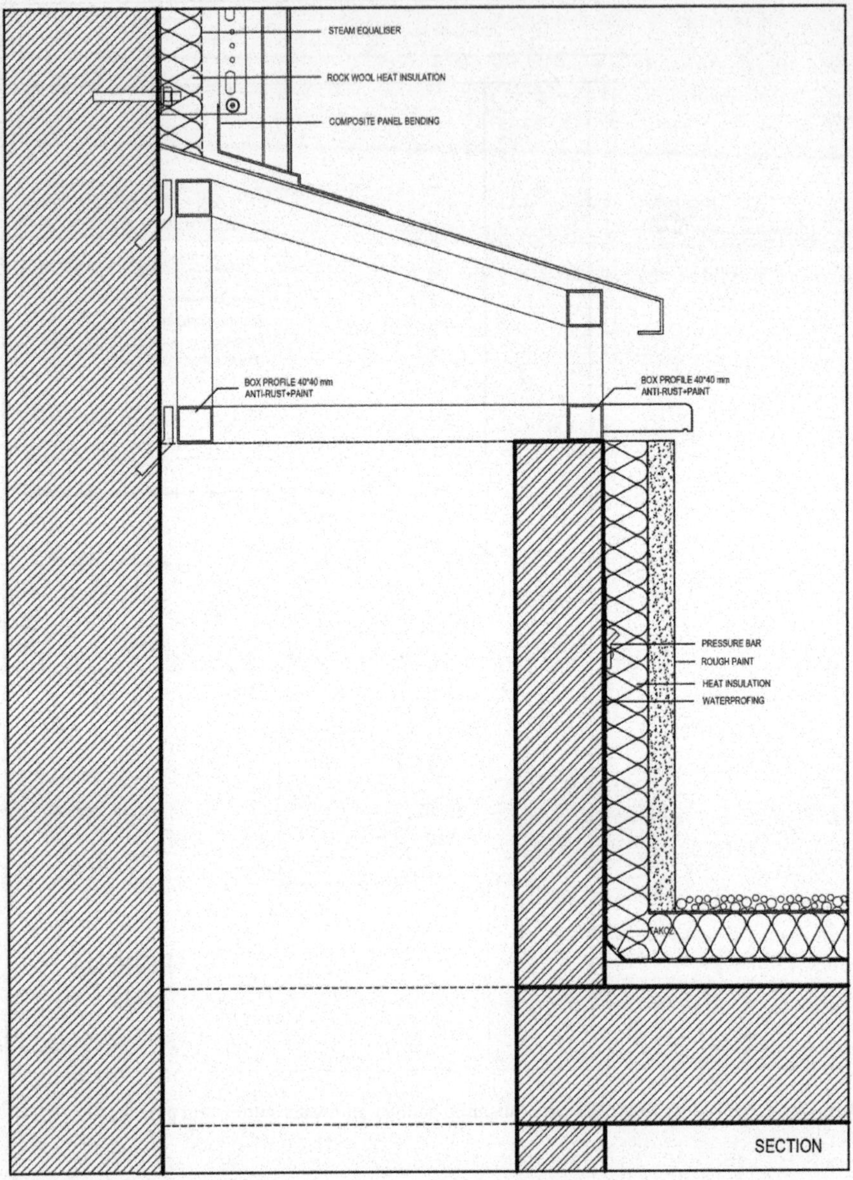

Fig. 30 A wall and chimney cap connection detail

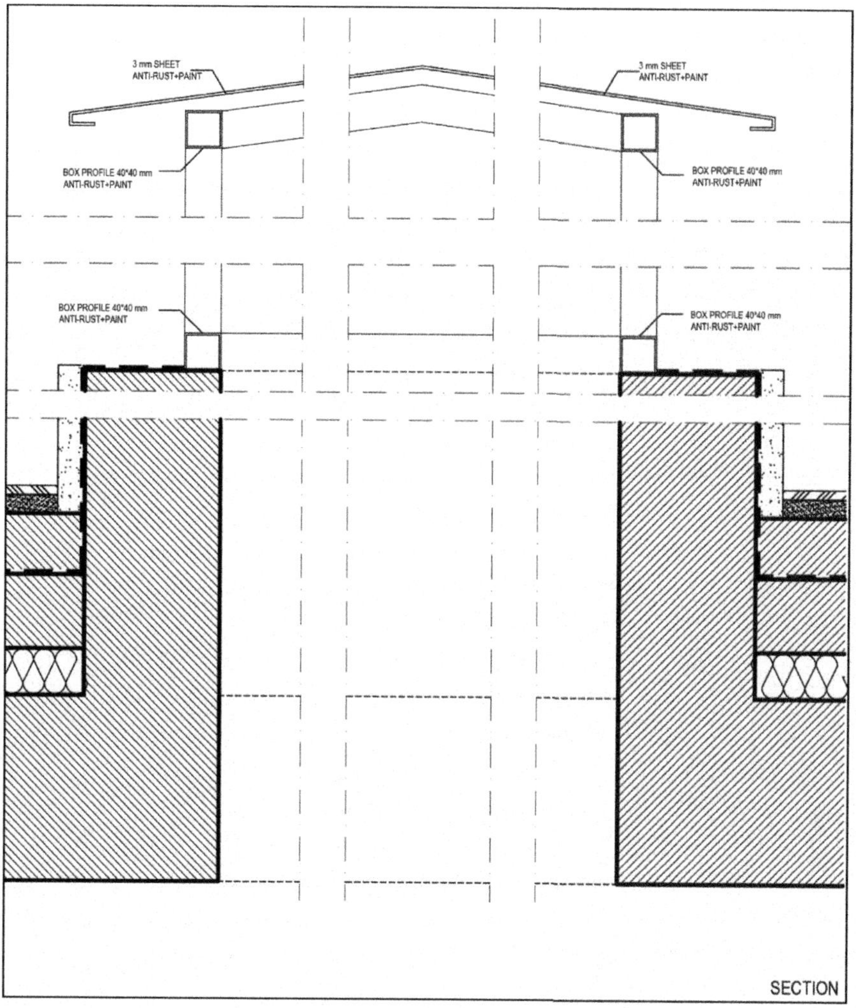

Fig. 31 A chimney cap connection detail

Reference

Özsarı Çelik (2023). Located in Antalya, all images were retrieved in 2023 with the company's founder's permission.

Chapter 5
Finishings and Divisions

Abstract Buildings and spaces are defined with horizontal and vertical components: walls, ceilings, and floors; their finishings are set for intended functions. This chapter explains the secondary components, including exterior and interior walls, glazing, ceilings, floor and coverings, and circulation systems. The materials, layers, and application details are presented with the conduction of figures and photographs.

1 Exterior Wall Cladding

Exterior walls are one of the major determinants of the appearance of a building. Together with the roof, exterior walls constitute the building's envelope, separating inside and outside and maintaining an acceptable interior environment. Exterior walls must:

- Prevent water infiltration from rain and snow.
- Control heat loss and heat gain
- Control air leakage and water vapor transmission
- Resist fire.
- Control sound transmission
- Accommodate movement due to thermal, moisture, and other causes.

1.1 Principles of Rainwater Infiltration Control

1.1.1 Gravity-Induced Infiltration

Gravity generally affects water penetration through an exposed horizontal surface of a wall and can be countered by providing a nominal slope in the surface. The purpose of the slope (generally 1:12) is to drain water away from vulnerable parts of a building. The coping over a roof parapet is generally sloped toward the roof so that the water drains on the roof. The slope is directed away from the building in exterior

R. Sarı and E. B. Çalışkan, *Building Construction Methods and Systems*,
https://doi.org/10.1007/978-3-031-50043-5_5

windowsills and thresholds under entrance doors. The critical point here is providing special details parallel with gravity direction and preventing water penetration into the vulnerable parts of the building layers. Coping and sills cover the edges of the wall layers.

1.1.2 Capillary-Induced Infiltration

Water molecules have cohesive bonds between them. Besides cohesive bonds, water molecules experience forces of attraction from the surface molecules with which they are in contact. These forces create adhesive bonds. The adhesive bond is stronger for most building surfaces than the cohesive bond. This is the main reason that instead of dropping vertically by gravity, a thin film of water can travel horizontally along the soffit. A drip mechanism, which consists of either a continuous groove or a vertical projection, counteracts this phenomenon (Figs. 1 and 2).

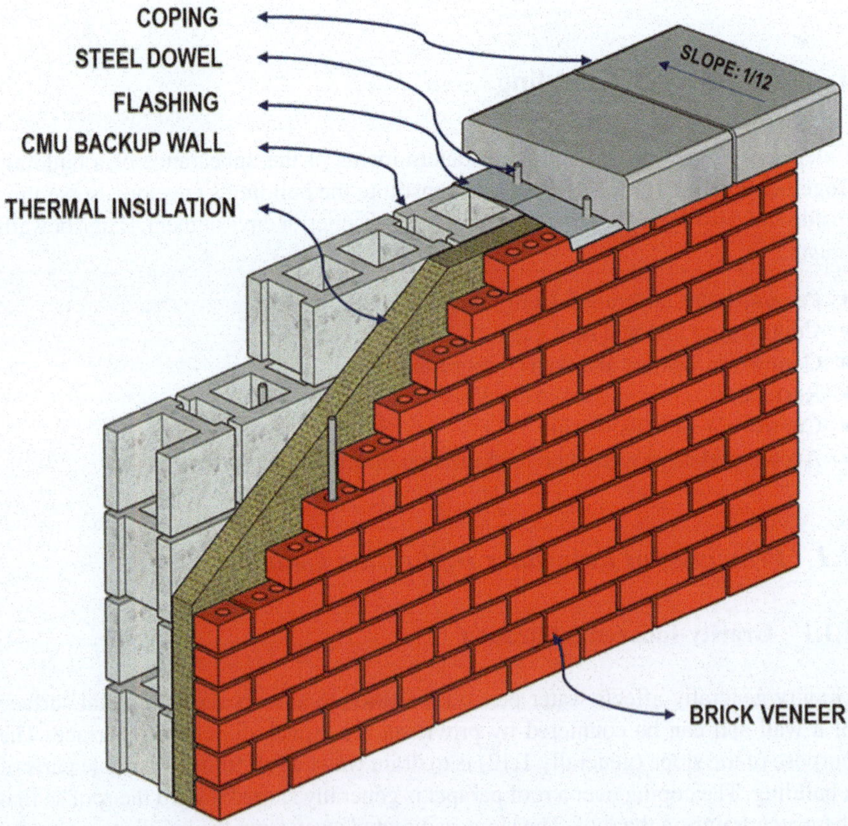

Fig. 1 3-D view of capillary-induced infiltration using concrete coping

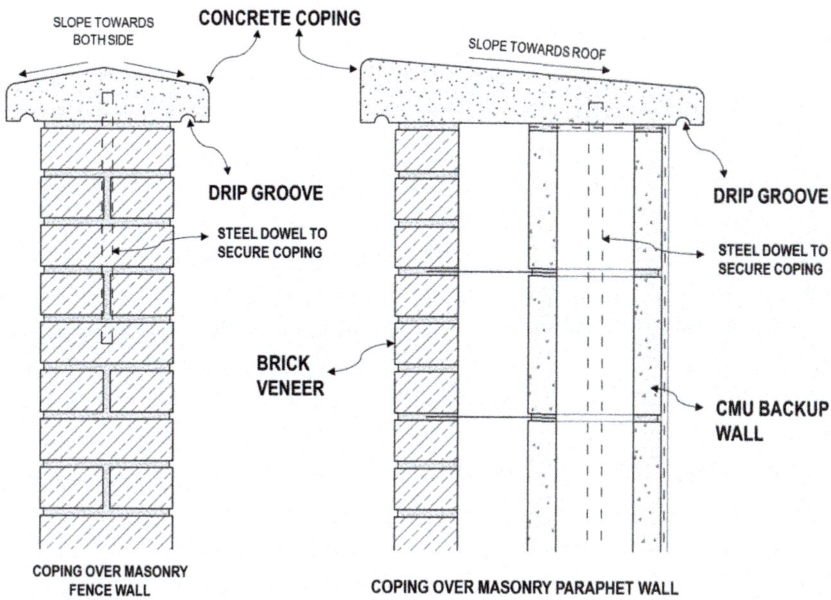

Fig. 2 Capillary-induced infiltration using concrete coping

1.2 Exterior Wall Cladding Systems

There are various cladding materials in the market. Although each material requires specific construction and application technics, dividing them into three categories regarding their implementation methods is possible. These are (i) curtain walls, (ii) veneer walls, and (iii) infill walls (Fig. 3).

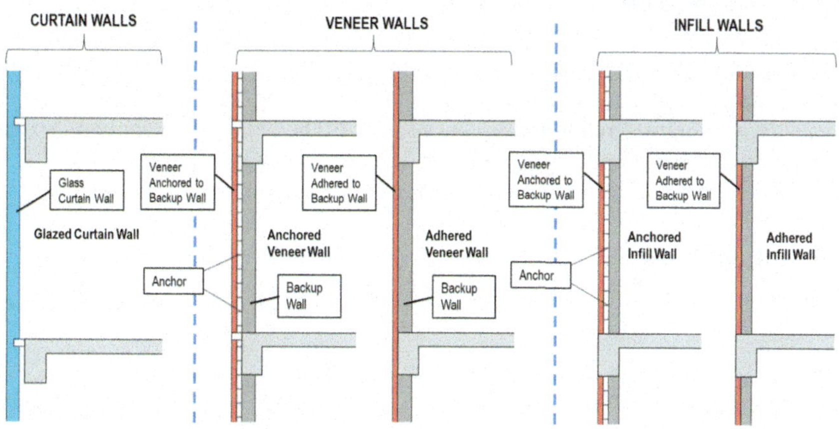

Fig. 3 Types of exterior wall cladding systems

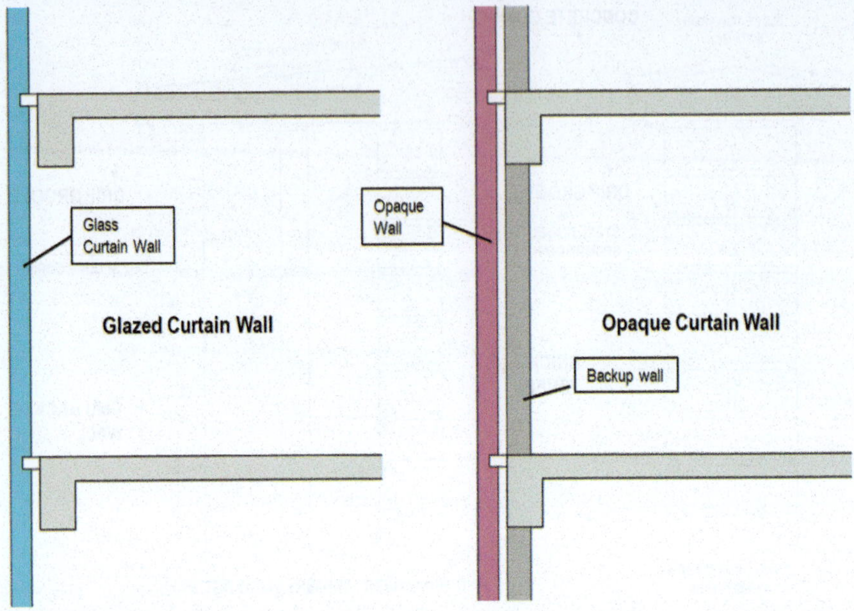

Fig. 4 Curtain wall system

1.2.1 Curtain Walls

Curtain calls can only be used with a frame structure. The cladding material can be transparent or opaque. There may be a backup wall in the background of the opaque curtain wall as an option (Fig. 4).

1.2.2 Veneer Walls

There are two types of veneer walls regarding the implementation in construction. Veneer walls are either adhered or anchored to the backup wall. The dead load of the veneer wall is transferred to the structural floor of the building by applying special details, as illustrated in Fig. 11.8 (Fig. 5).

1.2.3 Infill Walls

See Fig. 6.

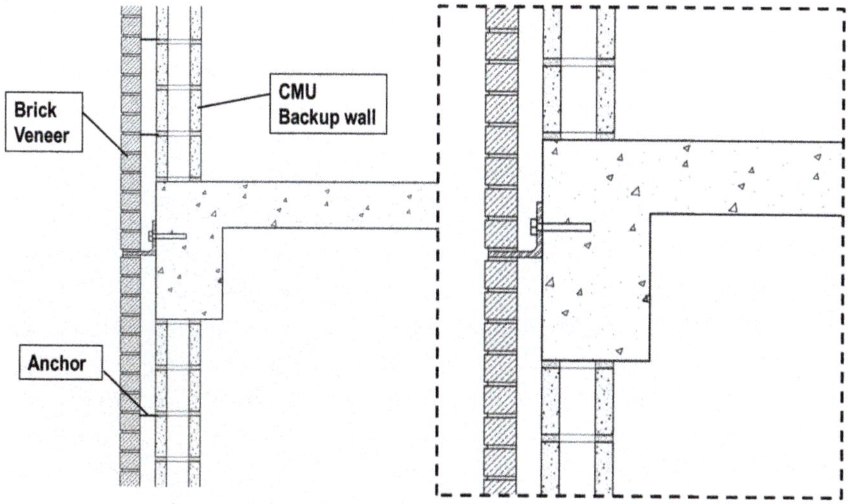

Fig. 5 Veneer wall system

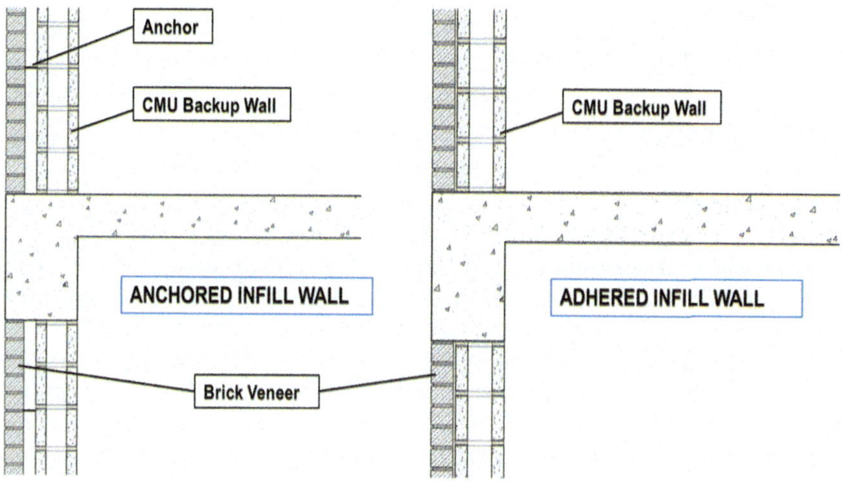

Fig. 6 Infill wall system

1.2.4 Brick Veneering

Brick veneer is by far the most popular system. Its popularity lies in its durability, fire resistance, and aesthetic appeal. The system requires almost no maintenance and can be used for buildings of all heights and complexity—from high rise to low rise, from rectilinear facades to intricate ones. A minimum air space is 2.5 cm for low-rise buildings and 5 cm for high-rise buildings (Fig. 7).

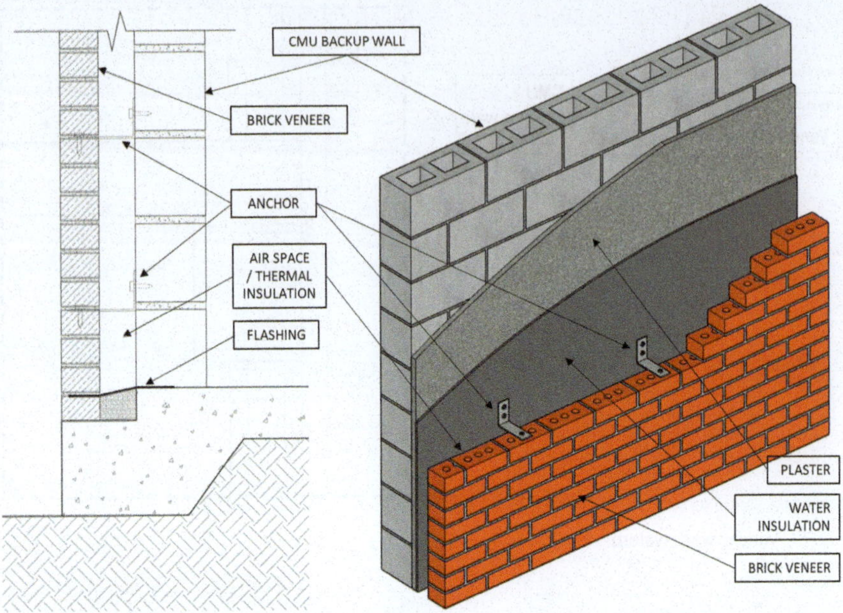

Fig. 7 Brick veneering application

1.2.5 Finishings

Stucco Cladding

Stucco consists of cementitious materials, sand, and water. It is generally used in exterior finishes, makes the surfaces more resistant to water and air filtration, and increases sound insulation and resistance to fire. Stucco should be applied if the ambient air temperature rises at least 5 C. Stucco can be used widely on (i) cold-formed steel stud walls, (ii) wood stud walls, (iii) masonry walls, and (iv) concrete walls.

Stucco is typically applied sequentially in two coats over a wall as the base and finish coats. In some situations, the base coat is applied in two layers: scratch and brown. The ingredients of the base coat are Portland cement, lime, sand, and water. Portland cement is the glue that bonds all mixed constituents, eventually curing into a strong and rigid surface. Lime imparts plasticity and cohesiveness to the mix. Plasticity implies that the mix can be spread easily, and cohesiveness implies that the mix will hold and not sag on a vertical surface during application (Figs. 8, 9, 10 and 11).

- **Control Joints and expansion joints**

Due to the existence of Portland cement, shrinkage is an inherent feature of a stucco surface, which leads to its cracking. Although stucco cracking cannot be

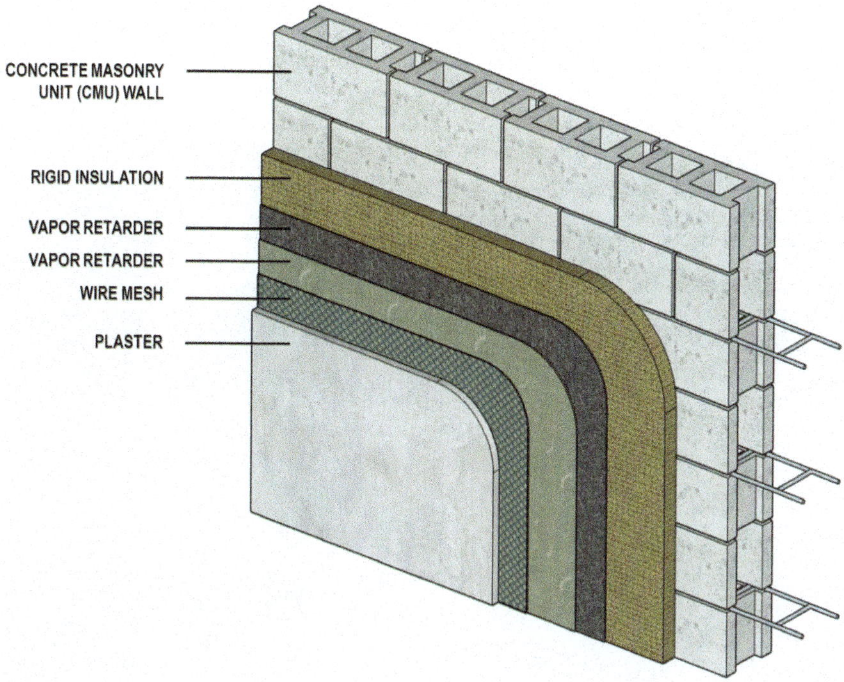

Fig. 8 Stucco cladding application in a Concrete Masonry Unit (CMU) wall

fully eliminated, it can be controlled by providing closely spaced control joints. In addition to providing control joints to control the shrinkage of the stucco finish, expansion joints are needed to respond to large movements in the building structure. Therefore, expansion joints should be provided at each floor level in the exterior wall of a multistory building to absorb the movement in the spandrel beam. Furthermore, expansion joints are utilized where there is a major change in building elevation or where a stucco-finished wall abuts a wall made of a different material.

Stone Cladding

Granite, marble, and limestone are commonly used for exterior cladding. Stone cladding can be field installed, slab by slab, at the construction site, or prefabricated into curtain wall panels. There are two methods available for stone cladding. These are (i) the standard set method, (ii) the vertical channel method, and (iii) thin stone cladding

- **Standard set method** (Fig. 12).
- **Vertical channel method** (Figs. 13, 14, 15, 16, 17, 18, 19, 20, 21, 22 and 23)

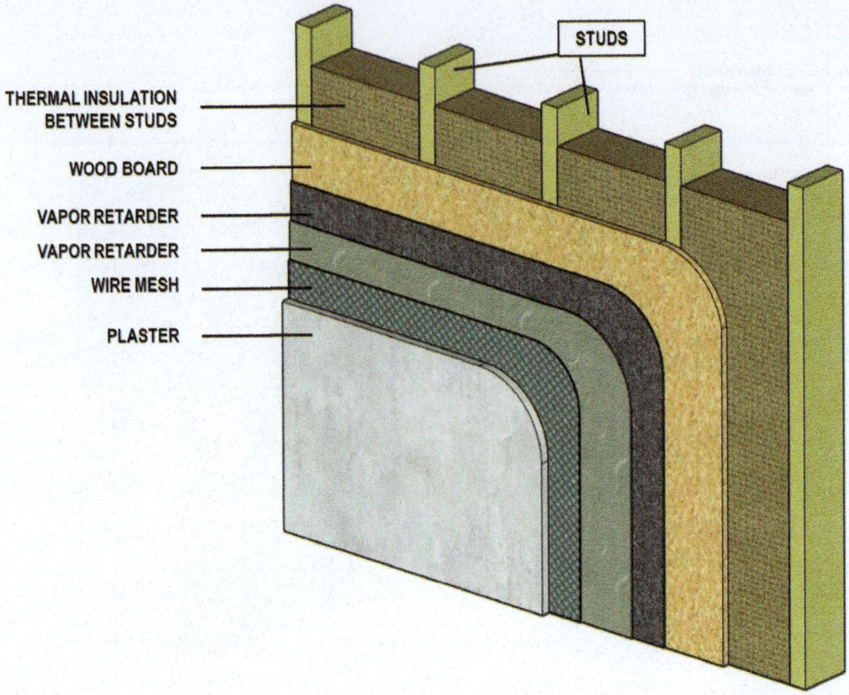

Fig. 9 Stucco cladding application in a wood wall

- **Thin stone cladding**

Another form of panelized stone cladding uses a thin (approx. 5 mm) stone veneer bonded to an aluminum honeycomb backing. This form of stone is even lighter than glass, having the same thicknesses. The panel's lightweight, high ductility and bending strength make it ideal for use in seismic areas.

Insulated Metal Panels

Another lightweight exterior cladding system consists of metal (typically steel) panels with 4 to 10 cm factory-installed polyurethane foam insulation between metal skins. The panels may be installed both horizontally and vertically. The panels can be integrated with windows and other openings with manufacturer-provided accessories and detailing assistance (Figs. 24 and 25).

Reference

Özsarı Çelik, located in Antalya, all images were retrieved in 2023 with the company's founder's permission.

Fig. 10 Rock wool thermal insulation application

2 Interior Walls

Interior walls are non-structural partition walls. The main usage is providing enclosures for the rooms and spaces within exterior enclosures. Thus, the expectations for interior walls are not as high as exterior walls. Interior walls can be divided into masonry, panel, and curtain walls.

Fig. 11 Styrofoam thermal insulation application

2.1 Masonry Interior Walls

Unlike exterior walls, masonry interior walls share the same construction methods. However, regarding the functions, interior walls do not require details related to environmental conditions. Furthermore, the width of the interior walls is not as thick as the exterior walls. Nominal interior wall thickness starts with 10 cm, 15 cm, 20 cm, and 25 cm. Increasing the interior walls' thickness causes a decrease in the rooms' net area. Thus, 10 cm and 15 cm widths are available for residential buildings, while 15, 20, and 25 cm walls are used for other types. Hollow Brick Units (HBU), Aerated Concrete Blocks (ACB), and Concrete Masonry Units (CMU) are the masonry wall materials used in general (Figs. 26, 27, 28, 29, 30, 31 and 32).

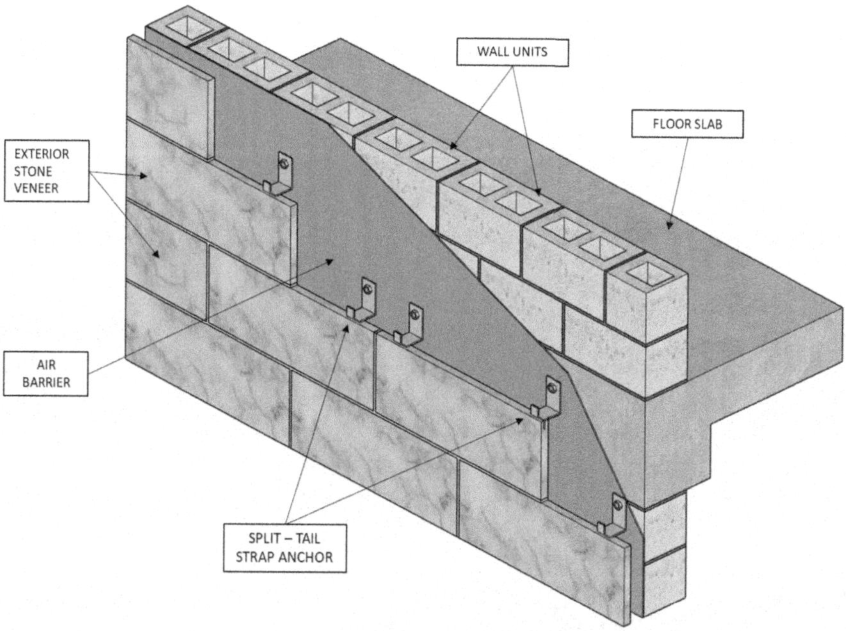

Fig. 12 3-Dimensional illustration of stone cladding by using the standard set method

2.2 *Interior Panel Walls*

Interior panel walls consist of studs and boarding materials installed at the construction area. The void among studs is sometimes filled with thermal insulation materials. The advantages of panel walls over masonry interior walls are that panel walls are lightweight structures requiring less material and labor, while masonry interior walls require special workmanship. The thickness of the panel wall width starts with 75 mm to 150 mm in general. The stud's depth and the boarding material's thickness affect the panel wall's total thickness (Figs. 33 and 34).

2.3 *Interior Curtain Walls*

Glass partition walls are applied in certain places. Besides providing transparent enclosure, glass partition walls are lightweight materials requiring less labor. They are made of tempered glass in general (Fig. 35).

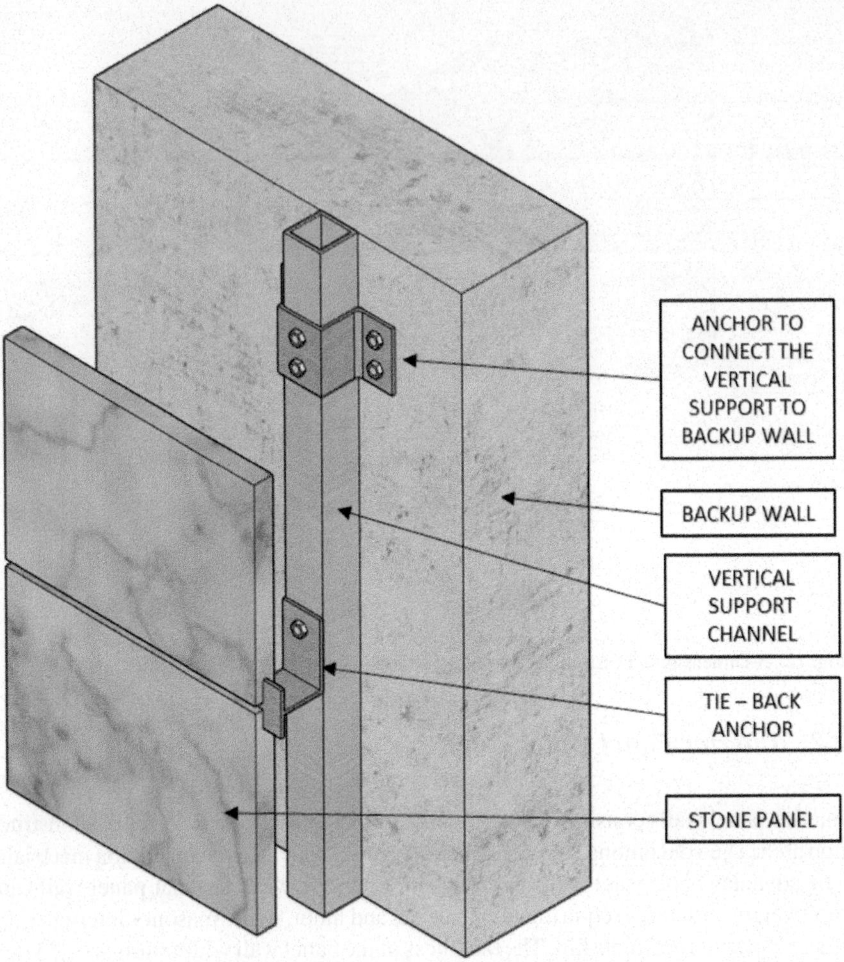

Fig. 13 3-Dimensional illustration of stone cladding by using the vertical channel method

3 Glass

3.1 Glass, Glazing, and Light Transmitting Plastics

In the market, commercially, annealed glass is available. Clear and tinted versions are generally used in the buildings. The basic features of Annealed Glass are:

- Brittle and very strong in compression
- Low tensile strength
- Low toughness
- Low thermal shock resistance Breaks due to sudden temperature changes.

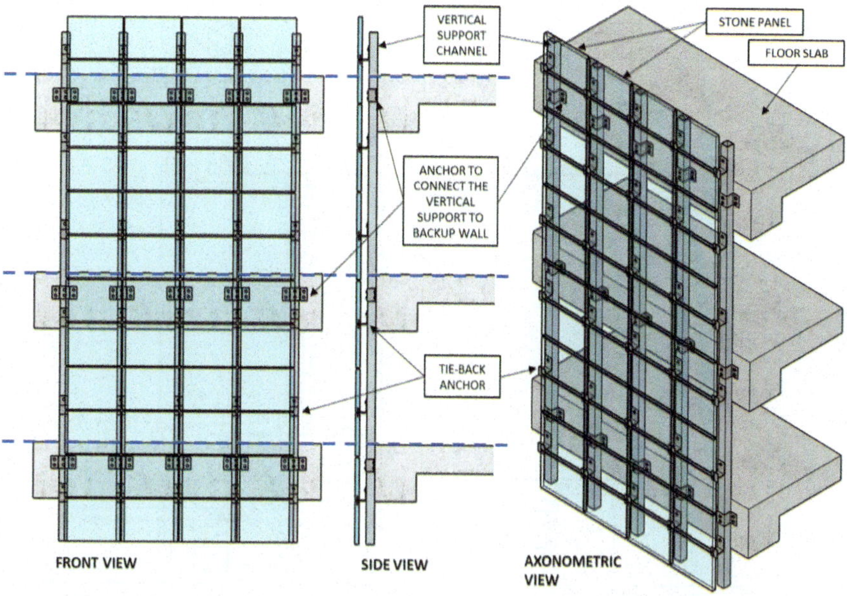

Fig. 14 Stone cladding illustration by use of vertical channel method in an elevation

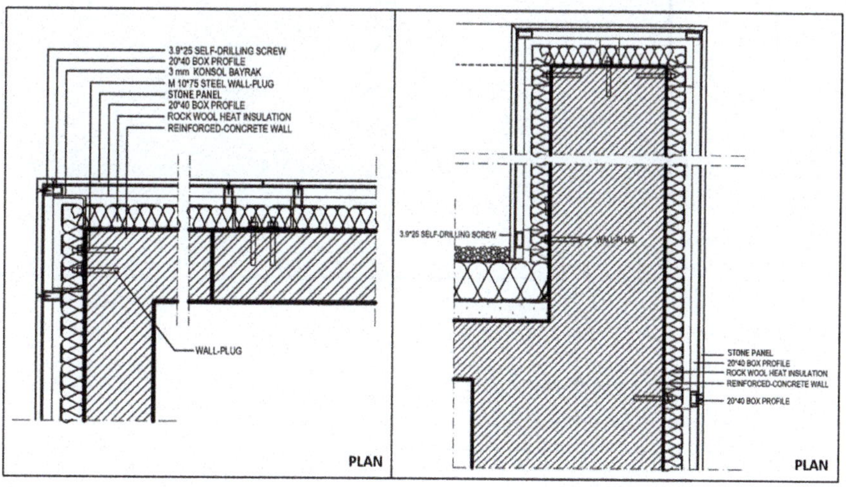

Fig. 15 Stone panel cladding detail

- If it breaks, it forms sharp fragments.

If stronger glass is needed, annealed glass is heated before use, and as a result of this heating process, the bending stress and temperature resistance of glass increase. There are three types of heat-modified glass:

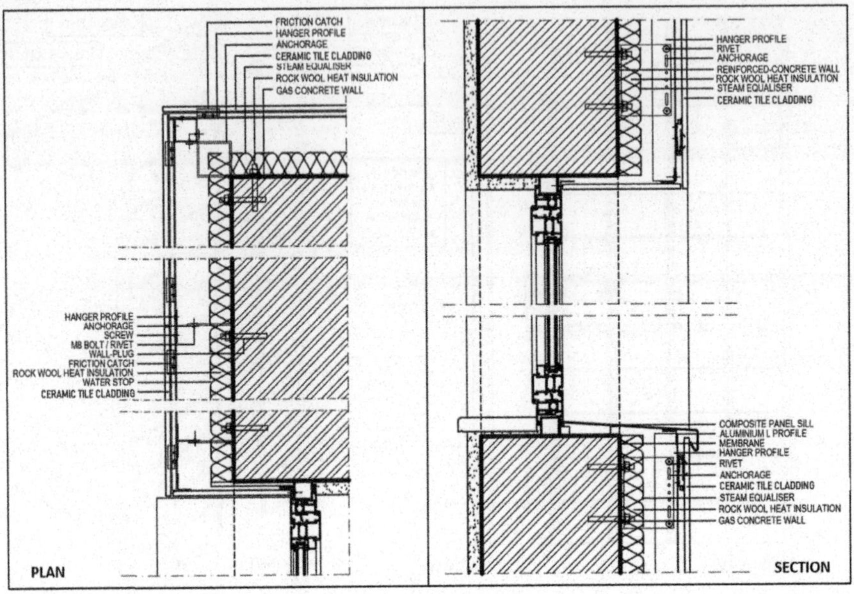

Fig. 16 Ceramic tile cladding

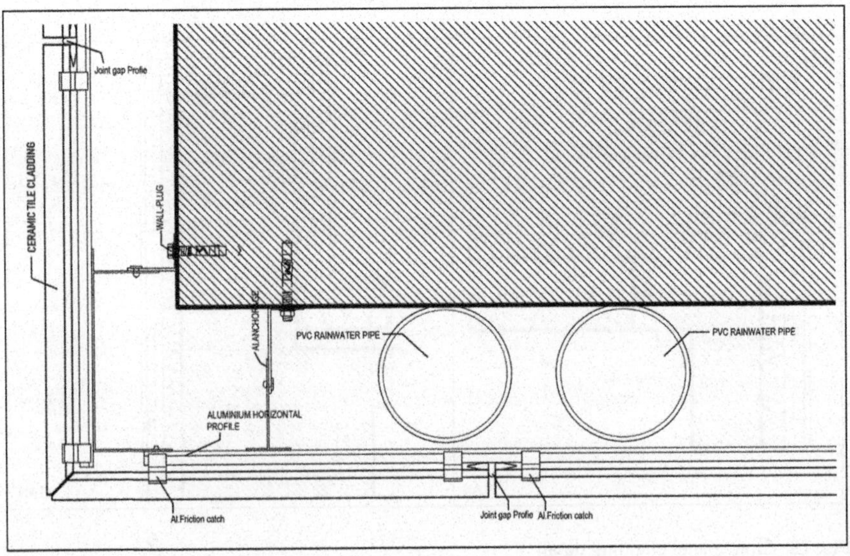

Fig. 5.17 Ceramic tile cladding corner detail

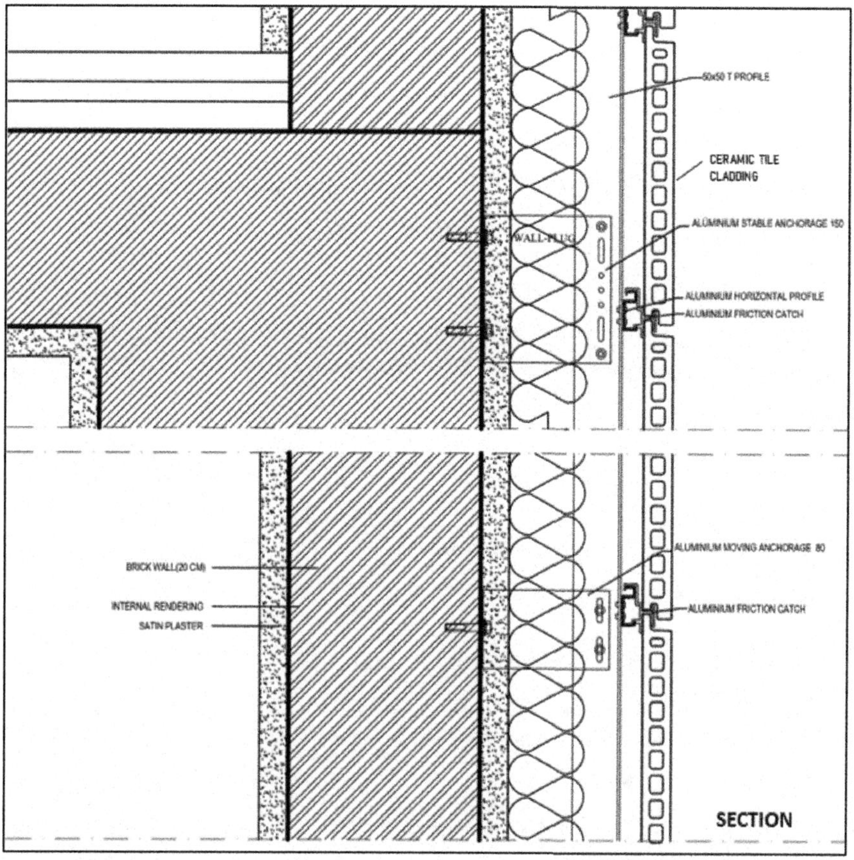

Fig. 18 Ceramic tile mechanical cladding detail

- Tempered glass, also referred to as fully tempered (FT) glass.
- Heat-soaked tempered glass
- Heat-strengthened (HT) glass

3.1.1 Tempered Glass

Tempering is the exact opposite of annealing. Tempering involves heating the glass below its softening point to approximately (700 C) and cooling (quenching) it by blowing a jet of cold air on all glass surfaces simultaneously. This causes the outer layers of the glass to harden quickly while the interior of the glass is still soft. As the interiors begin to cool, they tend to shrink but are prevented by the already-hardened outer glass surfaces. Consequently, the exterior of the glass comes under a state of compression, and the interior under a state of compensating tension. Tempered glass is approximately four times stronger than annealed glass in bending. It can

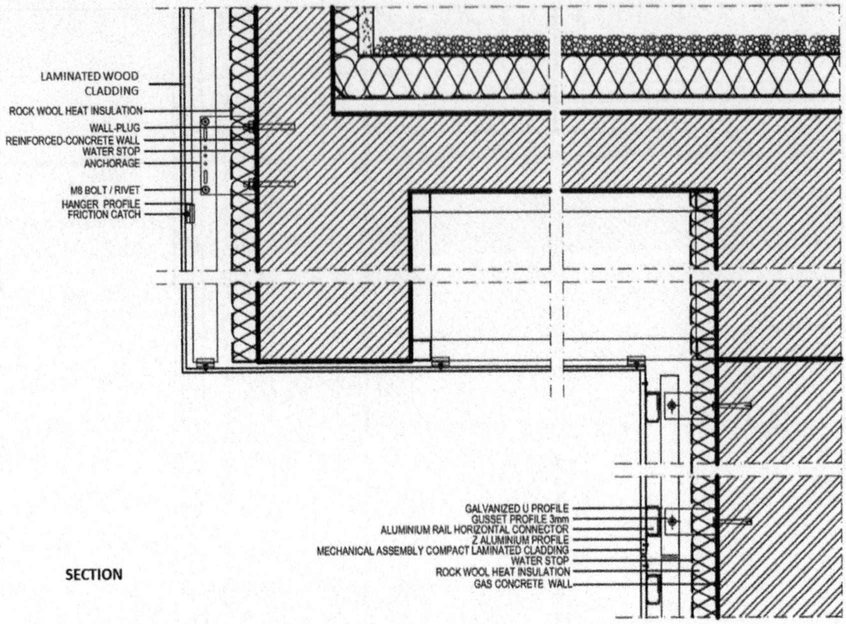

Fig. 19 Laminated wood cladding cantilever floor detail

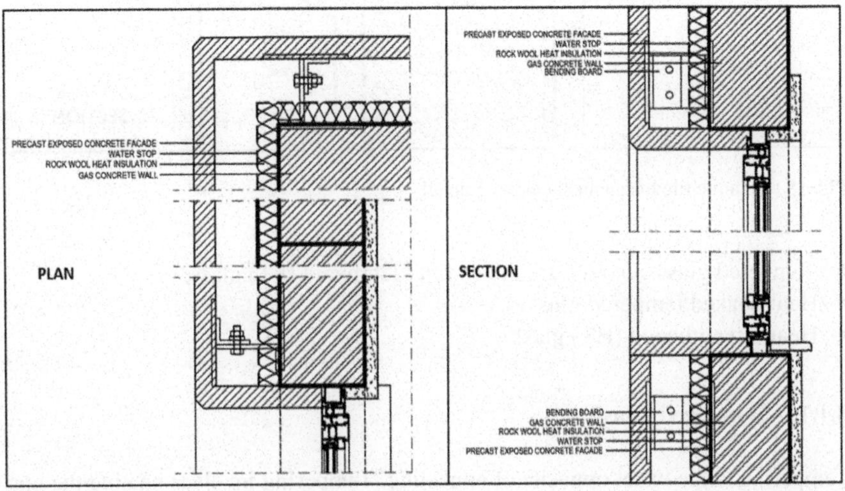

Fig. 20 Precast concrete cladding detail

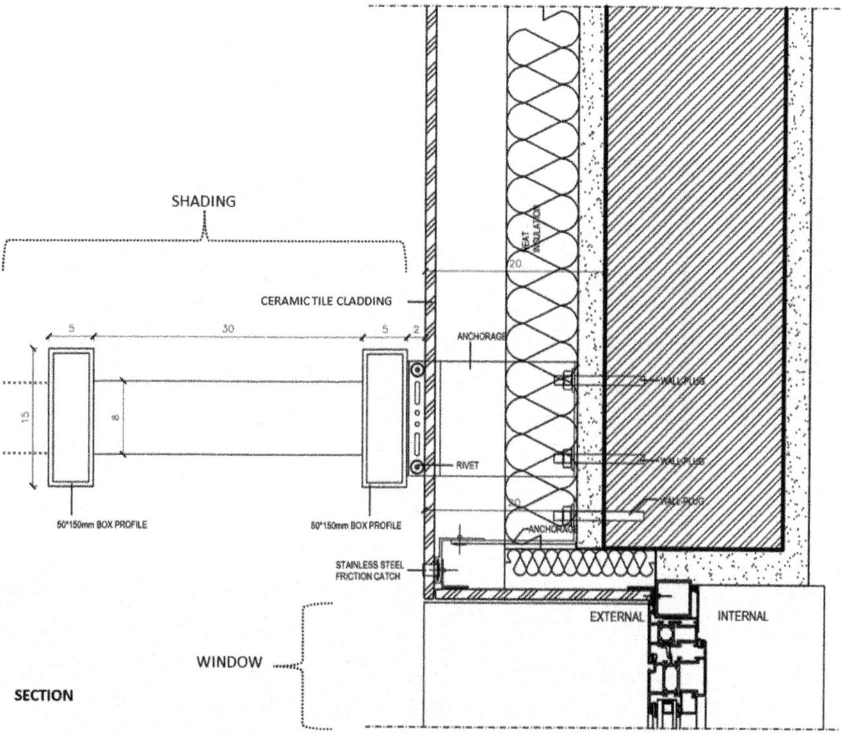

Fig. 21 Shading and ceramic tile cladding detail

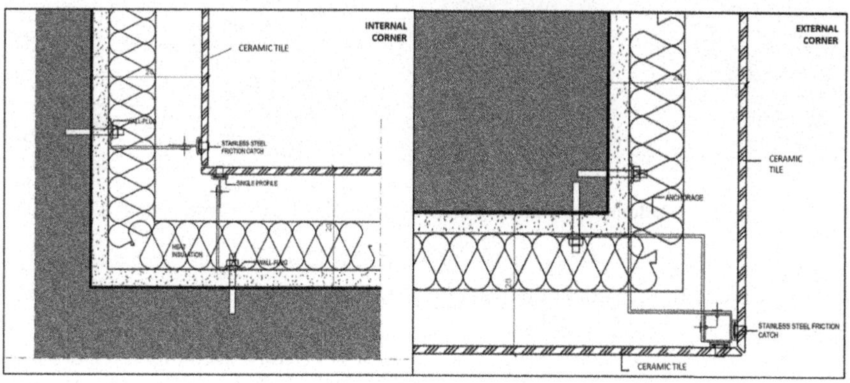

Fig. 22 Ceramic tile cladding internal and external corner details

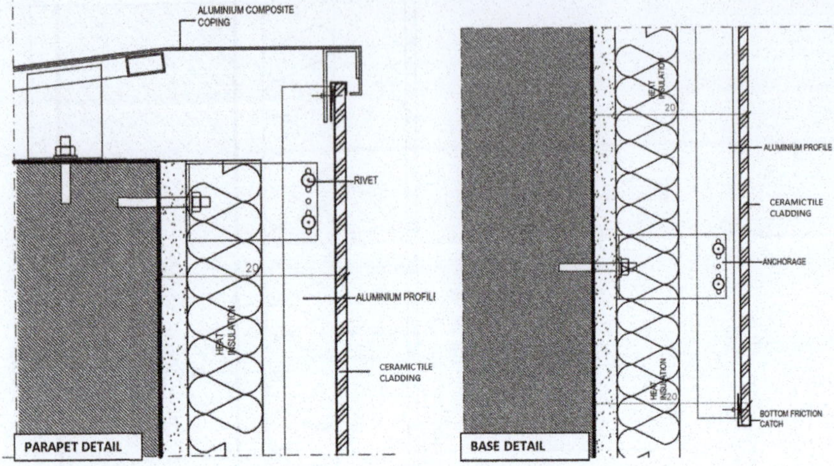

Fig. 23 Parapet and base connection details of ceramic tile exterior mechanical cladding

Fig. 24 Wall panel cladding application in a construction site. Image Credit: Özsarı Çelik (2023)

also withstand greater deflection than annealed glass of the same thickness and is far more resistant to impact and thermal stress.

Tempering must be done after the glass has been cut to size. A tempered glass sheet must not be cut, drilled, or edged. This is due to the fact that these processes release the locked-in stress, causing the glass to disintegrate abruptly. Abrasive blasting and etching may be done with some care. However, both of them reduce the thicknesses of the outer compressed layers, reducing the effectiveness of tempering. When tempered glass breaks into tiny square-edged, cubicle-shaped granules—a breakage pattern is usually called dicing, while annealed glass breaks into long, sharp-edged

Fig. 25 Flashing applications. Image Credit: Özsarı Çelik (2023)

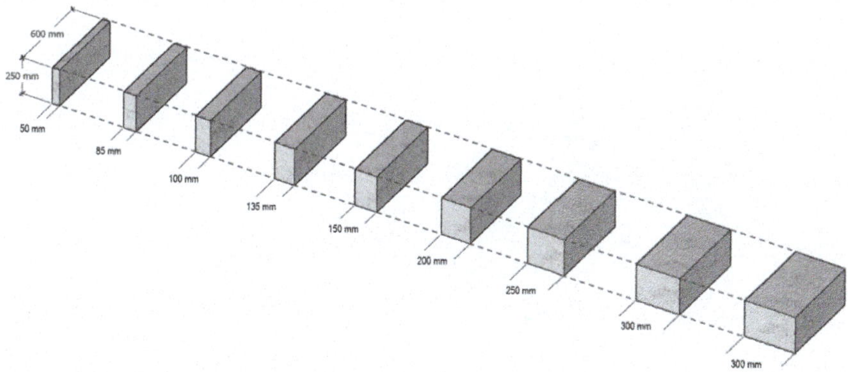

Fig. 26 Aerated Concrete Block Dimensions

pieces. Thus, tempered glass is used in hazardous locations to meet the safety glazing requirement.

3.1.2 Heat-Soaked Tempered Glass

Nickel Sulfide and certain other ingredients present in glass do not fully melt during glass manufacturing. When tempered glass is subjected to extreme thermal stresses or shock, these ingredients cause spontaneous breakage due to sudden expansion. Heat-soaked tempered glass is used when necessary to reduce or eliminate spontaneous breakage. In the heat-soaking process, tempered glass panes are subjected to cyclical heating and cooling to simulate long-term field conditions. Glass that does

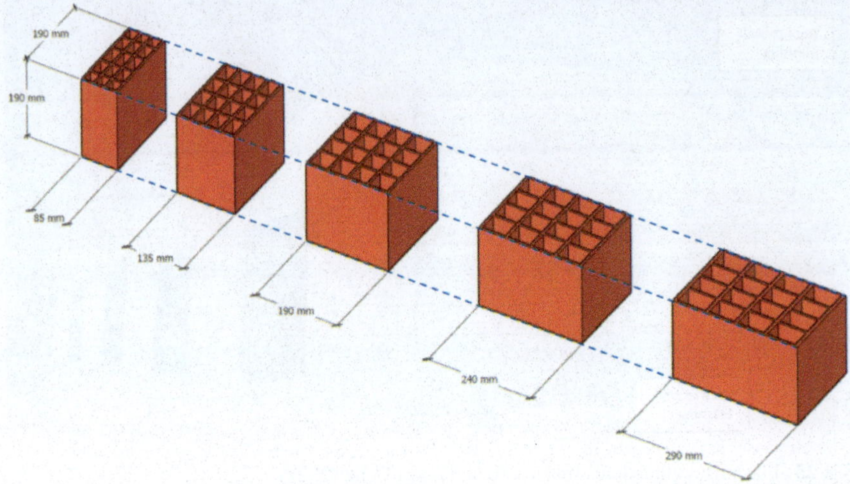

Fig. 27 Hollow Brick Unit dimensions

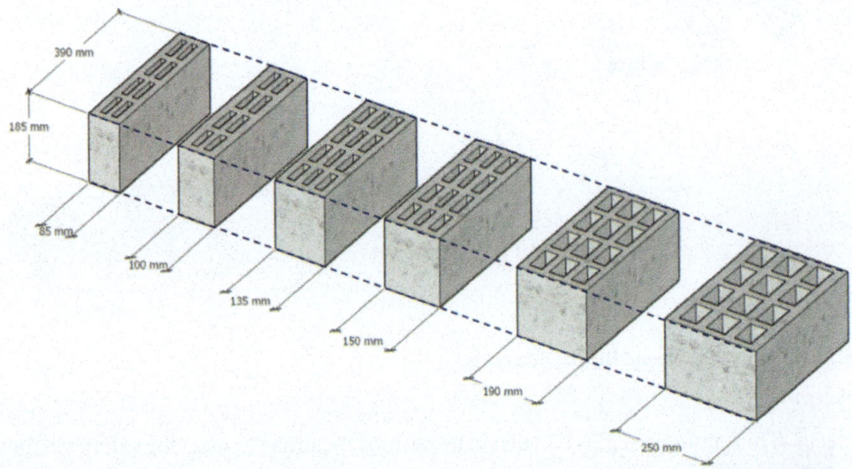

Fig. 28 Concrete Masonry Unit dimensions

not break in the process is safe against spontaneous breakage and is called heat-soaked tempered glass. Heat-soaked glass is expensive and is used only in projects where its usage is affordable.

Fig. 29 A CMU and CMU wall construction

Fig. 30 Concrete Masonry Unit (CMU) wall applications

Fig. 31 Brick interior wall applications

Fig. 32 Connection details of the perpendicular walls

3.1.3 Heat-Strengthened Glass

It is heat-treated exactly as tempered glass, but a much smaller volume of air is used during quenching so that the glass cools slowly. It is nearly twice as strong as annealed glass in bending. Heat-strengthened glass breaks into pieces that are sharper than tempered glass but blunter than those obtained from the breakage of annealed glass. In other words, heat-strengthened glass does not "dice" on breaking. Thus, heat-strengthened glass is not a safety glass. Furthermore, like tempered glass, heat-strengthened glass cannot be cut or drilled after heat treatment.

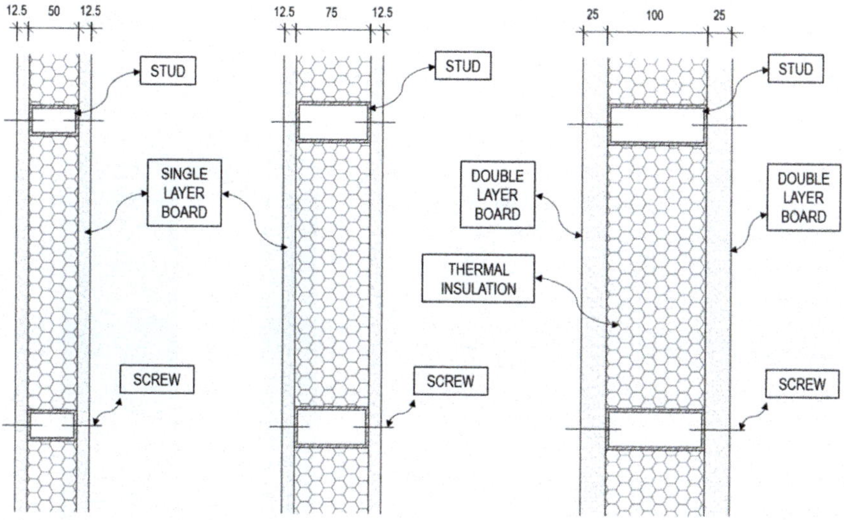

Fig. 33 Some examples of interior panel walls

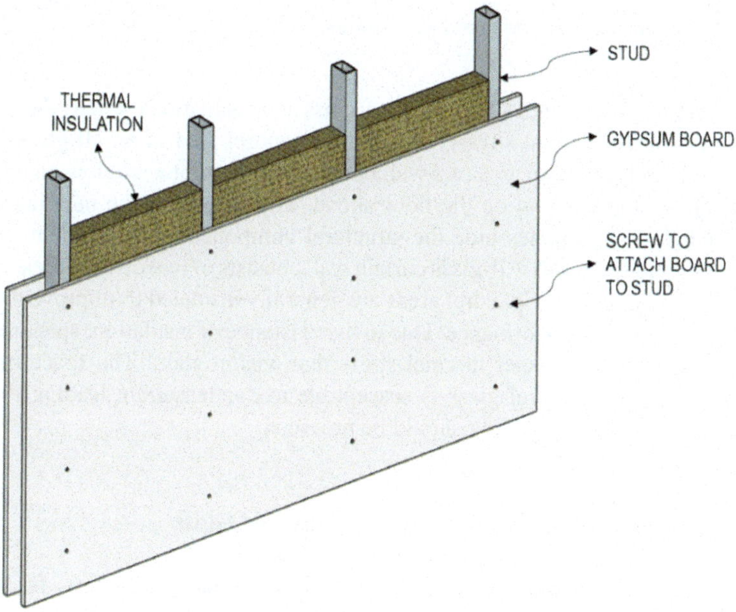

Fig. 34. 3-Dimensional illustration of an interior panel wall

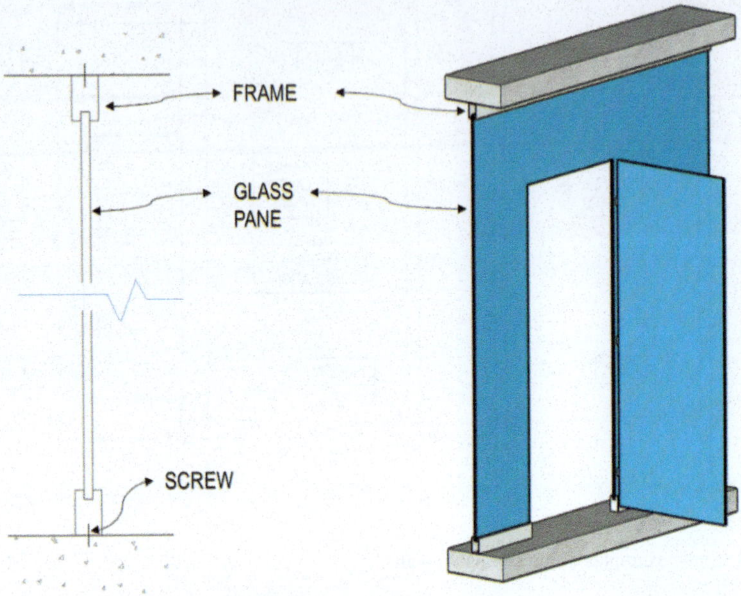

Fig. 35 Basic components of the glass panel partition wall

The primary use of heat-strengthened glass is in situations where the glass is subjected to large thermal stress, such as the spandrel area of an all-glass curtain wall. A spandrel area is an area of a wall between the head of a window on one floor and the sill of the window on the floor above. This is part of the building façade where spandrel glass panes hide the structural components behind them, such as edge beams and floor slab. All-glass curtain wall consists of two distinct areas: vision glass and spandrel glass. Spandrel areas are generally insulated to improve the glass curtain wall's energy performance. Due to the existence of insulation, spandrel glass is subjected to much greater thermal stress than vision glass. The space between the insulation and spandrel glass is susceptible to condensation, leading to metal corrosion. Thus, careful sealing should be provided.

3.1.4 Heat-Strengthened Glass Versus Tempered Glass

Tempered glass is nearly twice as strong as heat-strengthened glass. If it is not mandatory, however, heat-strengthened glass is preferable. Spontaneous breakage is less of a problem in heat-strengthened glass. When heat-strengthened glass breaks, it stays within the opening, similar to annealed glass. On the other hand, tempered glass fractures in small pieces and tends to evacuate the openings. There is generally less optical distortion in heat-strengthened glass.

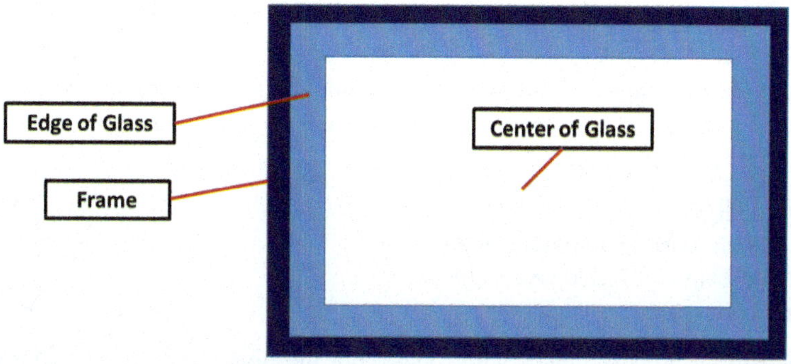

Fig. 36 There are three parts of typical glazing: the center, the edge of the glass, and the frame

3.1.5 Components of Glazing Units

Insulating Glass Unit (IGU)

A sealed assembly consisting of two glass sheets with an intervening space is called an insulating glass unit (IGU). There are three parts of typical glazing: the center, the edge of the glass, and the frame. A thermal break consists of an insulating connector (usually polyurethane) that connects the two parts of the aluminum frame (Fig. 36).

3.1.6 Glazing Types

Safety Glass

Two types of glass are considered to be safety glass:

- **Fully Tempered Glass**: On impact, tempered glass breaks into small, blunt, relatively harmless pieces.
- **Laminated Glass:** On impact, laminated glass is safe because it stays in place after breaking due to the plastic interlayer.

Laminated Glass

In its simplest form, laminated glass is made from two layers of glass laminated under heat and pressure to a plastic interlayer so that all three layers are fused. In the event of an impact, broken glass tends to remain bonded to the interlayer, minimizing the hazard of shattered glass. In architectural applications, laminated glass is the choice for skylights, sloped and overhead glazing, zoos, and aquariums. Depending on the PVB laminations (interlayer material), it is possible to derive hurricane-resistant and blast-resistant glass.

Fire-Resistant Glass

Ordinary glass (annealed, tempered, or laminated) is not fire-resistant. When subjected to a typical building fire, it shatters in 2 or 3 min. The type of glass used as a fire-resistant glass are:

- Wired Glass
- Intumescent Multi-laminate Glass
- Intumescent Gel-Filled Glass Units
- Wired Glass:

The rolling process makes wired glass, not the float glass process. During the rolling process, welded wire mesh is embedded in the middle of the glass thickness so that the resulting product is a steel wire-reinforced glass. Wired glass is not impact-resistant. The bending strength is only half of the annealed glass. On impact, wired glass breaks into sharp pieces, and the broken wires will generally project out of the glass, causing injury to a person.

- Intumescent Multi-laminate Glass:

Until 2 h, intumescent multi-laminate glass can resist fire.

- Intumescent Gel-Filled Glass Units

Intumescent gel-filled glass is an IGU in which the intervening space is filled with a clear gel. In the case of fire, the gel absorbs the heat.

Plastic Glazing

Clear plastic sheets (referred to as light-transmitting plastics) have gained popularity as an alternative glazing material due to the following reasons:

- Plastic can be bent to curves far more easily than glass.
- Plastic is several times stronger than glass of the same thickness and more impact-resistant.
- Plastic glazing is lighter than glass.

The disadvantages of plastic glazing far outweigh its advantages:

- Thermal expansion capacity is higher of plastic than glass. Thus, if the movement of plastic is restricted, it will visibly bow out in the direction of higher temperatures. In order to control expansion and shrinkage, complex framing details are required, which further adds cost.
- Humidity causes the same problem as thermal expansion.
- Plastic yellows with age, reducing its clarity and light transmission.
- Plastic is a combustible material.

Anatomy of a Glazing Pocket

Important components of a glazing pocket are:

- Setting blocks
- Edge blocks and (in seismic zones) corner blocks
- Glazing seals
- Weep holes
- SETTING BLOCK

Setting blocks provide a hard but resilient, bearing glass support. Setting block lift glass from the bottom of the glazing pocket to allow water to drain out from underneath and to prevent the glass from coming in direct contact with water.

- EDGE BLOCK

Edge blocks separate the vertical edges of glass from vertical frame members.

3.2 *Windows and Doors*

3.2.1 Window Styles

Fixed Window

The fixed window does not let the opening of the window sash. Thus, the glass is directly held by the window frame. The fixed window does not allow any ventilation opportunity. However, the air leakage ratio is low compared to operable window types.

Operable Window

- HUNG WINDOW

Hung type may be increased depending on the operation. However, all hung types have sashes that can slide vertically over a fixed glass lite. Opening area can be provided up to 50% of the window sash.

- CASEMENT WINDOW

A casement window consists of one or two operable sashes that close on each other with or without a center mullion. The sash closes on the frame with pressure, providing a compression seal causing less air leakage than the hung window.

- SLIDING WINDOW

A sliding window has one sash that slides horizontally over a fixed sash. Alternatively, both sashes may slide concerning each other. Like the hung window, the sliding window can provide an opening area of up to 50% of the window sash.

- AWNING WINDOW

An awning window is similar to a casement window but provides a degree of rain protection when the window is partially open. It can provide up to 100% opening area in the window sash.

- HOPPER WINDOW

A hopper window is similar to an awning window but opens inward at the top. Similar to the awning window, the hopper window can provide up to 100% opening area in the window sash.

- PIVOTING WINDOW

A pivoting window may be pivoted at the center or off the center. It allows easy window cleaning from the inside and provides up to 100% opening area (Fig. 37).

3.2.2 Window Frame Materials

- WOOD

Wood is the oldest window material. The following problems are why wood is not used as window material: swelling, shrinkage, fungal decay, termite vulnerability, and the need for periodic staining or painting.

- STEEL

Before introducing Aluminum as a frame material, steel was the popular frame material. Although it does not include swelling, shrinkage, fungal decay, and termite vulnerability problems that wood frames represent, steel requires periodic and frequent painting to prevent corrosion. However, there is no match for steel in terms of strength among window framing materials.

- ALUMINUM

Aluminum is the most widely used material for window and curtain wall frames. Unlike steel and wood, Aluminum frames require almost no maintenance. Furthermore, it provides easy and adaptable usage on framing. Moreover, Aluminum frames do not represent any problems wood window frames face.

- VINYL OR PVC

Similar to aluminum frames, Vinyl frames require almost no maintenance. Furthermore, the initial construction cost of Vinyl frames is cheaper than Aluminum frames. The main problem of Vinyl frames is the high coefficient of thermal expansion. Furthermore, the surface color of the frames is changing.

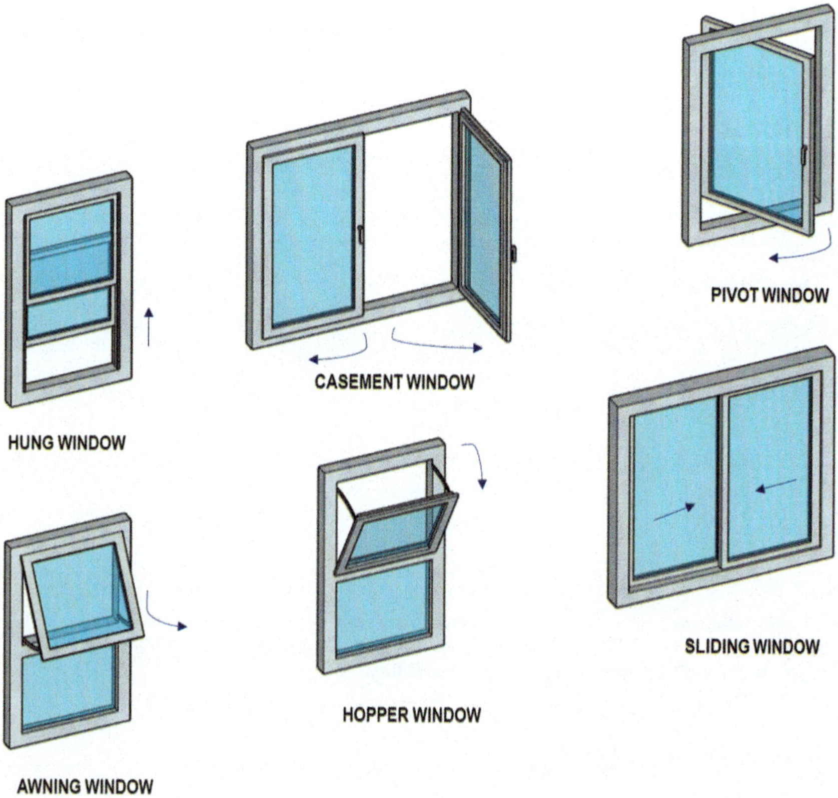

Fig. 37 Commonly used window types and opening directions

3.2.3 Classification of Doors

Material Based Classification

- WOOD
- METAL
- FIBER-GLASS PLASTIC (FGP)

Operability Based Classification

- SINGLE-LEAF HINGED DOOR
- DOUBLE-LEAF HINGED DOOR
- SLIDING POCKET DOOR
- BYPASS SLIDING DOOR
- SIDE-HINGED FOLDING DOOR

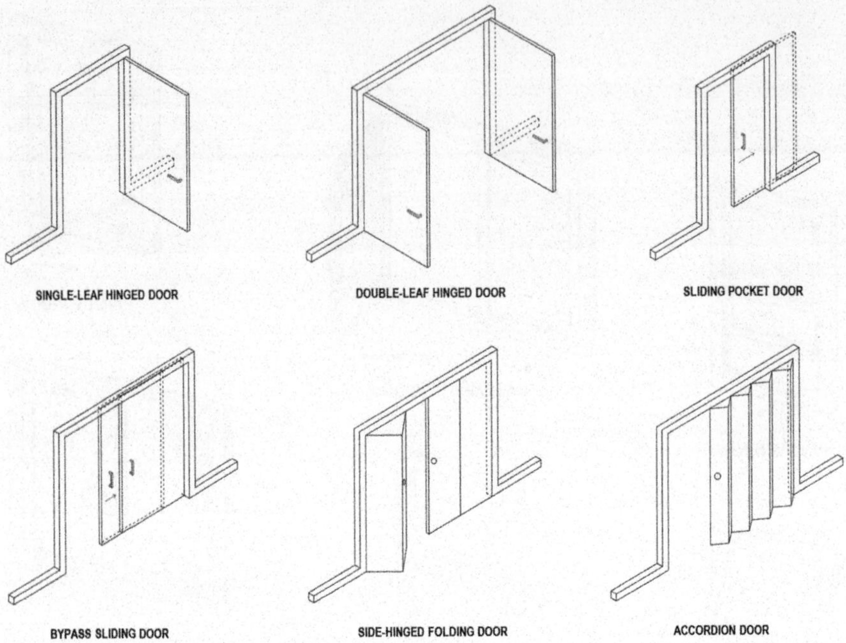

Fig. 38 Various door types regarding their operability

- ACCORDION DOOR (Fig. 38)

Style Based Classification

- FLUSH DOOR
- LOUVERED FLUSH DOOR
- PANEL DOOR
- FRENCH DOOR
- DIVIDED FRENCH DOOR
- DUTCH DOOR (Fig. 39)

3.2.4 Fire-Rated Doors and Windows

According to fire resistance time, they are classified as 3-h, 90-min, 60-min, or 20-min rated doors. A glazed window with 45-min fire protection is called a fire window.

3.2.5 Window Terminology

See Figs. 40, 41, 42 and 43.

Fig. 39 Various door types regarding their style

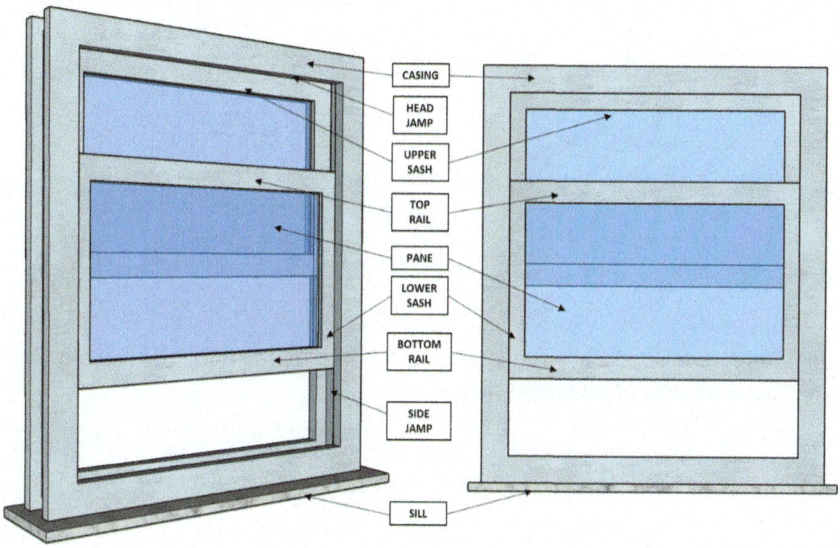

Fig. 40 Hung window terminology

3.2.6 Door Terminology

See Figs. 44, 45, 46, 47, 48 and 49.

3.3 Glass Aluminum Curtain Wall

Transparency, luminosity, and elegance are the reasons for the popularity of glass walls in modern architecture. In glass walls, the glass panes (also called liters) are

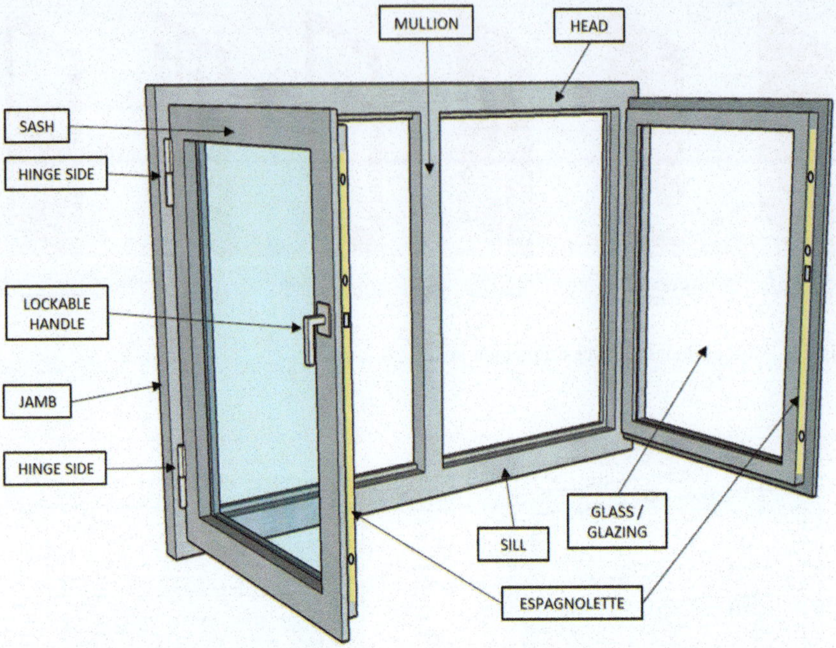

Fig. 41 Casement window terminology

held within vertical and horizontal aluminum framing members. Thus, they share some characteristics of aluminum windows.

3.3.1 Glass-Aluminum Curtain Wall Types

Stick-Built Systems

The curtain wall is installed piece by piece at the site. Generally, the mullions are installed first, followed by rails. Subsequently, the glass panes are installed within the mullion-rail framework. The anchorage of the wall to the structural frame is through the mullions. The mullions may span from floor to floor or over two floors. Expansion joints accommodate thermal expansion and contraction of mullions in mullions. Its disadvantages include longer on-site assembly time and more on-site labor than the other systems (Fig. 50).

Unitized Systems

The unitized system comprises shop-fabricated, preassembled, and generally pre-glazed framed wall units. The units are designed so that adjacent units' vertical

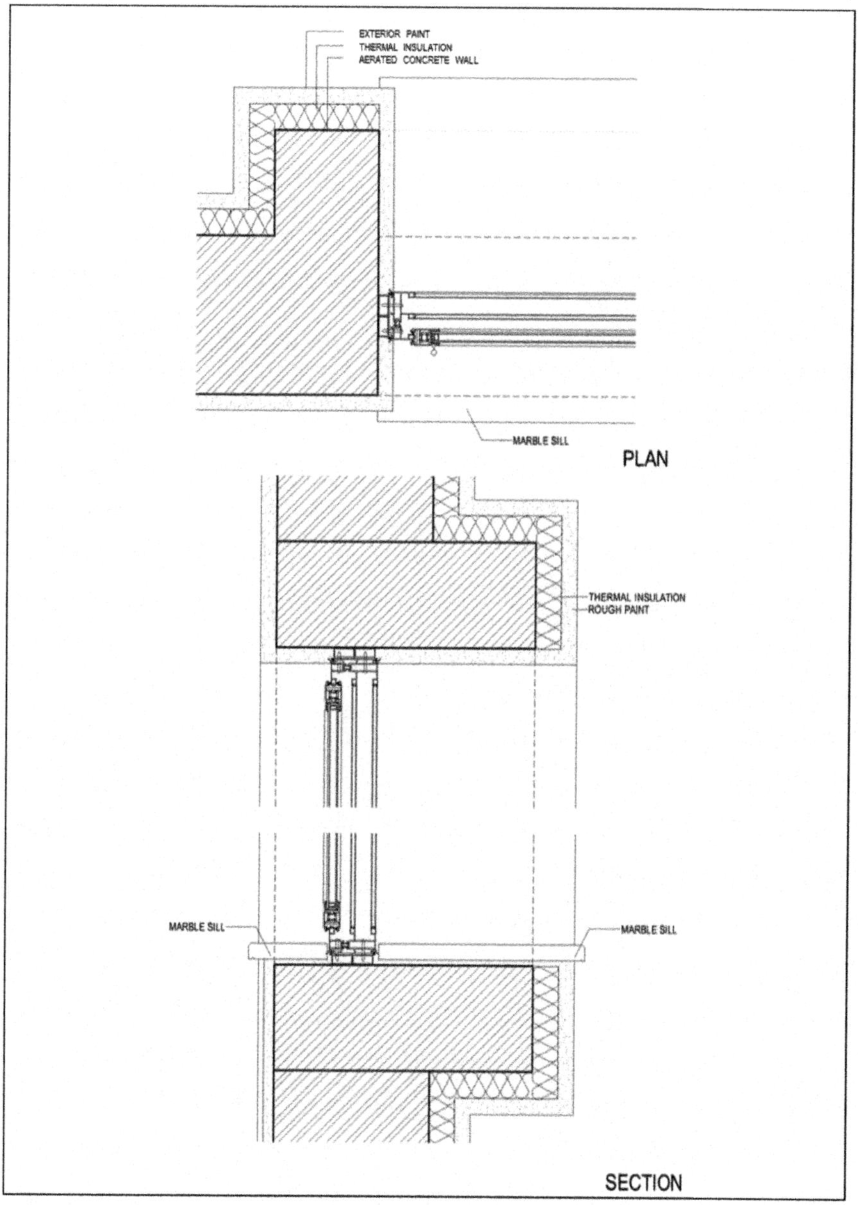

Fig. 42 Aluminum Frame Sliding Window Detail

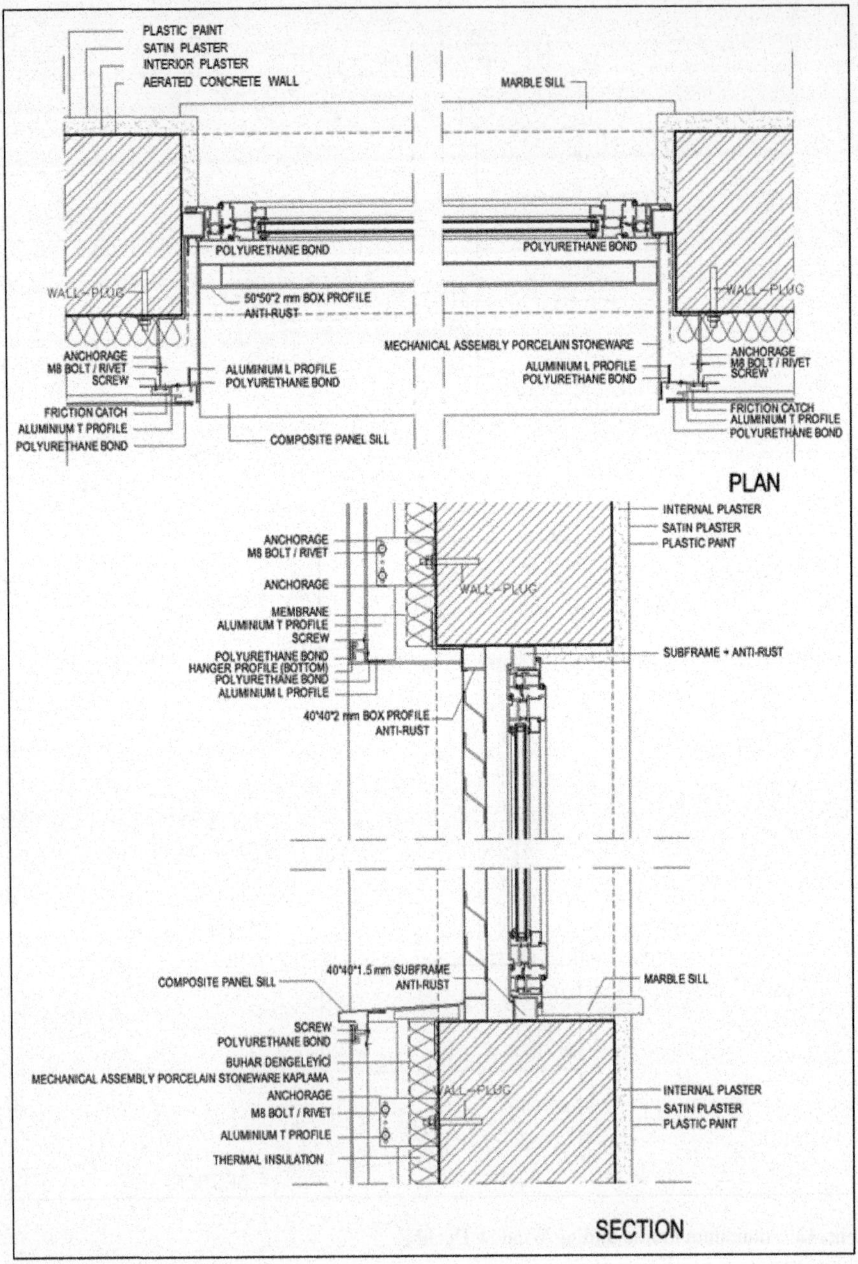

Fig. 43 A window detail

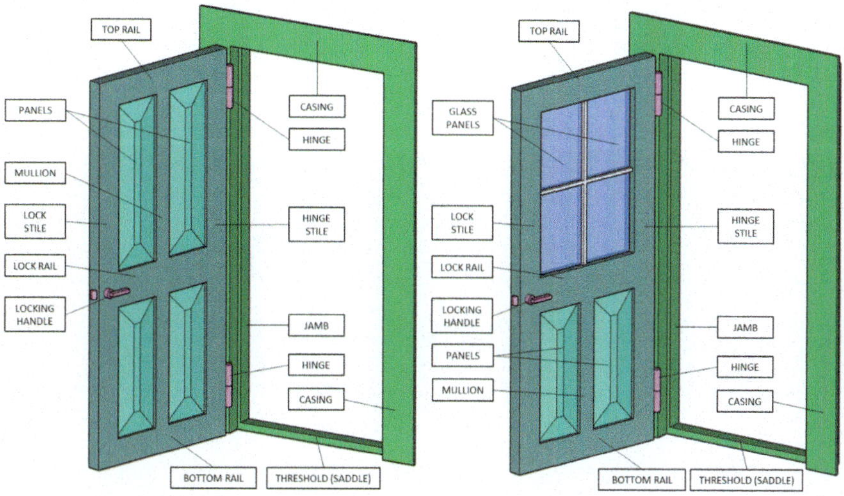

Fig. 44 Door terminology

and horizontal members interlock to form common mullions and rails. The units may be one or two stories high. They are anchored to the building's structural frame, essentially like the mullions in the stick system. This system's advantage is its greater quality control resulting from shop fabrication. Its disadvantages are the greater shipping cost because of the added bulk from assembled units, the need for greater protection of units during transformation, and a lower degree of field adjustment (Fig. 51).

Unit and Mullion Systems

The unit and mullion system combines the advantages of the stick and unitized systems. It is constructed by first installing the mullions; factory-assembled units are placed between them. Because the system is a compromise between the stick and unitized systems, it has the advantages and disadvantages of both its transportation cost is lower than that of the unitized system but greater than that of the stick system. A greater degree of site adjustability is available in the unit and mullion system, but it is less than that of the stick system (Fig. 52).

Panel Systems

The panel system consists of preassembled (and sometimes preglazed) homogeneous sheet metal panels with glass infills that generally span from floor to floor. The curtain wall's appearance is more integrated and comprehensive than a grid pattern of horizontal and vertical elements. The panels can be formed by stamping or casting.

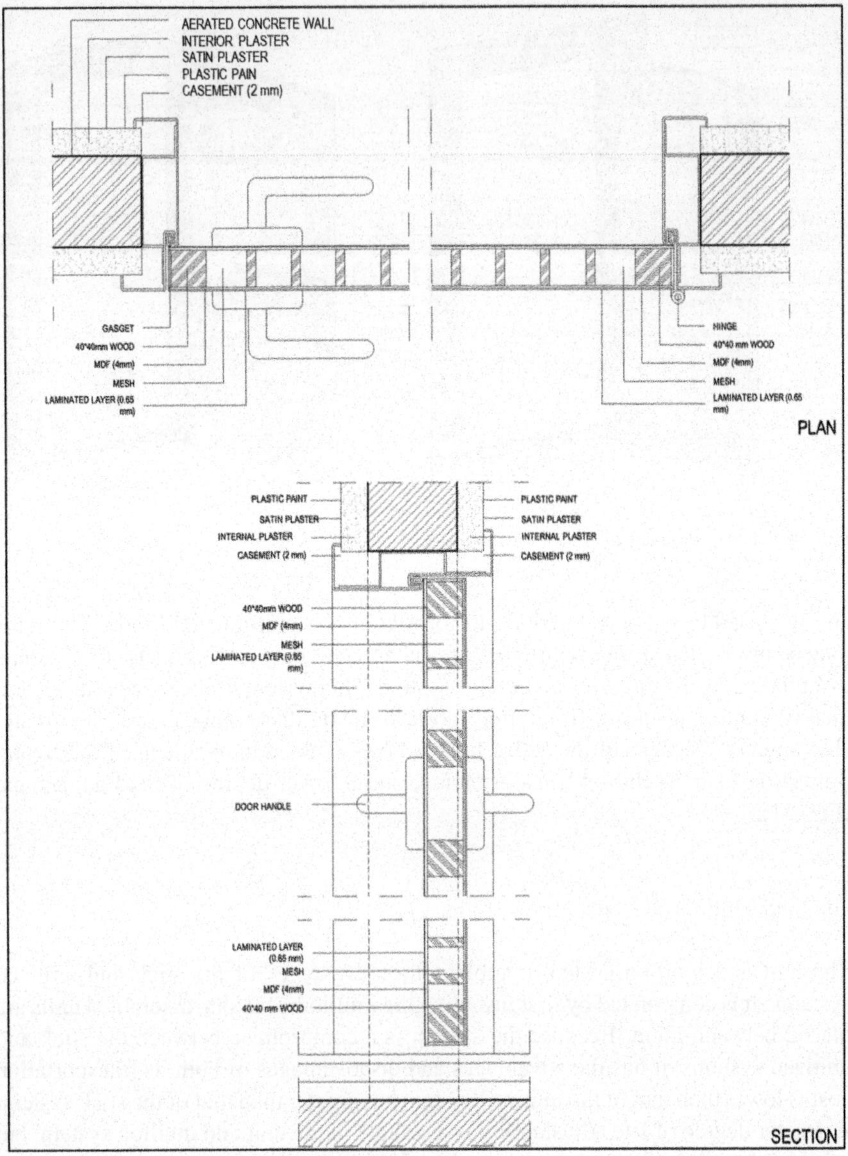

Fig. 45 A wood door detail

The casting system is economical only where many identical panels are needed (Fig. 53).

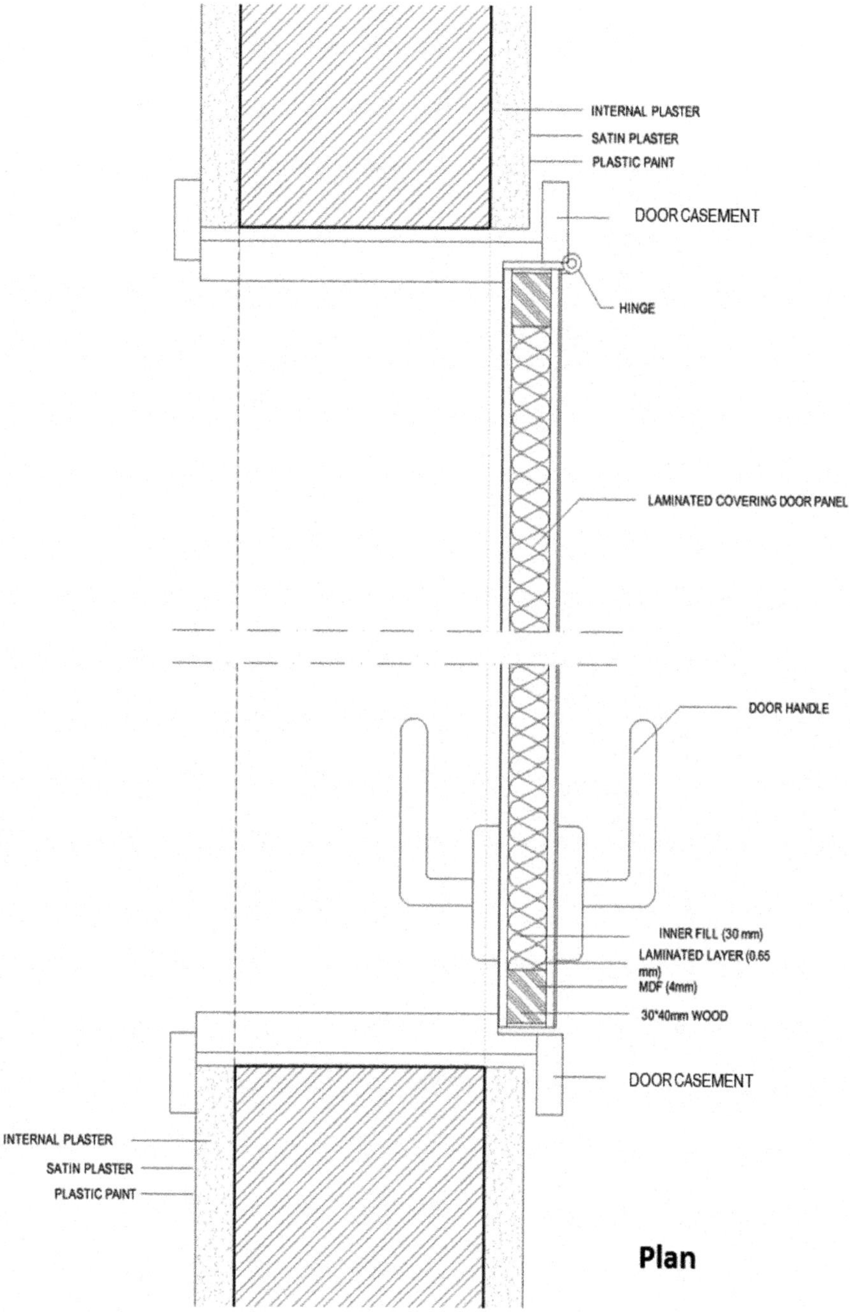

Fig. 46 A laminated wood door detail

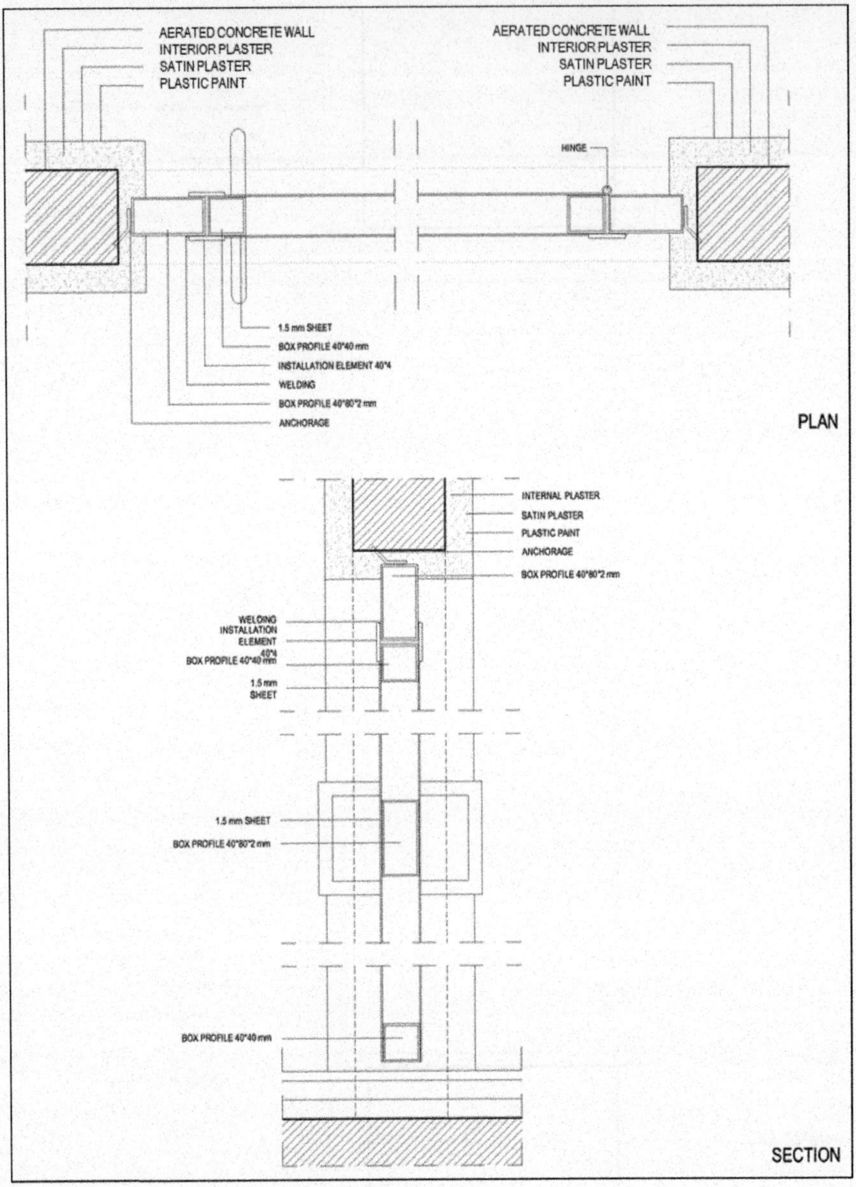

Fig. 47 Single leaf steel door detail

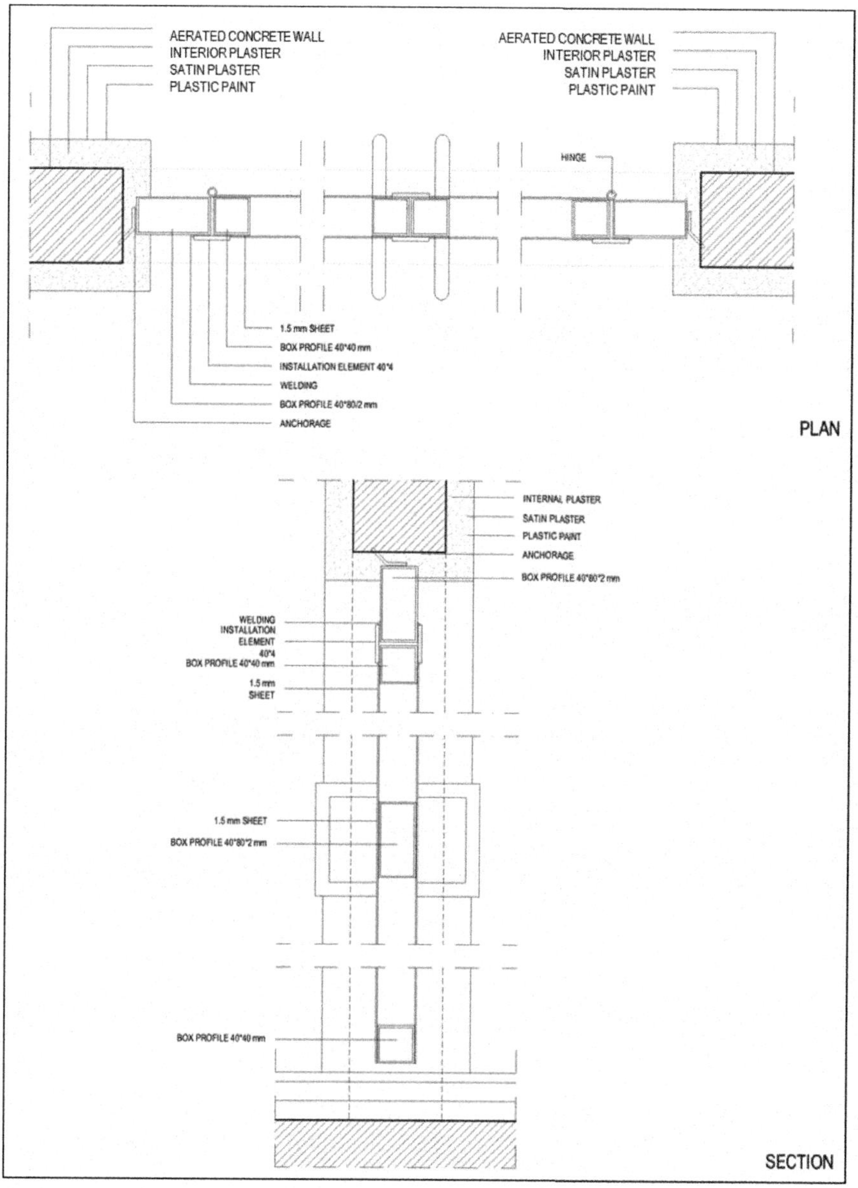

Fig. 48 Double leaf steel door detail

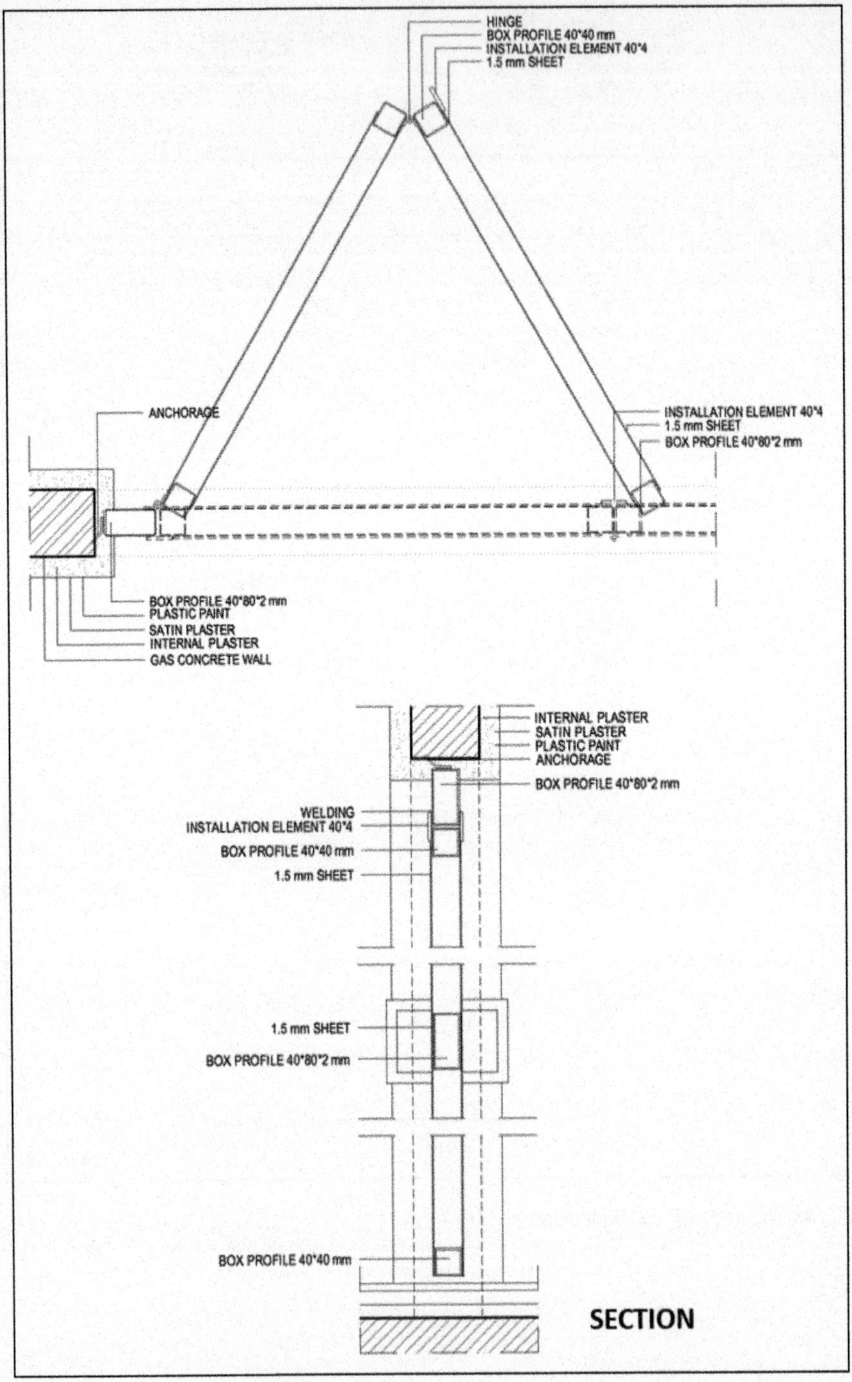

Fig. 49 Steel folding door detail

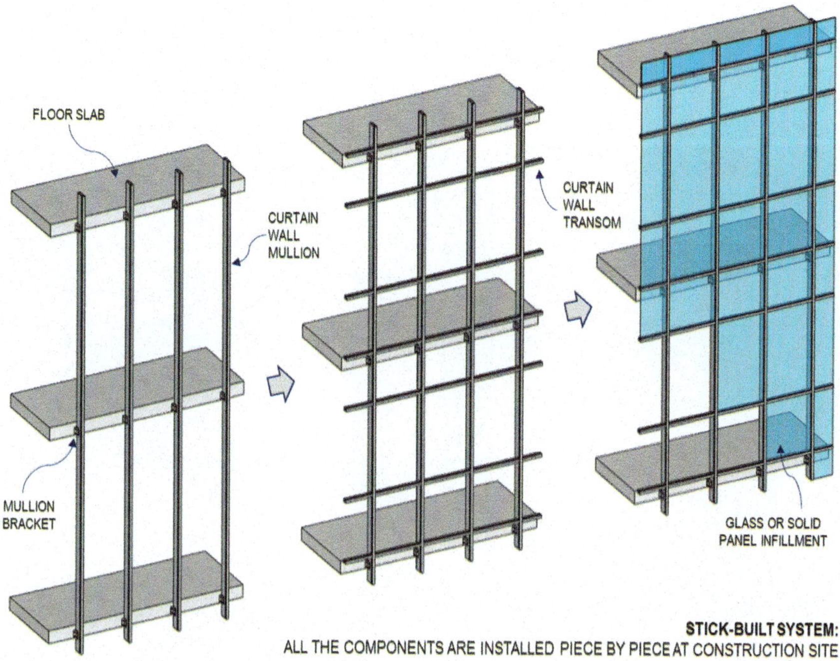

Fig. 50 Stick-built system

Column Cover and Spandrel System

This system consists of separate column covers connected to spandrel covers that generally span from column to column. Infill glazing units may be preassembled or assembled at the site like a stick-built system. The system provides an independent expression of the structural system rather than concealing it behind a wall (Fig. 54).

Comparison of the Systems

See Fig. 55.

3.3.2 Examples of Some Curtain Systems

See Figs. 56, 57, 58, 59, 60, 61, 62, 63, 64, 65, 66, 67, 68, 69, 70 and 71.

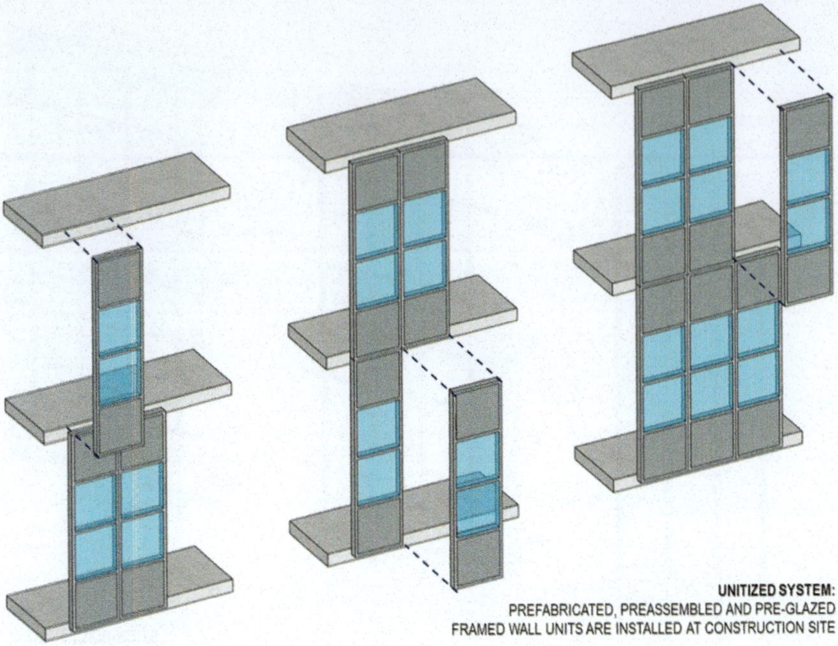

Fig. 51 Unitized system

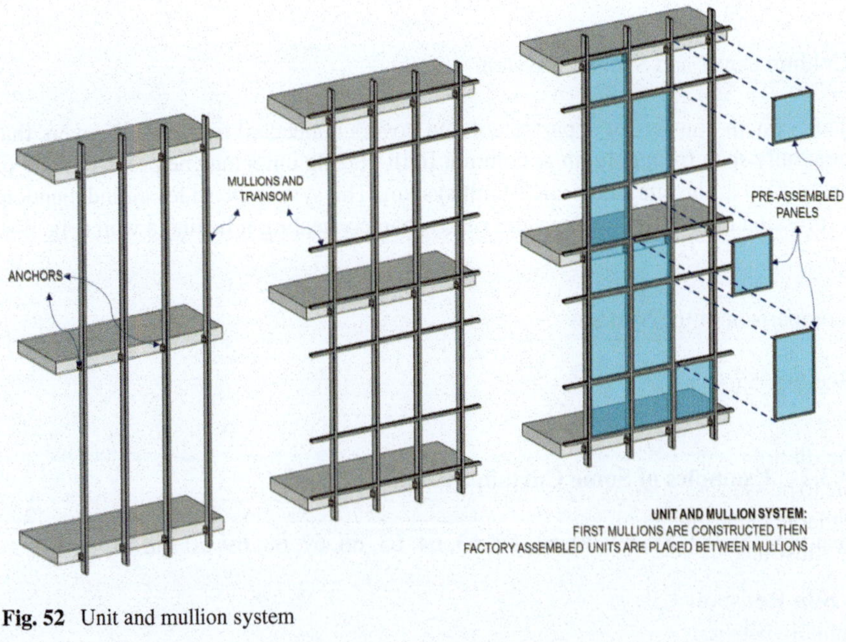

Fig. 52 Unit and mullion system

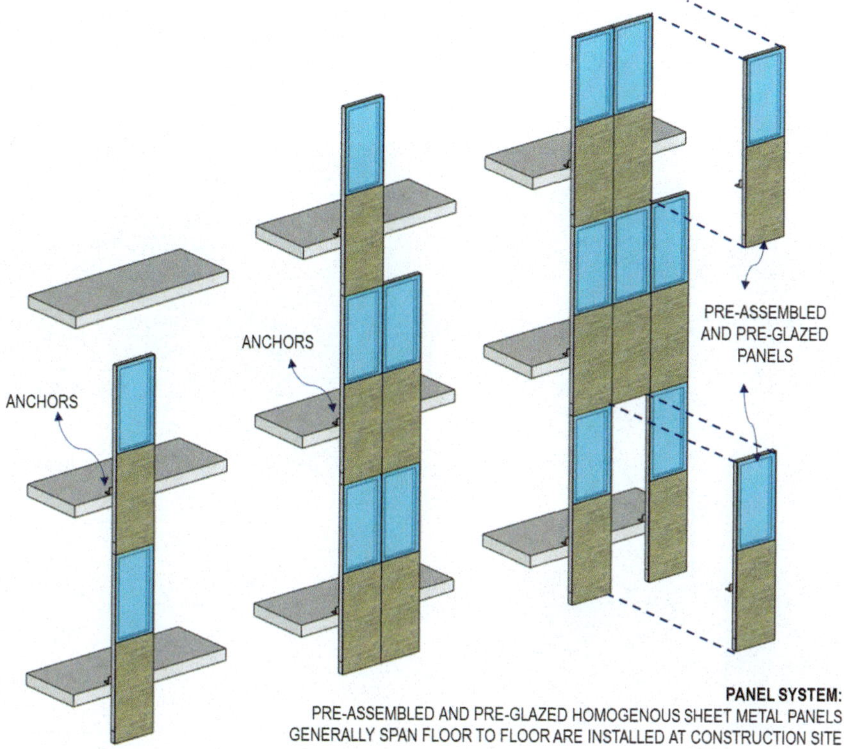

Fig. 53 Panel system

4 Floor Coverings

A floor consists of multiple layers overlapping with certain requirements. These layers are represented as (i) subfloor, (ii) floor slab/deck, and (iii) ceiling (Fig. 72).

4.1 Subfloor

The subfloor is the top surface of the structural floor. There are various materials available to be used as subfloors. However, certain application techniques and methods are available regarding the material category. In the following sections, certain methods are illustrated. The selection criteria for the subfloor are listed as follows:

- **Slip Resistance**: Prevention of slips is important. Otherwise, certain injuries can likely occur.

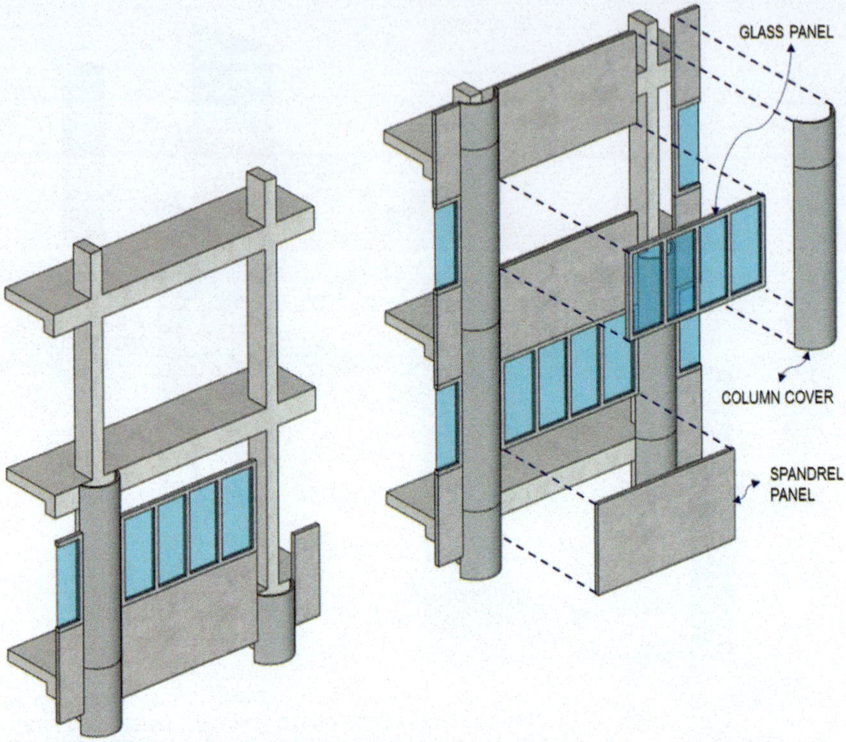

Fig. 54 Column cover and spandrel panel system

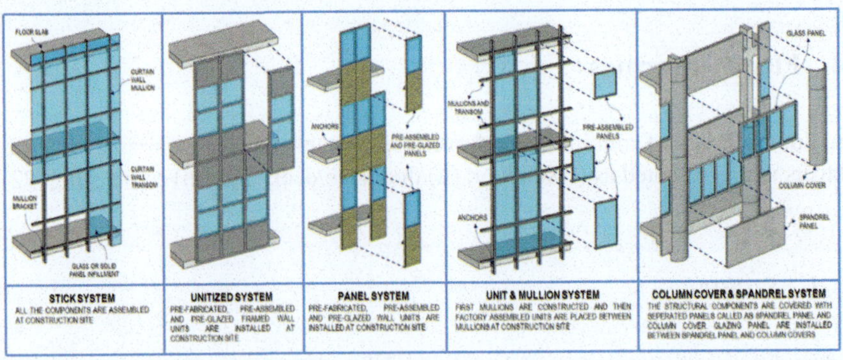

Fig. 55 Comparison of the glass aluminum curtain wall types

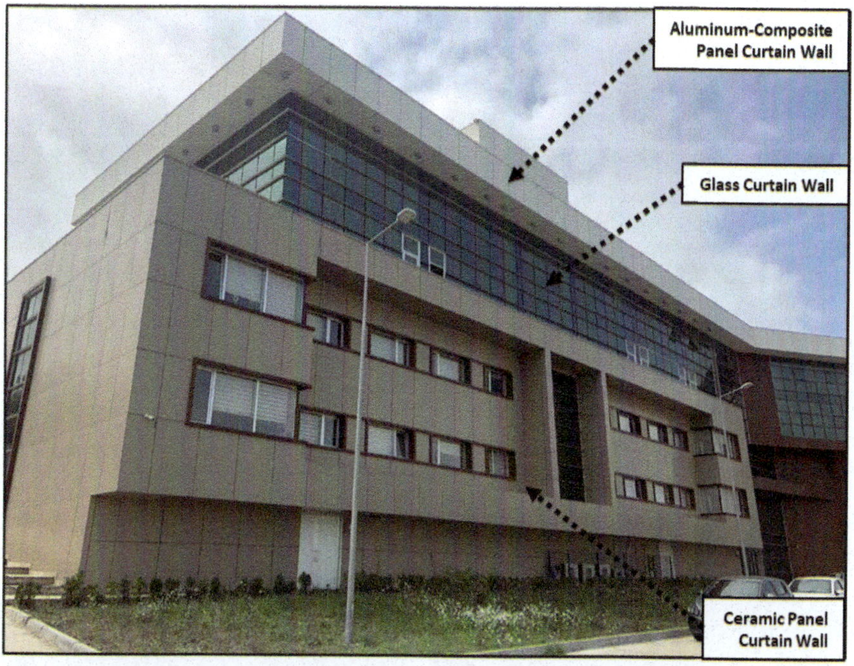

Fig. 56 Aluminum-composite, glass, and ceramic panel curtain wall application

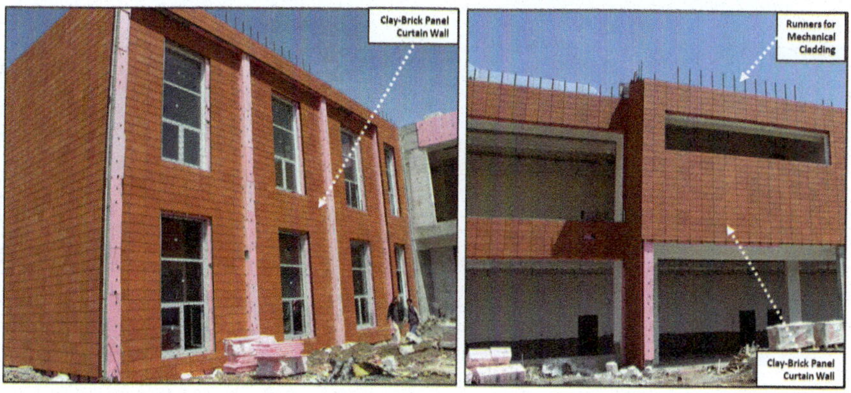

Fig. 57 Clay-brick panel curtain wall application

- **Durability and Longevity**: The service life and performance of the selected material affect not only the construction but also the maintenance cost of the project.
- **Flammability**: Regarding the codes and regulations, selected material is needed to provide certain resistance to fire for a while.

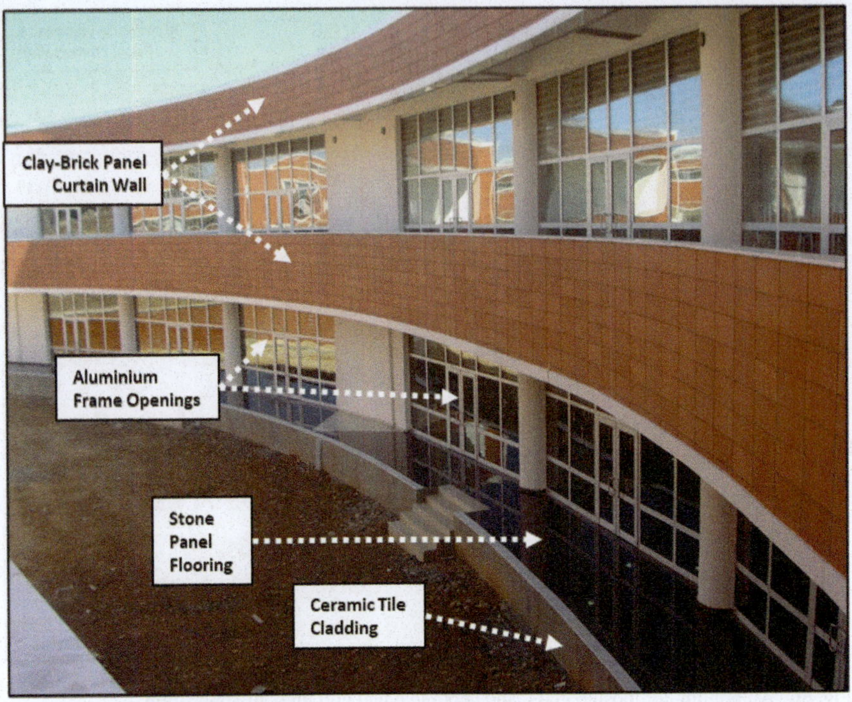

Fig. 58 Various finishing applications in a building's exterior surface

Fig. 59 Application layers and details for a ceramic panel

- **Sound**: Sound absorption and conductivity are factors influencing subfloor performance.
- **Hygienic Quality**: Hygienic performance is important, especially in hospitals and kitchens.
- **Walking Comfort**: The selected material is required to comfort the occupants.
- **Sustainability**: Production and use of the material is required to be sustainable.

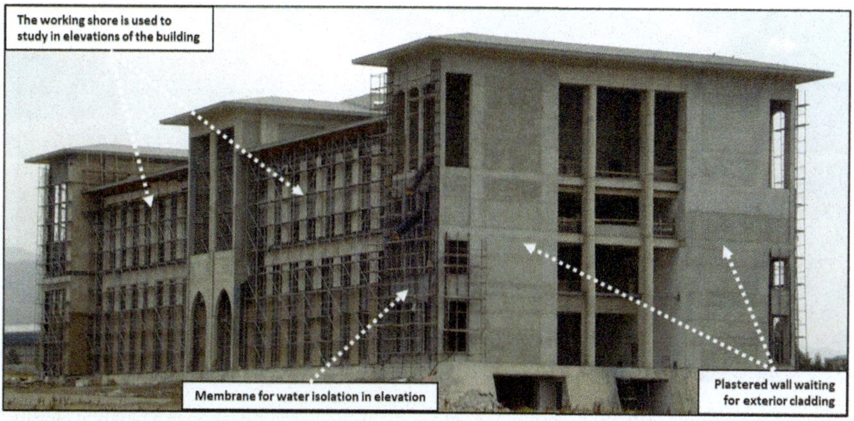

Fig. 60 The working shore is installed to study the finishings in the elevation of the building

Fig. 61 The stone panel is installed

- **Maintenance Requirements**: Low maintenance requirement is the preferable option for material selection (Figs. 73, 74 and 75).

Fig. 62 The thermal insulation, membrane, and vertical channel application in a building elevation

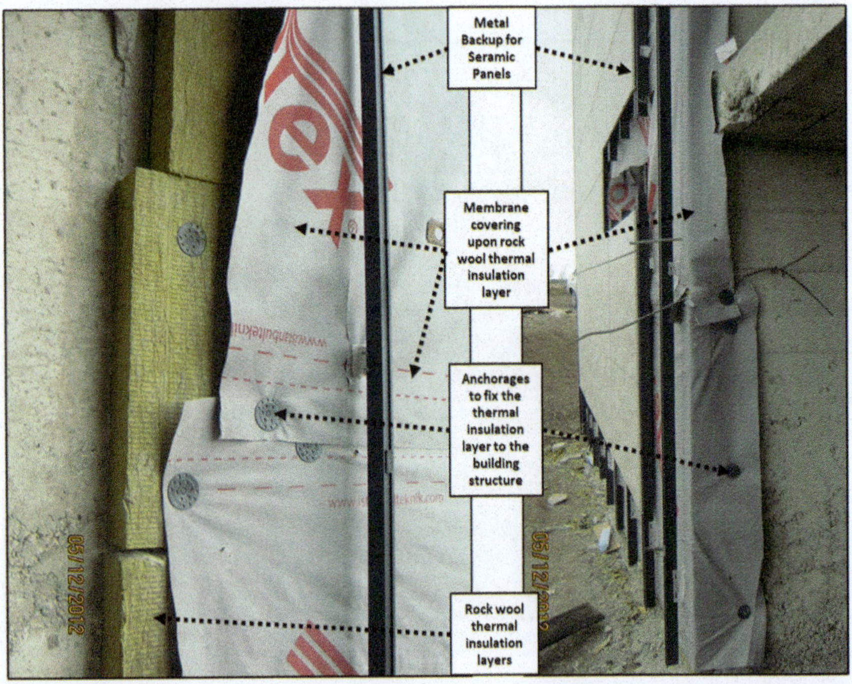

Fig. 63 The images of thermal insulation, membrane, vertical channel, and stone panel application

4.1.1 Stone and Tile Flooring

The following physical properties of the tiles have an impact on the selection of appropriate tiles for particular purposes:

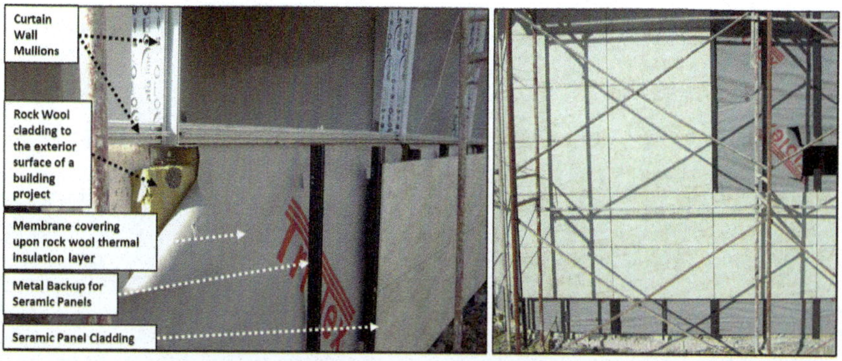

Fig. 64 Ceramic panel cladding application layers in a building elevation

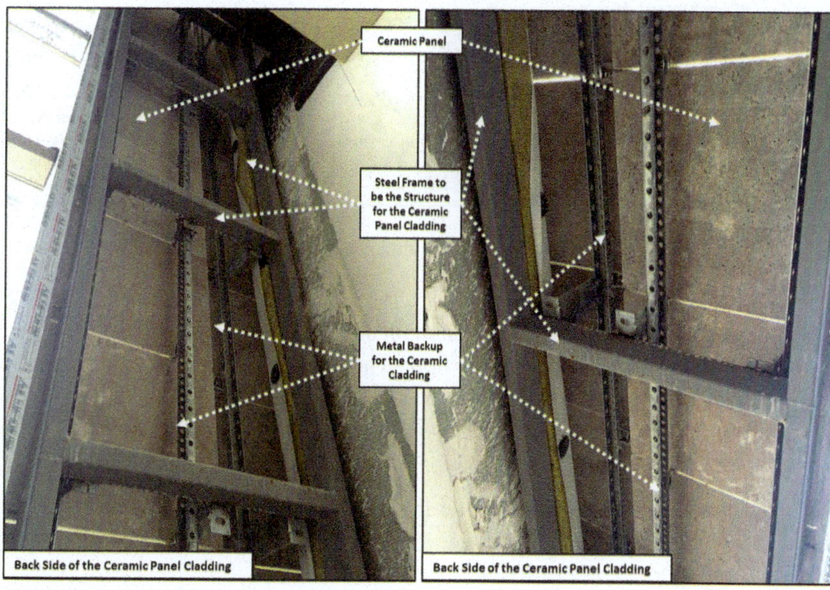

Fig. 65 Metal backup application for ceramic panel cladding

- **Quality and Uniformity**: Due to being fired products, facial and structural defects in ceramic tiles may develop that disqualify them for installation.
- **Shapes and Dimensions**: Tight control over dimensional variation allows grout joints, whereas looser control permits wider joints.
- **Warpage**: If tiles are not uniformly flat, they will not lay flat on the subfloor and will break under heavier loads because the underlying mortar will not adequately support portions of the tile
- **Water Absorption**: Tiles do not absorb atmospheric moisture, as wood does; however, they absorb water when in direct contact.

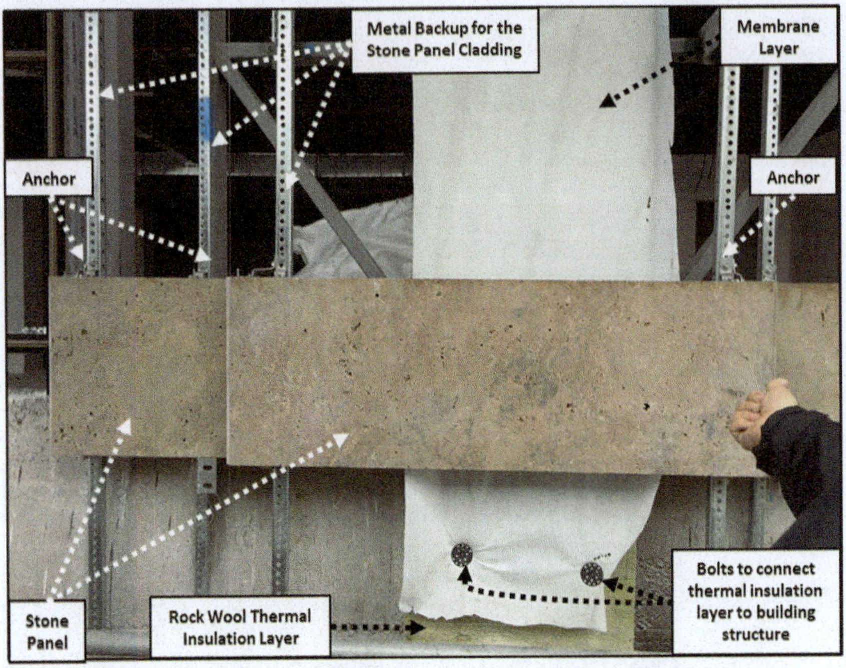

Fig. 66 Stone panel cladding illustrating application layers

- **Breaking strength**: Tiles should maintain their integrity when subjected to loads or when objects are dropped on the floor. The breaking strength of a tile is a good indication of how well it will perform.
- **Abrasion Hardness**: Tiles should be hard enough to resist the abrasion that will occur over their service life.

Regarding the setting methods, tile flooring is divided into two categories: (i) thick set and (ii) thin set.

Thick Set (or Thick Bed)

A thick-set requires a 5–7.5 cm thick mortar bed. This method is necessary in the following circumstances.

- The floor tiles are large, generally more than 30 cm × 30 cm,
- The floor slopes to floor drains,
- There is excessive variation in the thickness of tiles, as is generally the case with natural stone panels,
- The subfloor has surface irregularities.

Fig. 67 Plaster, thermal insulation, and membrane cladding application

Thin-Set (or Thin Bed)

This is more popular than a thick set due to requiring less material and labor. The mortar bed is generally 3 mm thick and consists of polymer-based adhesives. This method is used in the following circumstances:

- The tiles are small, generally less than 30 cm × 30 cm,
- No slope to floor drains is required,
- The tile thickness is relatively uniform,
- The subfloor does not have excessive surface irregularities (Fig. 76).

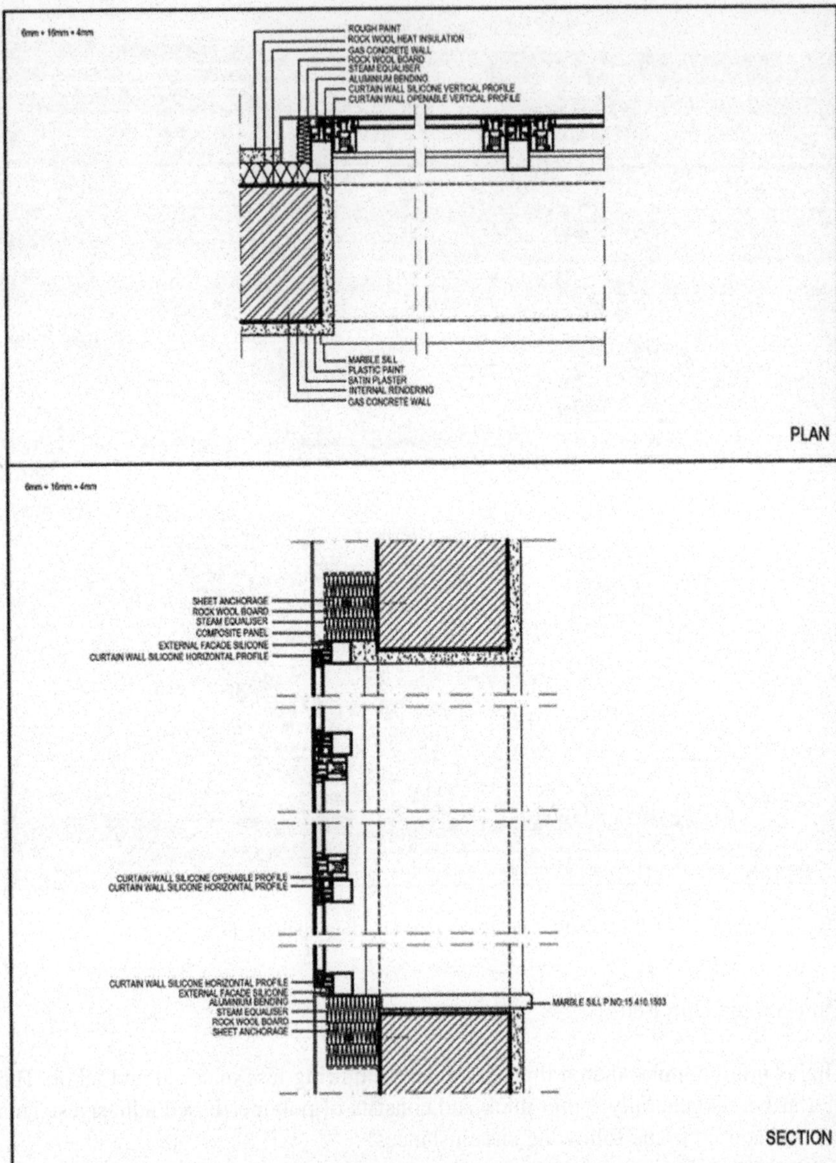

Fig. 68 A glass curtain wall connection detail

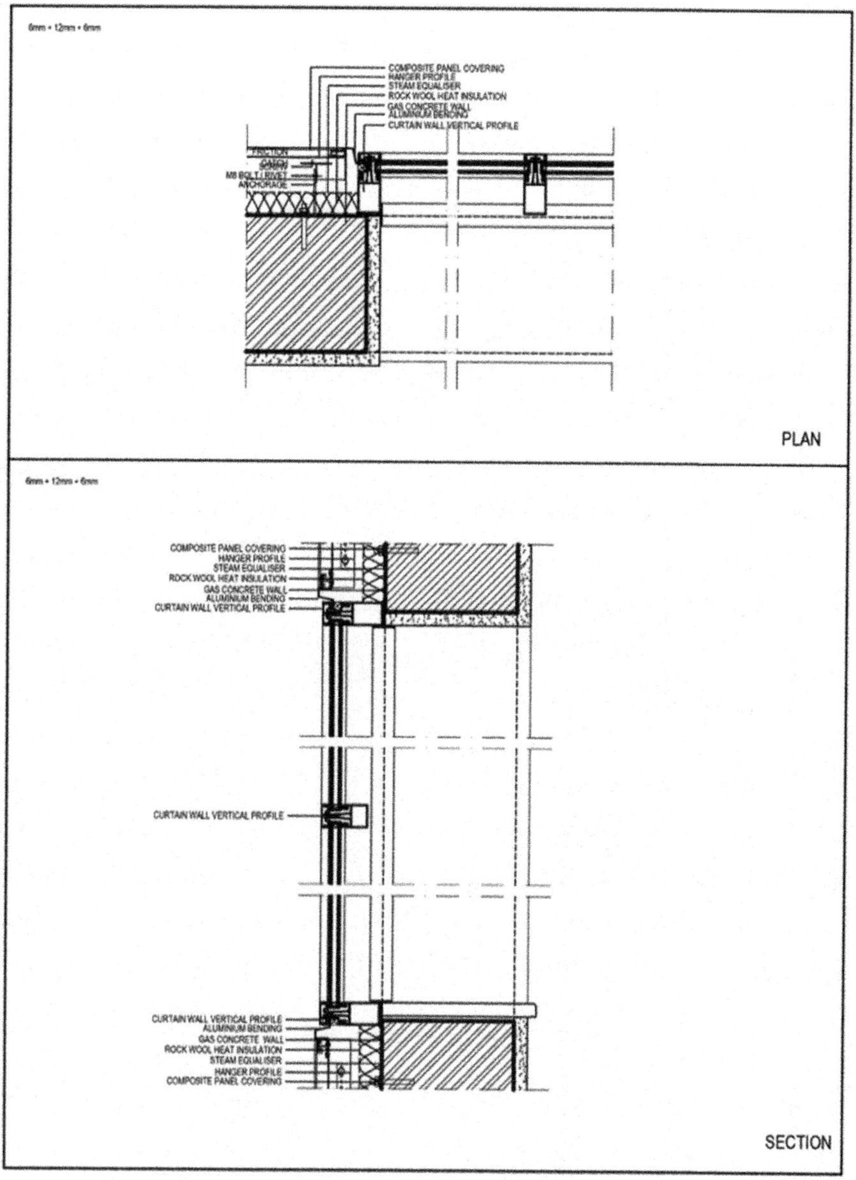

Fig. 69 A glass curtain wall connection detail

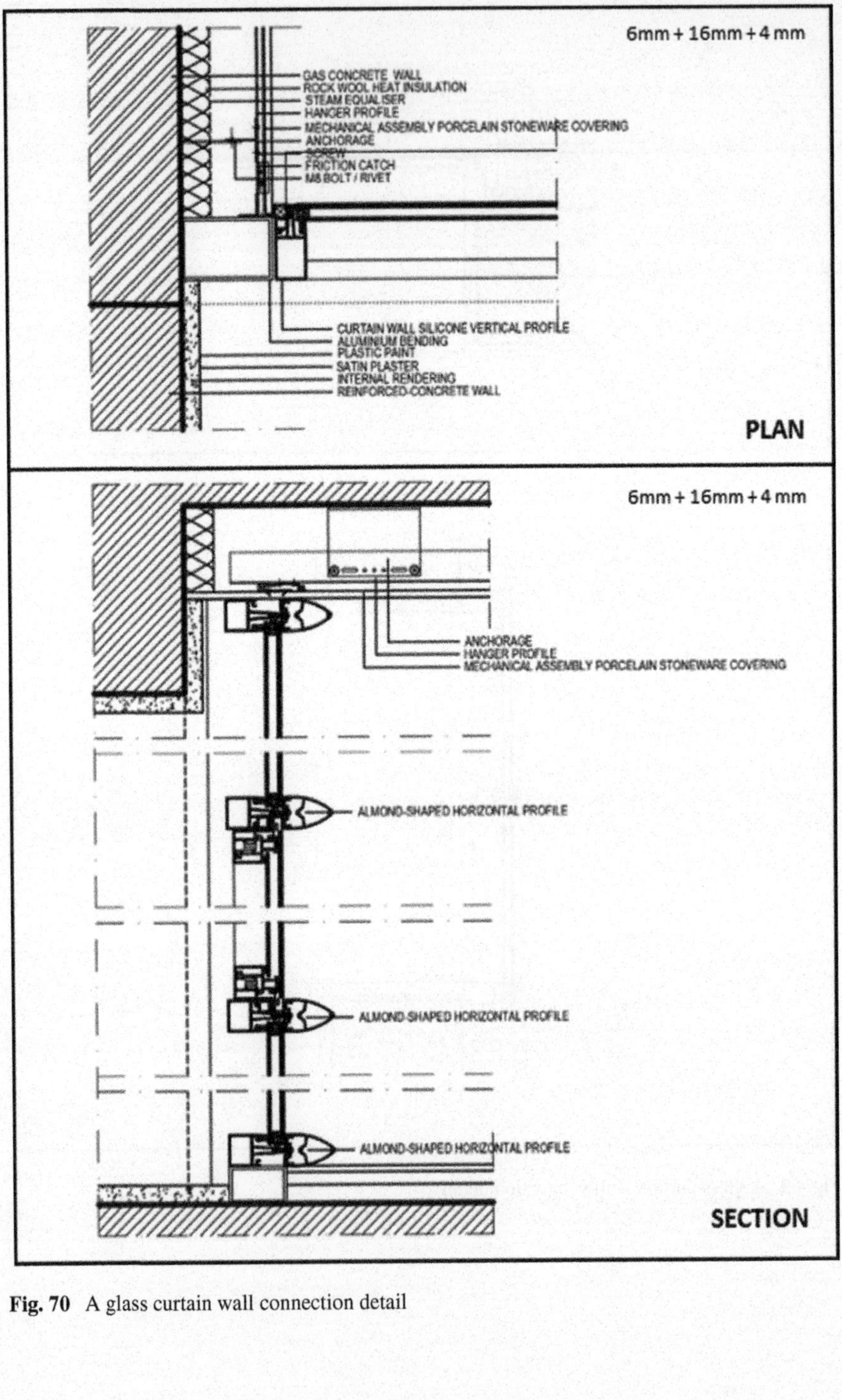

Fig. 70 A glass curtain wall connection detail

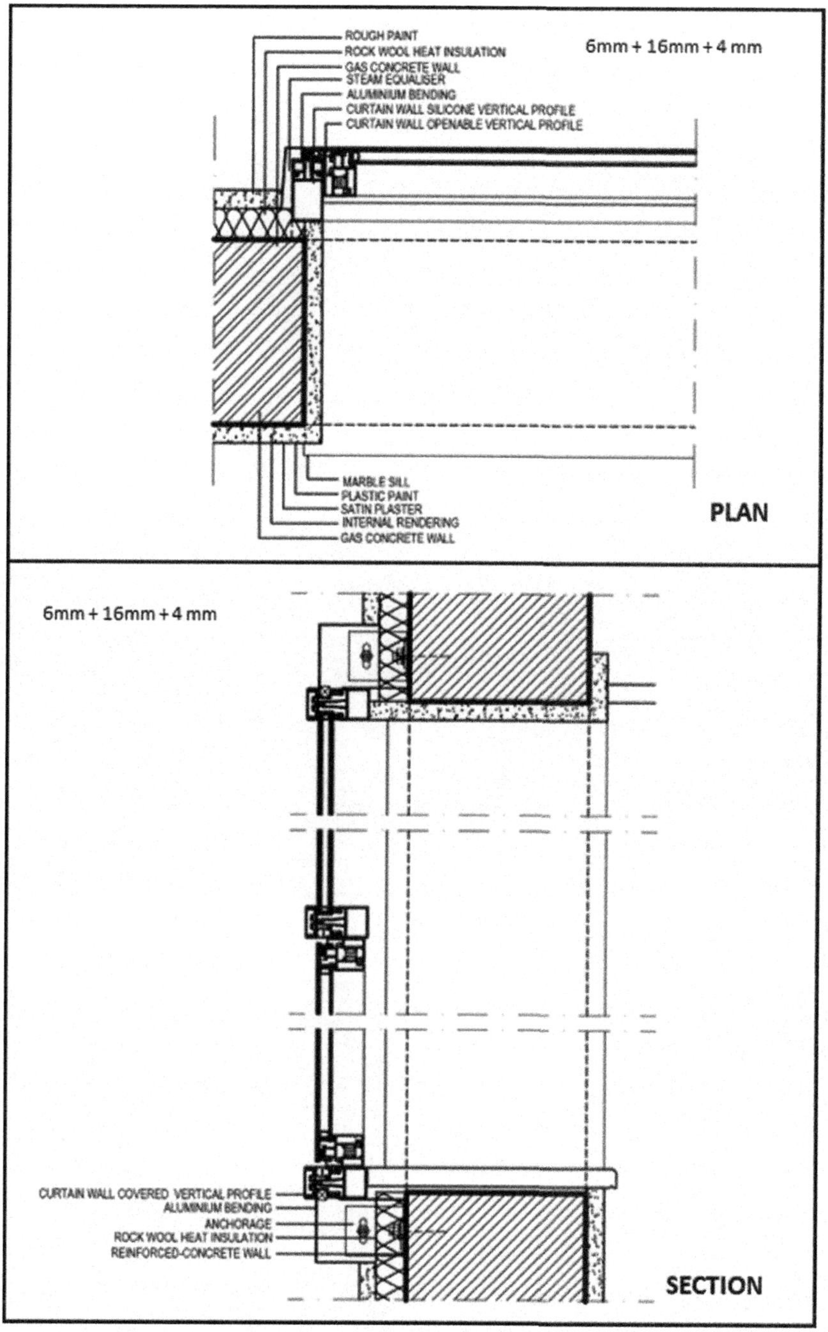

Fig. 71 A glass curtain wall connection detail

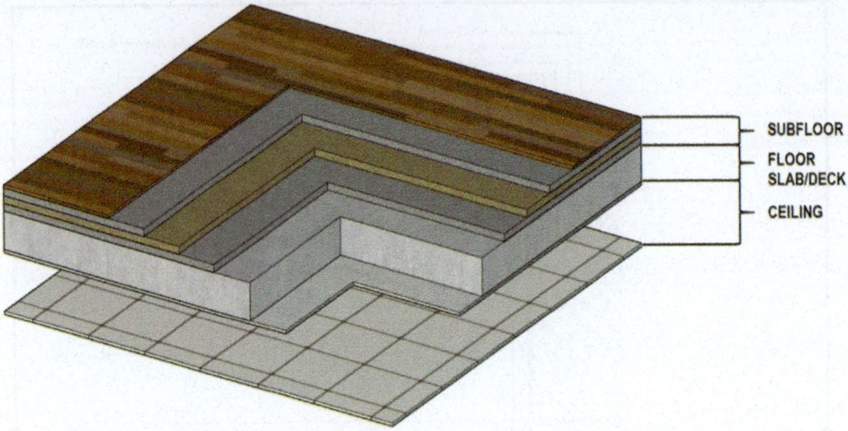

Fig. 5.72 Layers of a typical floor

Fig. 73 The screed application over the concrete slab

Fig. 74 Screed application upon concrete slab

Fig. 75 The plumbing pipes and electrical cables remain within the screed layer

4.1.2 Setting Materials

- **Organic adhesives** are generally ready-to-use liquid or powdered water-emulsion latex products cured by evaporation. They are typically for light-duty, interior-use-only installations and unsuitable for high temperatures.
- **Cement mortars consist** of mixtures of Portland cement, sand, water, and water-retentive additives and are for general-duty installations.
- **Water-Cleanable Epoxies** An epoxy resin and a hardener are suitable for heavy-duty installations, high-temperature conditions, and specific functions.
- **Furan Resin Mortars** consist of furan resin, a carbon or silica filler powder, and an acid catalyst for chemical resistance.

4.1.3 Grout

- **Sand—portland cement grouts**: They are used for joints greater than 3 mm wide, whereas unsanded cement grouts contain water-retentive additives and are for joints up to 3 mm wide.
- **Polymer-modified cement grouts**: They tend to perform better than Portland cement grouts. These grouts possess increased color stability, good flexural and

Fig. 76 Ceramic floor tile application as a thin set

bond strengths, stain resistance, and lower moisture absorption, so they resist frost damage.

- **Water Cleanable Epoxy and Furan Resin Grouts**: They are essentially the same as the mortars and are used with appropriate mortar.

4.1.4 Stone Panel Flooring

Stone panels for flooring, also known as dimension stone, are natural Stones that have been selected and fabricated (cut and trimmed) to specific shapes and sizes, with or without mechanical dressing of one or more surfaces. Natural Stones used for panel flooring include granite, marble, limestone, slate, and other quartz-based Stones such as sandstone, bluestone, and quartzite. The panels must be installed over a thick-set mortar bed due to not being uniformly thick. Most stones are more than adequate for use as a floor covering because they can resist abrasion, wear, and absorption. When used in exterior applications, damage is possible due to water permeability, inelasticity, or low compressive strength.

Fig. 77 Installation of stone panels at ground surfaces with thickset

Regarding the finishing surface, stone panels are categorized as polished, honed, and thermal:

- **Polished**: Finished to a reflective sheen and resistant to wear (granites more so than marbles), a polished finish can be stretched and dulled by abrasive materials such as sand on the shoes of people who walk over it
- **Honed**: Finished to a uniformly matte sheen, a honed finish can be used to mask wear
- **Thermal**: Exposure to an open flame essentially burns off the immediate surface, leaving a slightly roughened surface with improved slip resistance.
- **Installation of Stone Panels**

The thick bed method is used. Because the backs of stone panels are usually ungauged, the panels are larger than tiles, and there is a slight variation in thickness among panels. Thus, the thin-set installation method cannot be used. Because stone panel flooring is a rigid assembly, it will likely develop cracks if the subfloor deflects excessively. Thus, a cleavage membrane is needed so the setting bed does not bond to the subfloor. If the stone panel flooring is installed on a concrete slab on the ground that is not subject to deflection, a cleavage membrane is not needed (Figs. 77 and 78).

4.1.5 Terrazzo Flooring

Terrazzo flooring is a common flooring type. Like concrete, a terrazzo binding matrix is mixed with several aggregates and placed, wet and plastic, in its final location. However, unlike concrete, the exposed surface is ground and polished after it cures to expose the binding matrix and aggregates, thus revealing a smooth and colorful finish. A binding matrix is the material that holds the aggregate chips in position.

Fig. 78 Granite stair panels are applied as a thick set to the surface

Nowadays, the common matrix material is a resinous epoxy, including polyester, polyacrylate-modified cement, or polyurethane. Marble chips are the most common aggregate; however, granite, onyx, travertine, or glass chips are also used, and mother-of-pearl is especially favored. Metal dividers and control strips are used to control cracking and to create designs. Cementitious terrazzo requires closely spaced strips to control cracking. Resinous terrazzo does not crack, so the strips are used to create decorative designs. There are various terrazzo flooring types. These are (i) Cementitious Terrazzo, (ii) Sand-Cushion Cementitious Terrazzo, (iii) Bonded Cementitious Terrazzo, (iv) Structural Terrazzo, (v) Rustic Terrazzo, (vi) Epoxy Terrazzo and (vii) Precast Terrazzo (Fig. 79).

4.1.6 Carpet and Carpet Tile Flooring

Two basic types of carpet are available: Rolled goods or tiles. Generally used in commercial and worship buildings. Carpet flooring has rubber backings and can be easily changed. Fibers are the basic component of carpets. Three fibers are used to make the yarn: nylon, polypropylene, and wool. Fibers are grouped variously to make yarn, which is then used to construct the exposed face of the carpet.

Carpet Installation

Some rolled goods carpets require a cushion between the subfloor and the carpet to achieve their desired performance. The most commonly used installation methods are:

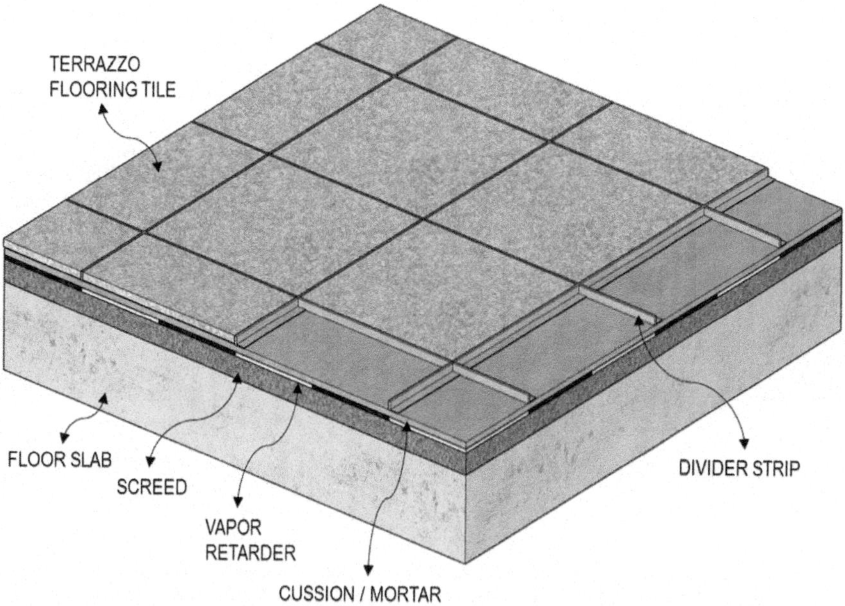

Fig. 79 Application of terrazzo flooring tile

- **The stretch-in method**: The carpet is placed over an independent cushion, stretched (placed in tension), and then securely hooked to tackle nail strips at the perimeter of the room
- **Direct glue-down:** The carpet tiles are attached by a compatible adhesive to the subfloor.
- **Double glue-down** is the same as direct glue-down, but the carpet cushion is also glued to the subfloor this time (Figs. 80, 81, 82 and 83).

4.1.7 Wood Flooring

Wood flooring is valued in residential applications for the warmth and beauty it brings to a room. Two types of material are used: Solid wood flooring and engineered wood flooring. Solid wood flooring: The same wood species is used throughout the piece. Engineered wood flooring: like plywood, it is a combination of a surface veneer, usually hardwood, which is laminated to one or more plies of a wood veneer from a less expensive species that provides dimensional stability and added strength. Controlling the moisture content is very important in wood flooring. Changes in temperature and humidity easily cause the wood to expand and shrink; thus, enough expansion space must be provided in wood flooring.

Three types of sophisticated wood flooring installations are:

- Floating System:

Fig. 80 The workers are rolling the adhesive to the surface

Fig. 81 The carpet tile and trim application

Resilient materials such as neoprene, rubber, or polyvinyl chloride sheets may be adhered to the substrate, or a foam underlayment may be used to isolate the wood flooring from the subfloor (Fig. 84).

- Fixed-wood-sleeper and fixed-metal sleeper systems.

In fixed-wood sleeper and metal tubes, wood strips, channels, or other shapes are used in fixed-metal sleeper systems to create a grid on the subfloor (Fig. 85).

Fig. 82 The carpet tiles are cladded by the workers at the construction site

4.1.8 Resilient Flooring

Resilient flooring is typically used for utilitarian purposes in rooms that require regular wet cleaning and where there is a probability that water will be on the floor. Two forms of resilient flooring are available: tiles and sheets (in roll format). Many resilient flooring products are chemical resistant. Resilient flooring types are (I) solid vinyl tiles, (ii) vinyl composition tiles, (iii) rubber tiles, (iv) sheet vinyl, (v) linoleum tiles and sheet, and (vi) static control resilient flooring.

Solid Vinyl Tiles

Composed of three primary ingredients—pigments, fillers, a vinyl chloride polymer or copolymer binder—and other modifying resins, plasticizers, and stabilizers. The thickness is approximately 2.5 mm, and sizes range from 30 to 90 cm square (Fig. 86).

Fig. 83 Various carpet tile applications

Vinyl Composition Tiles

Similar to solid vinyl tiles, vinyl composition tiles are less expensive. The thickness is usually 3 mm, and the size is usually 30 cm square (Fig. 87).

Rubber Tiles

Rubber tiles are manufactured from a vulcanized compound of natural or synthetic rubber, pigments, fillers, and plasticizers. Thicknesses range from 4.5 mm to 8 mm, and sizes range from 30 to 60 cm square. Rubber tiles are commonly used for stair treads and riser covers (Figs. 88 and 89).

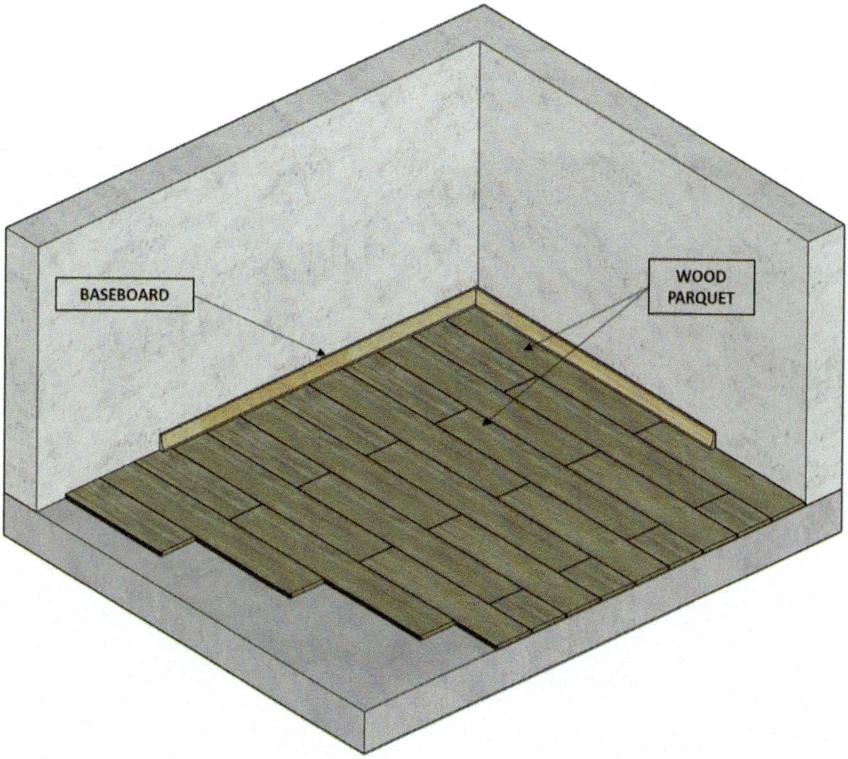

Fig. 84 Floating system

Sheet Vinyl

Like solid vinyl and vinyl composition tiles, sheet vinyl is ideally suited for where floors are regularly or frequently cleaned with water and cleaners. It provides minimal joints, and the sheets can be bent up the Wall to form an integral, flash cove base. The joint is either chemically bonded or welded with hot air, making the joint watertight and giving a seamless appearance. The thickness is approximately 2.5 mm, and the widths are 180 cm and 360 cm.

Linoleum Tiles and Sheet

Linoleum is the original resilient sheet on which all other resilient floorings have been patterned. Invented in England in 1860, linoleum consists of a binding agent and a filler. This mixture is then solidified, bonded, and keyed to a fibrous backing such as burlap (jute). The joints of sheets and tiles can be made watertight by sealing them with heat-welded rods of linoleum. Thicknesses range from 2 mm to 3.5 mm, and sheets are as wide as 200 cm (Fig. 90).

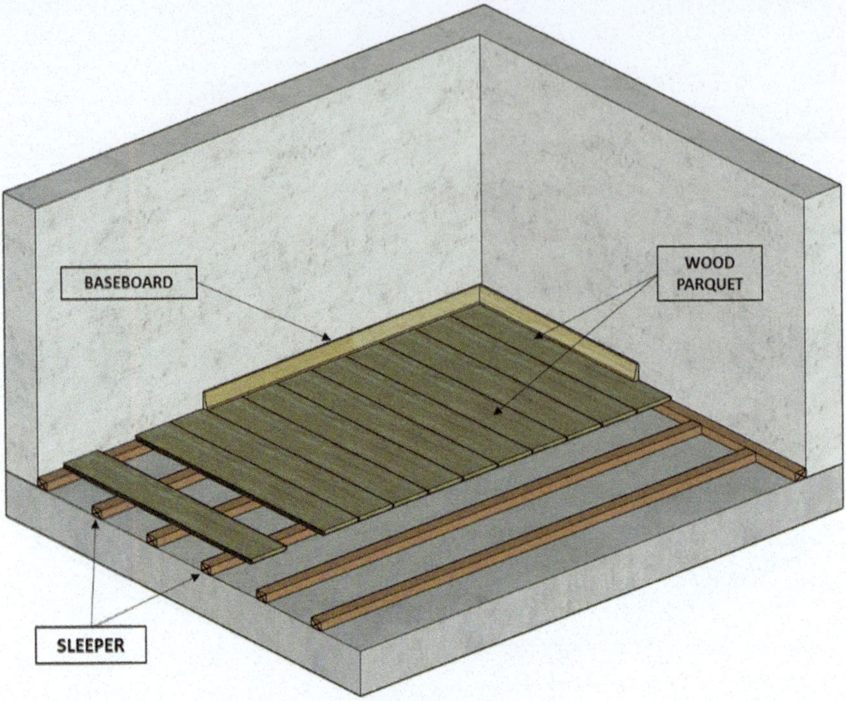

Fig. 85 Fixed-wood-sleeper system

Fig. 86 Solid vinyl tile application

Fig. 87 Vinyl composition tile application

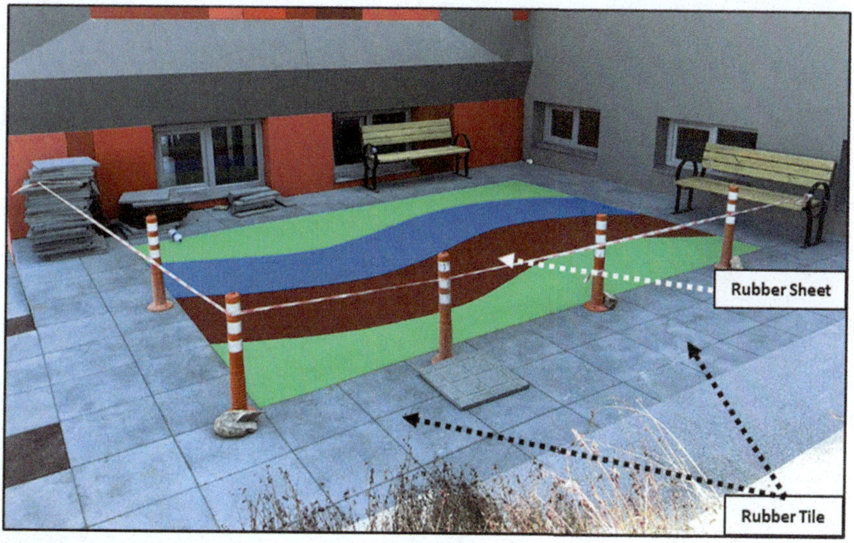

Fig. 88 Rubber sheet and tile application

Static Control Resilient Flooring

There are situations in which floor coverings are required to control electronic discharge, such as sensitive manufacturing environments, computer and electronic rooms, and explosive environments. A static-control floor covering system is vital in resisting the electrostatic charges generated by people, furniture casters, and equipment movements. Static-control resilient flooring directs electrical charge to a reliable grounding source. These systems comprise floor-covering products with conductivity elements within the material body, requiring static-control adhesive and grounding strips.

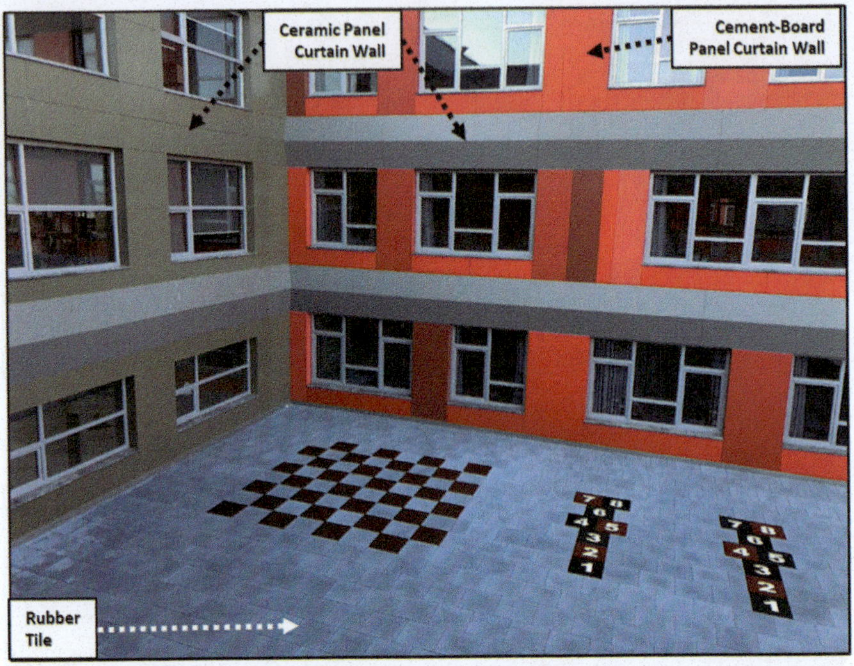

Fig. 89 Rubber tile, ceramic, and cement-board panel curtain wall application

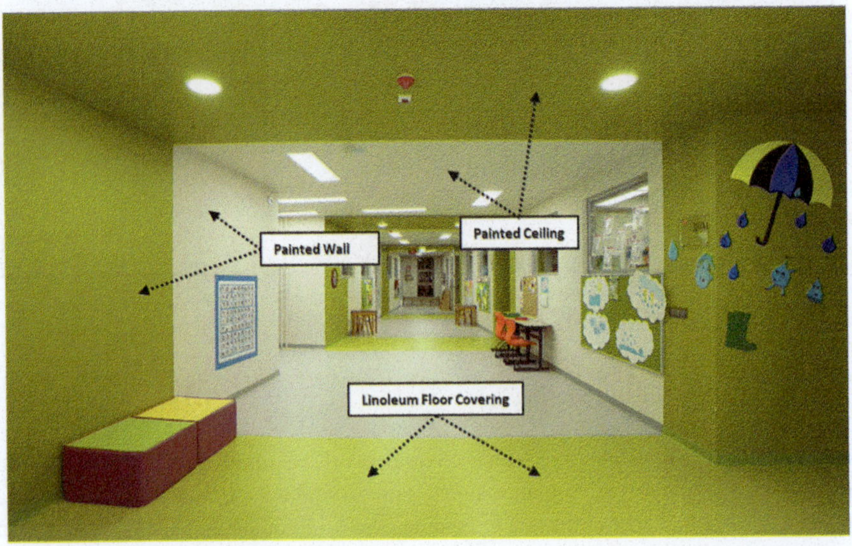

Fig. 90 Linoleum sheet flooring application

4.1.9 Other Floor Covering Materials

Brick and concrete are also widely used as floor covering. Full-size or split-face bricks can be used for interior and exterior flooring coverings as thin-set or thick-set mortared for interior floors with or without grouted joints. Concrete is also used as being dyed or stained. The surface may be ground, honed, and polished using abrasive media with finishes ranging from matt to reflective sheens.

4.1.10 Underlayment

An underlayment is a thin material that is adhered to or applied to the subfloor before the final floor covering is installed. It either provides protection or prepares the subfloor to receive the flooring. There are three types of underlayments: membrane, fill, and solid.

Membrane underlayment can perform three functions; some products may simultaneously perform two or all three.

- **Waterproofing**: These products prevent water absorption. Product types include liquid-applied components, plastic sheets (PVC chlorinated polyethylene, polyethylene), and self-adhering bituminous sheets.
- **Crack Isolation**: Because subfloor structures expand, contract, and deflect very slightly throughout the life of a building, it is sometimes necessary to provide crack-isolation membranes. They are designed to either eliminate the transfer of cracks or greatly reduce the likelihood of cracks. They are fungus and microorganism-resistant and have good shear strength, point-load, and crack resistance.
- **Sound Reduction**: These membranes mitigate or reduce the noise between floors, such as between two residential units.

There are two types of underlayment. These are:

Fill Underlayment: It can be troweled over irregular surfaces, providing a smooth surface. They are also used for leveling or resurfacing a subfloor. They are typically hydraulic cement-based and gypsum-based materials.

Solid Underlayment: It is used when the subfloor's elevation has to be raised slightly, or the subfloor is significantly damaged. Common materials include medium-density fiberboard, cement fiberboards, and cement backer units (Fig. 91).

4.1.11 Accessories as Wall Base and Moldings

Resilient accessories are typically manufactured of PVC. This material can be softened by heating, shaped to fit, and hardened. The wall base is a narrow strip of material installed at the base of the Wall adjoining the floor covering. Molding covers small gaps between floor coverings and transitions from one covering to another.

Fig. 91 Liquid waterproofing material is applied to a wet core

5 Ceilings

Selection criteria for ceiling finish materials are explained as follows:

- **Aesthetic Expectations**: Ceiling finish materials, heights, and profiles can create inviting environments and influence how light interacts with the building's interior. Ceilings introduce a sense of scale and proportion to an interior space.
- **Concealing the Building's Utilities Overhead** (mechanical and electrical equipment and components): The most common reason for a ceiling is to conceal the building's structure and overhead utilities.
- **Wind Loading**: Wind can impose an uplift load on the exterior soffit surface; therefore, wind-uplift resistance becomes an important consideration.
- **Volume of Occupied Space**: Ceiling height is related directly to the space required to be heated and cooled.
- **Humidity**: The ceiling adjacent to openings in exterior walls may be subjected to a higher humidity than the ceiling in the remainder of the building.
- **Flammability**: Building codes sometimes require ceilings to resist fire propagation and spread.

- **Seismic Activity**: In seismically active areas, building codes require products, materials, and equipment installed overhead to resist the movements caused by an earthquake. During seismic activity, support structures and ceiling finish materials must not fall, which would cause injuries to people trying to vacate the building.
- **Sound Absorption**: Ceiling finish materials and their application method can help absorb sound generated within a room or space.
- **Sustainability**: Many ceiling products contain recycled content and can make contributions to sustainability goals
- **Antimicrobial Resistance**: In addition to normal maintenance, some applications may require ceiling to be resistant to bacterial growth and mold and mildew development.
- **Light Reflectance**: A ceiling can reflect or diffuse light from other sources to distribute it uniformly throughout a space.
- **Maintenance**: Some occupancies, such as healthcare facilities, require ceiling finishes that can be regularly cleaned and scrubbed to remove possible contaminants.

The types of ceilings are (i) no ceiling finish, (ii) ceiling attached to the building structure, and (iii) suspended ceiling.

5.1 No Ceiling Finish

Building structure and utilities are exposed. Thus, carefully design, detailing, and coordinating all exposed elements are required to produce a satisfactory result.

5.2 Ceiling Attached to the Building Structure

In residential and light-commercial buildings, lightweight ceiling finish materials are attached directly to the building structure. This is the most economical approach in residential applications, especially when few mechanical, plumbing, or electrical systems are accommodated within the plenum. A ceiling attached directly to the building structure does not usually require another structure, as is needed for a suspended ceiling. The most commonly used material is gypsum board. It is either nailed to the underside of the wood framing or screwed to light-gauge steel framing (Figs. 92 and 93).

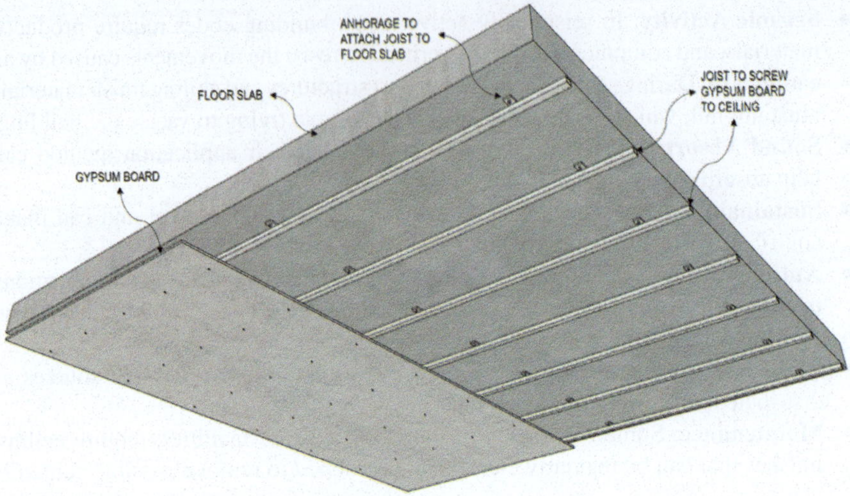

Fig. 92 Ceiling attached to the building structure

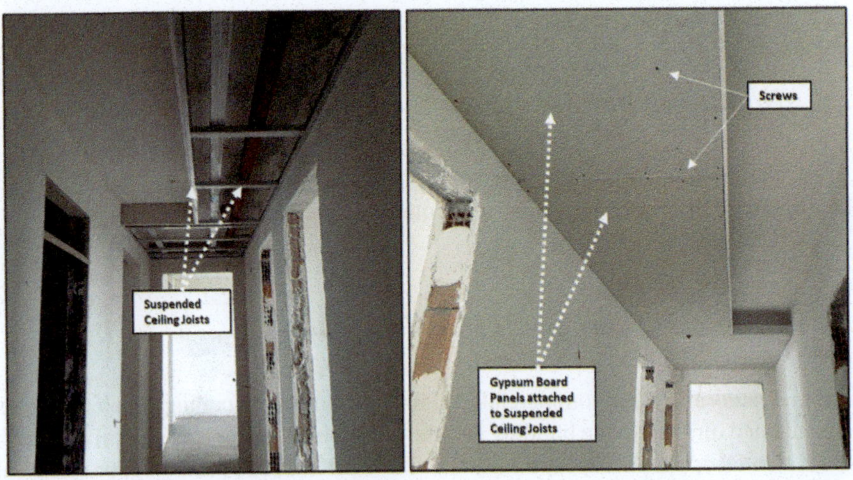

Fig. 93 Gypsum board panel application

5.3 *Suspended Ceiling*

A lightweight metal grid structure is suspended from the building's overhead structure to support the ceiling finish products and establish the ceiling height. This grid structure consists of main runners, suspended from above by steel wire and interconnecting cross runners spaced uniformly to support the ceiling finish products (Fig. 94).

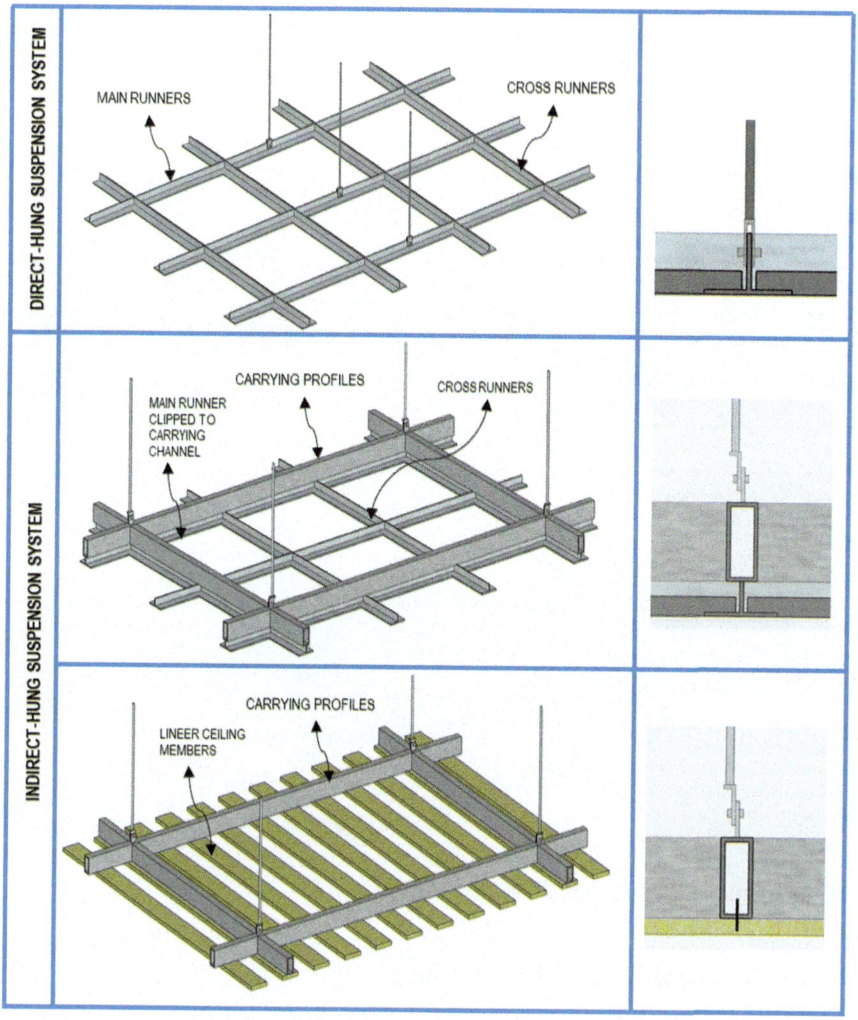

Fig. 94 Direct and indirect hung suspension systems are used for suspended ceiling construction

5.3.1 Suspended Acoustical Ceiling

The most common commercial ceiling consists of modular acoustical panels in a direct-hung suspension system called an inverted-tee grid. The acoustical panels are typically 60 × 60 cm or 60 × 120 cm and are laid on the horizontal legs of inverted tees. This grid structure is composed of corrosion-resistant steel or extruded aluminum inverted tees. Acoustical panels and tiles comprise various materials, including mineral wool, mineral fibers, fiberglass, and perlite (Figs. 95, 96, 97, 98, 99 and 100).

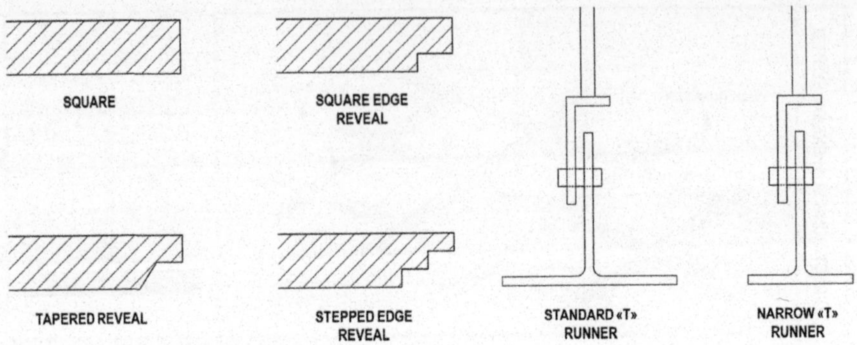

Fig. 95 Acoustical panel cut and runner types

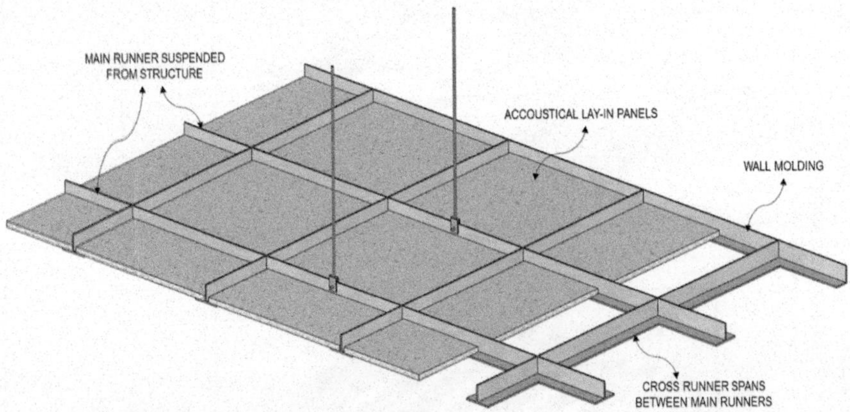

Fig. 96 Application of suspended acoustical ceiling

5.3.2 Suspended Gypsum Board Ceiling

The second most common commercial ceiling consists of gypsum board sheets that are screw-attached to an inverted-tee grid structure (Figs. 101, 102, 103 and 104).

5.4 Some Applications for Finishings

See Figs. 105, 106, 107, 108, 109, 110, 111, 112, 113, 114, 115, 116 and 117.

Fig. 97 A suspended ceiling application in a construction site

Fig. 98 A suspended ceiling application in a construction site illustrating wall moldings

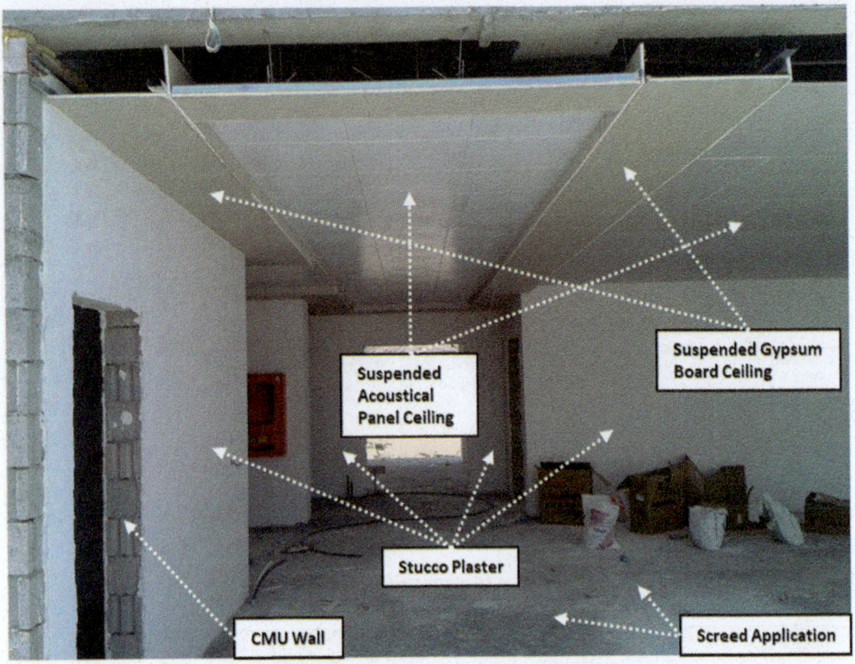

Fig. 99 Various finishing applications

Fig. 100 An acoustical and gypsum board suspended ceiling application

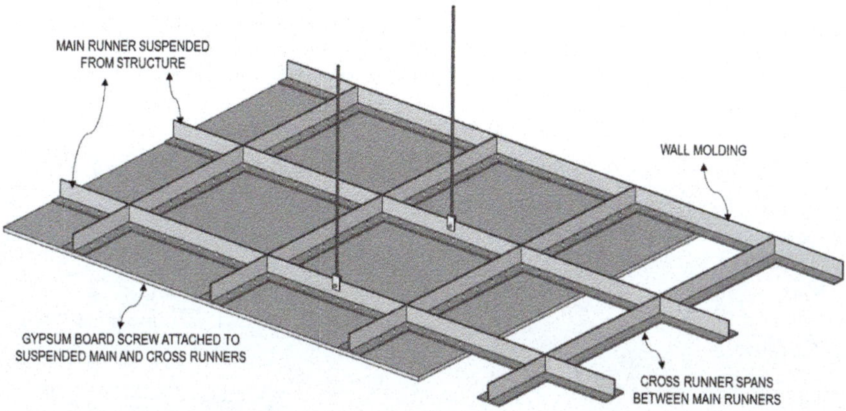

Fig. 101 Application of suspended gypsum board ceiling

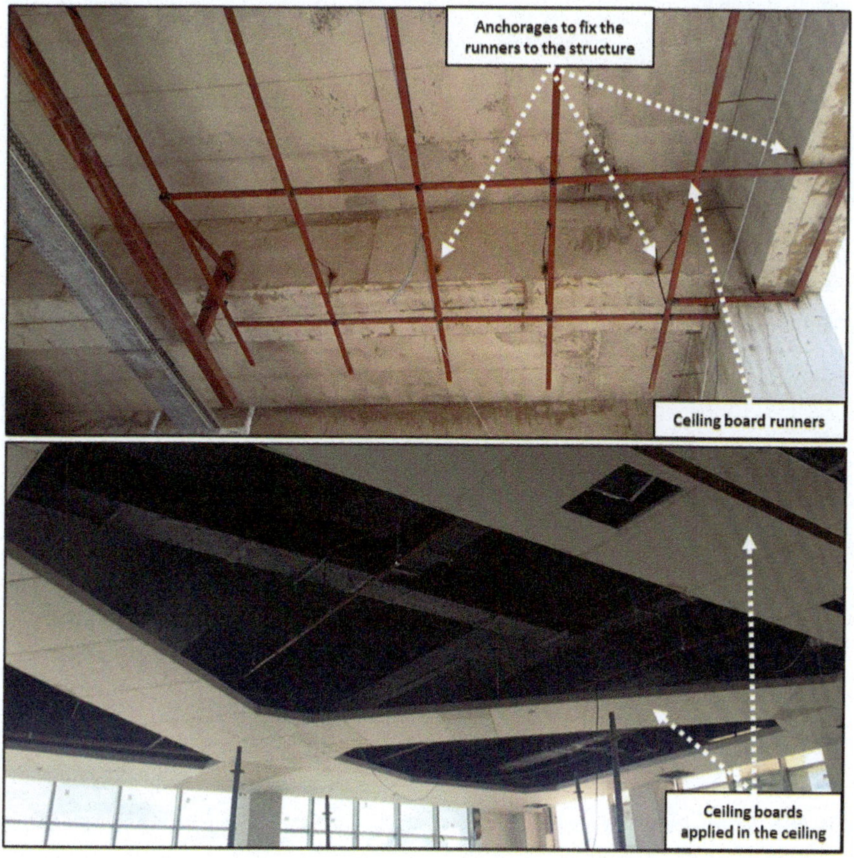

Fig. 102 Suspended gypsum board ceiling application in a construction site representing backup frames

Fig. 103 Suspended gypsum board and laminated wood panel applications

Fig. 104 Suspended ceiling application in a construction site representing the use of mobile shores for the installation

Fig. 105 Ceramic floor tile, gypsum board ceiling, and painted surface applications

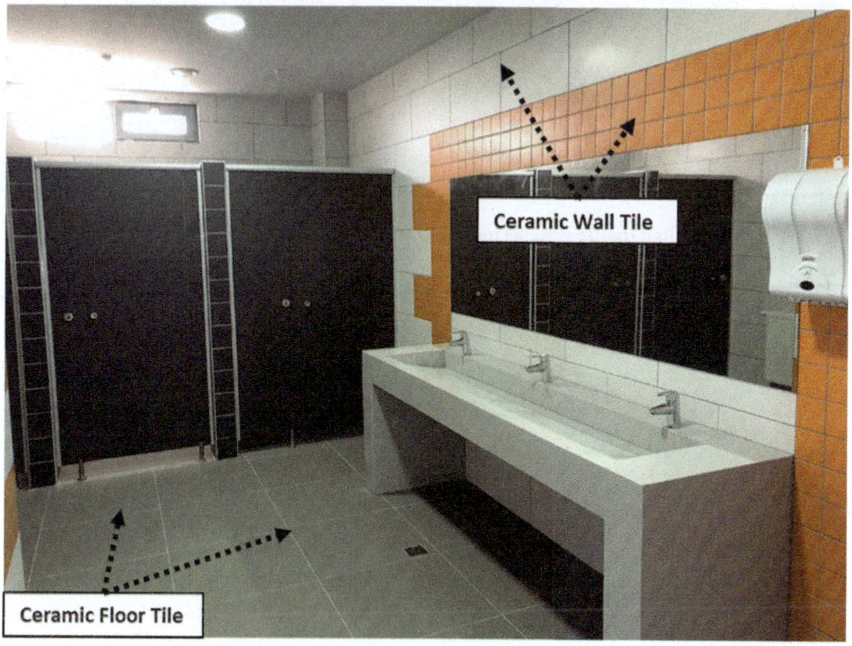

Fig. 106 Ceramic floor and wall tile application in a WC

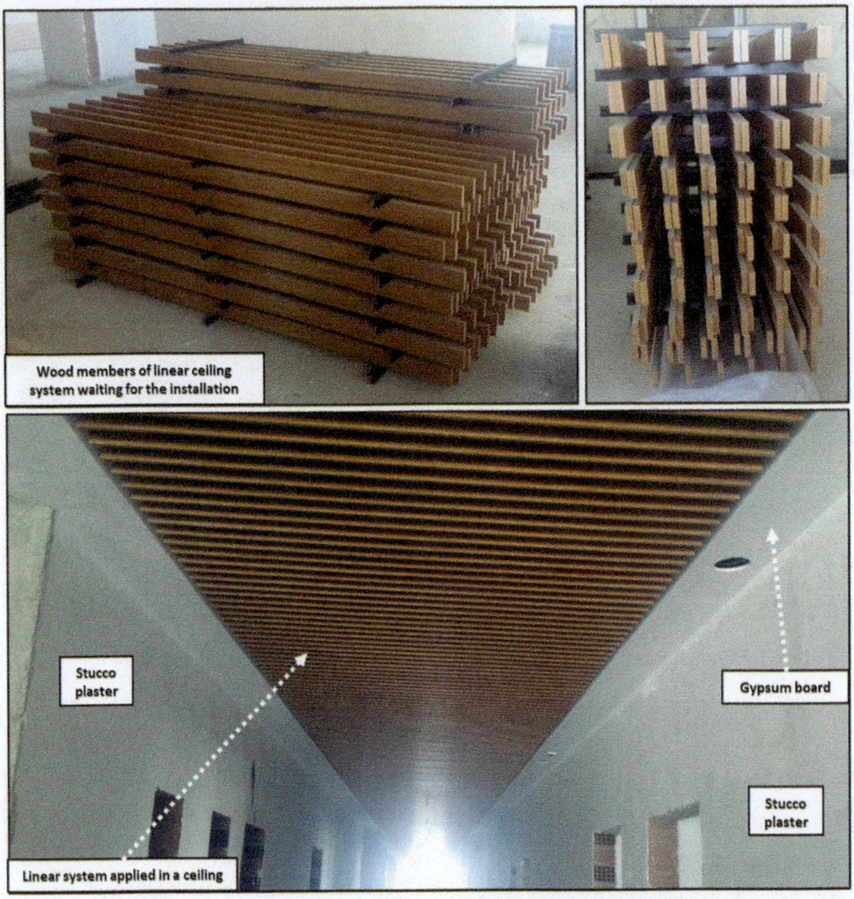

Fig. 107 Linear ceiling (baffle system) members and application in a construction site

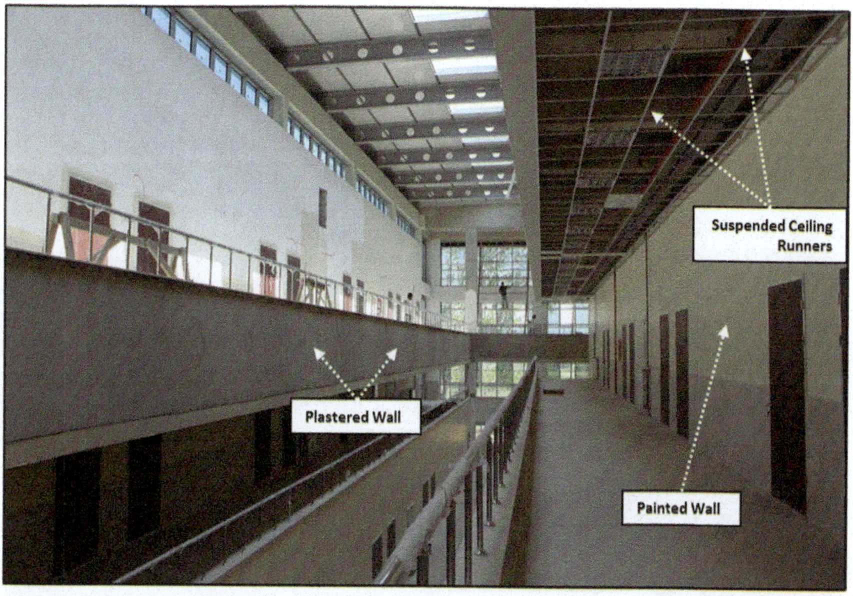

Fig. 108 A plastered wall to be ready for painting, painted wall, and suspended ceiling application

Fig. 109 Steel roof joists, joist girders, and painted wall application

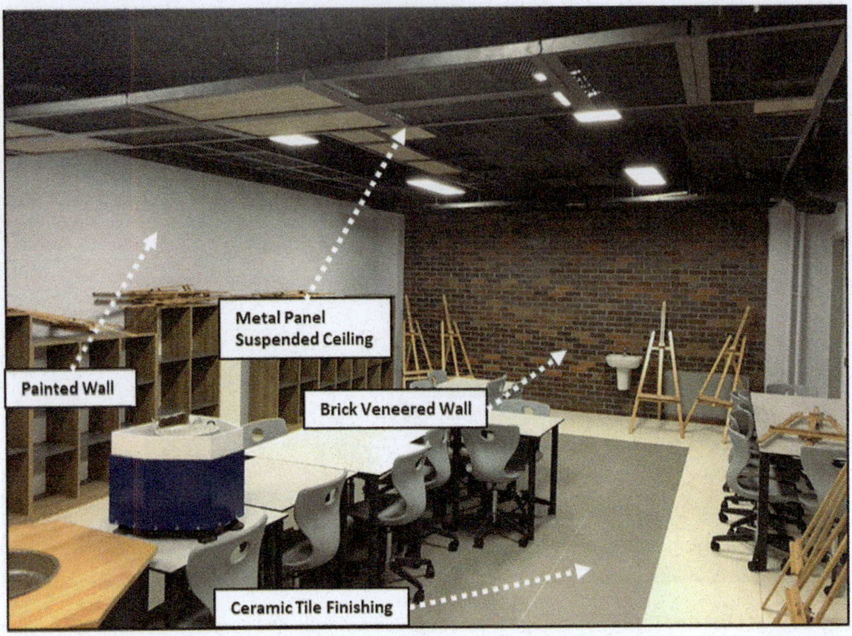

Fig. 110 Metal panel suspended ceiling, painted wall, brick veneered wall, and ceramic tile finishing application

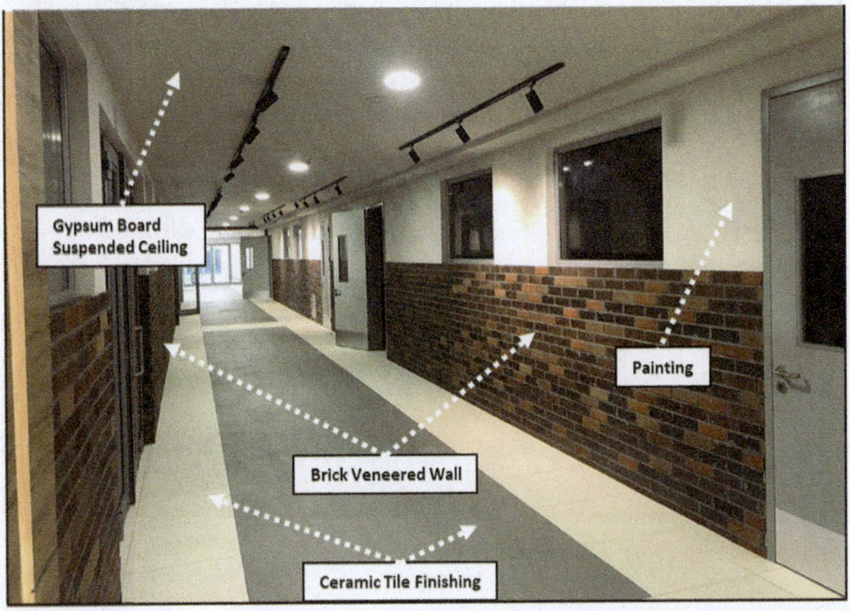

Fig. 111 Gypsum board ceiling, painted wall, brick veneered wall, and ceramic tile flooring application

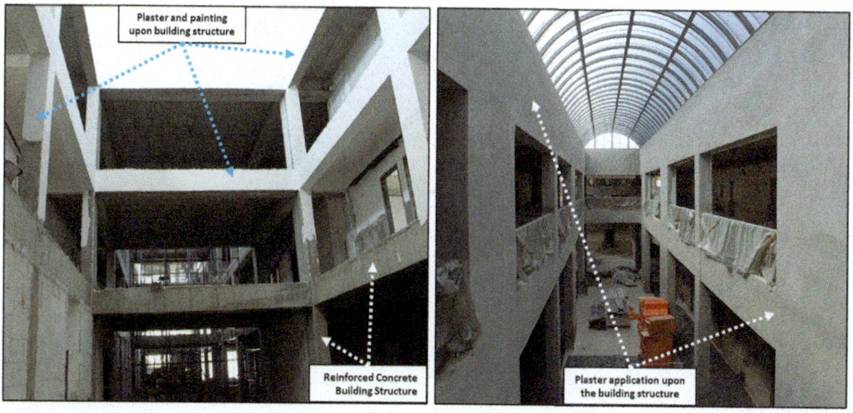

Fig. 112 Plaster and painting application upon the building structure

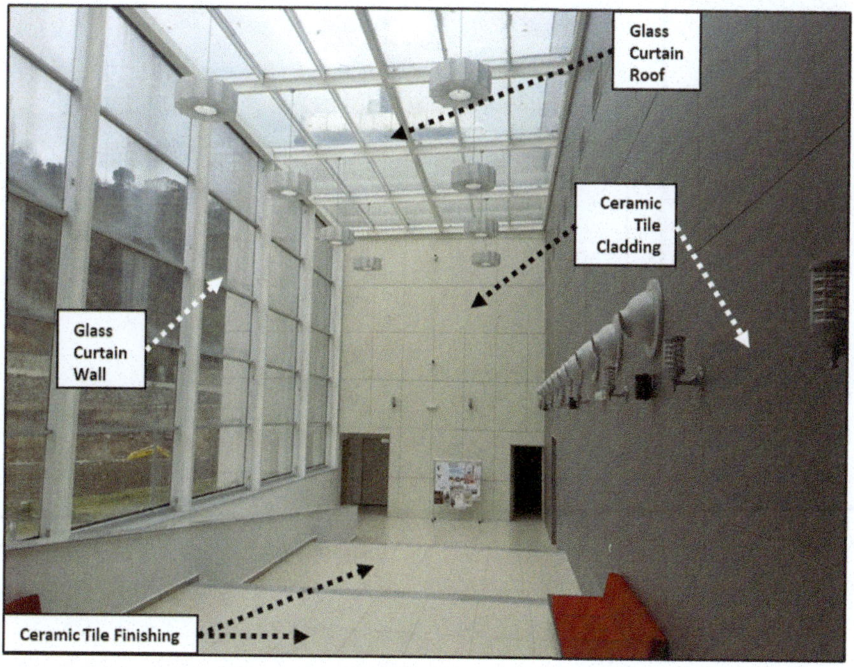

Fig. 113 Glass curtain wall and roof application

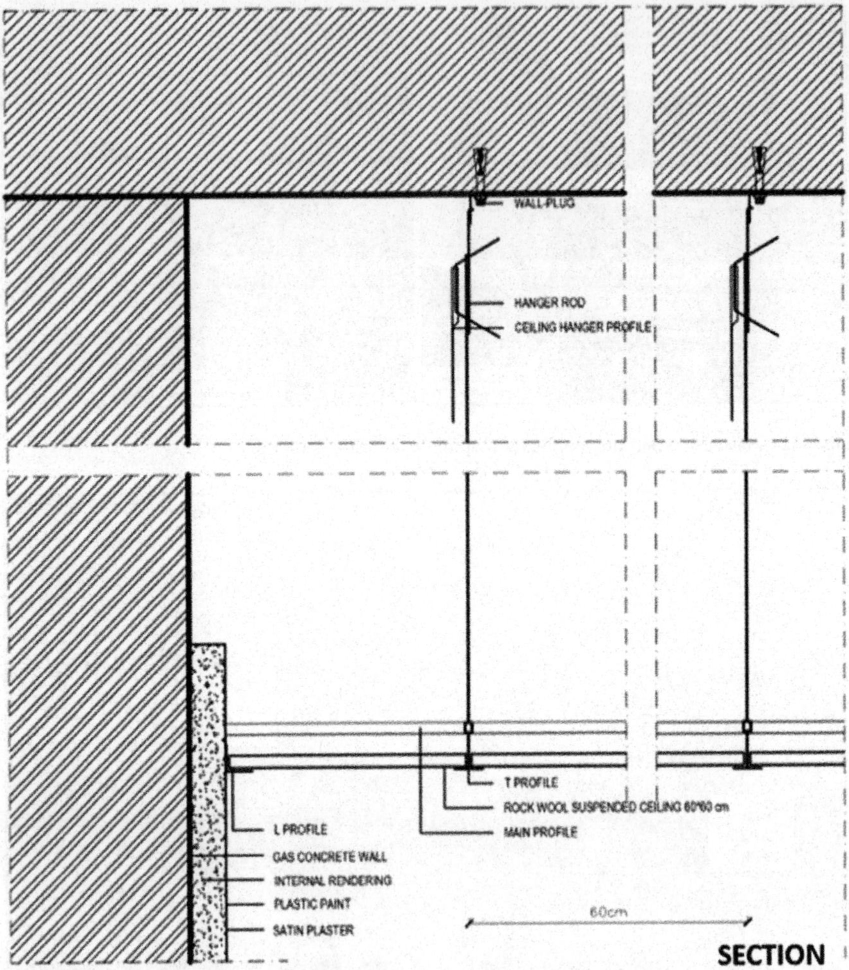

Fig. 114 Rock wool suspended ceiling and wall connection detail

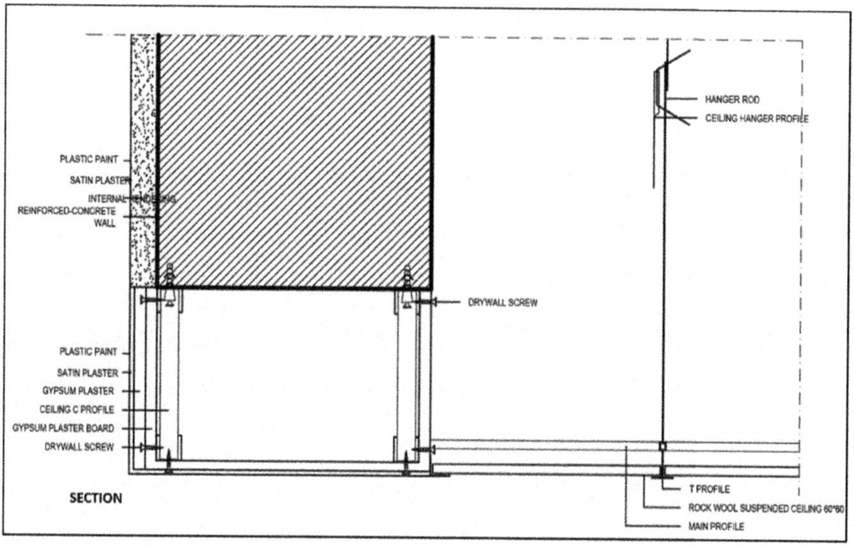

Fig. 115 Rock wool suspended ceiling and gypsum board connection detail

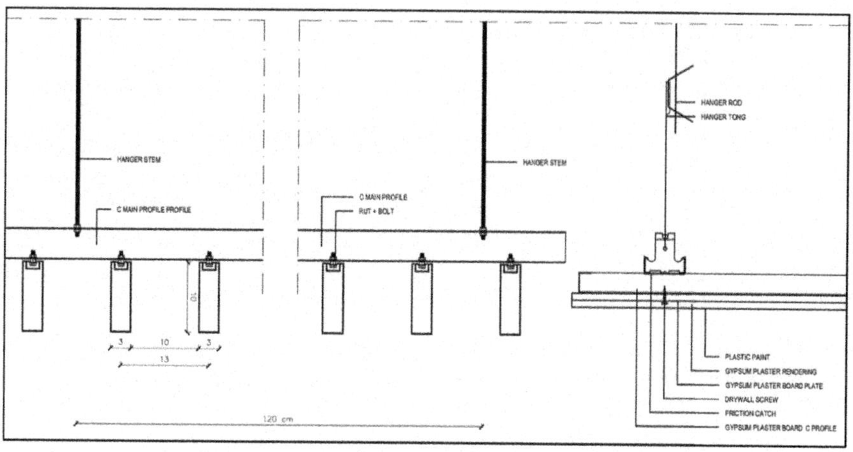

Fig. 116 Linear and gypsum board suspended ceiling

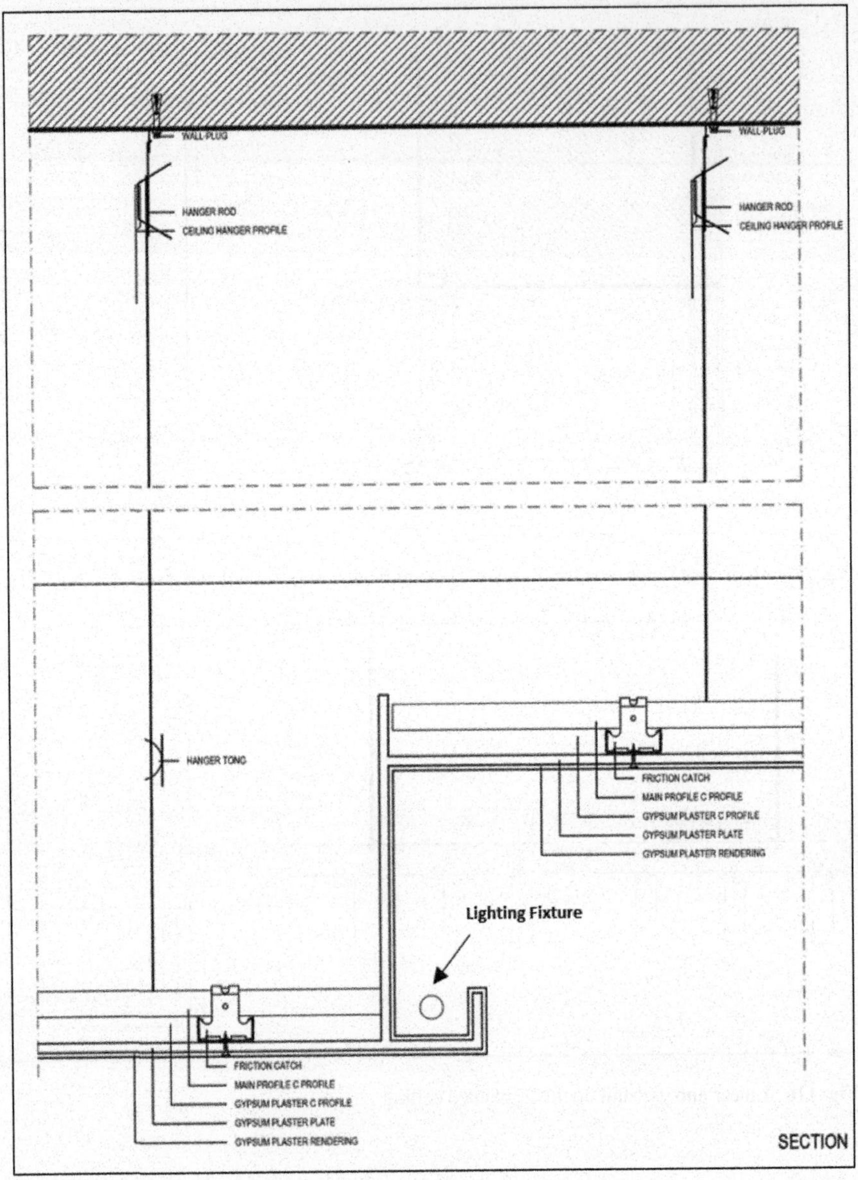

Fig. 117 Suspended gypsum board and lighting channel detail

Chapter 6
Emerging Technologies in Building Construction

Abstract The construction methods and materials are always in the change and improvement process regarding technological innovations and trends. This chapter introduces and underlines additive manufacturing and 3D printing in construction with governing principles and preliminary examples from the world.

1 Additive Manufacturing and Construction

1.1 Additive Manufacturing

Additive Manufacturing (AM) creates an object by building the form layer by layer at a time. AM has started to be conceptualized and initiated since the 1980s. The first concept was called rapid prototyping because it allows the creation of a scale model of the final object rapidly with low cost and without the typical setup process (Linke, 2017). A product design created in a Computer-Aided Design (CAD) environment is translated into a layer-by-layer framework so that AM machines can read and print the design using printable materials. Seven categories were described by the American Society for Testing and Materials (ASTM, 2012) regarding the AM. These are:

- **VAT Photopolymerization:**

VAT Photopolymerization is "an additive manufacturing process in which liquid photopolymer in a vat is selectively cured by light-activated polymerization" (ASTM, 2012). The method builds the object layer by layer within a liquid photopolymer resin (AMRG, 2023). An ultraviolet (UV) light is used to cure or harden the resin. Although the laser altitude remains constant, the object base is moved vertically, allowing the object to rise layer by layer, as depicted in Fig. 1.

- **Material Jetting:**

Material Jetting is "an additive manufacturing process in which droplets of build material are selectively deposited" (ASTM, 2012). The material jetting method is

R. Sarı and E. B. Çalışkan, *Building Construction Methods and Systems*,
https://doi.org/10.1007/978-3-031-50043-5_6

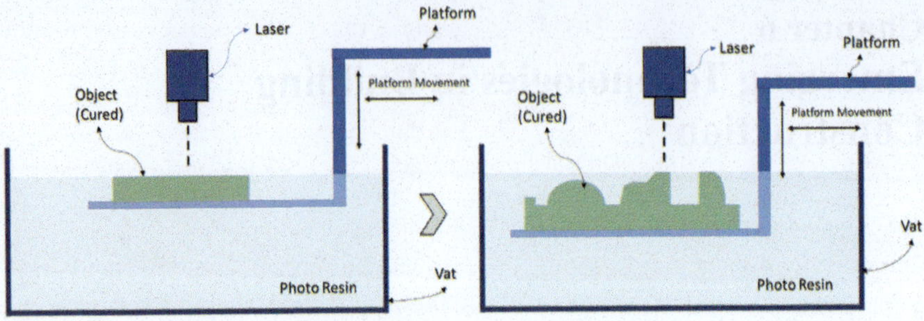

Fig. 1 Vat photopolymerization printing process

similar to a 2-dimensional inkjet printer. Material is jetted onto the build surface, where the nozzle follows the path of printing horizontally to raise the object layer by layer (AMRG, 2023). The material used for material jetting is limited due to the deposition and drop of material during the printing. Support material could be needed to hold the structure during printing, and support material could be removed after completion (Fig. 2).

- **Binder Jetting:**

Binder Jetting is "an additive manufacturing process in which a liquid bonding agent is selectively deposited to join powder materials" (ASTM, 2012). The binder jetting

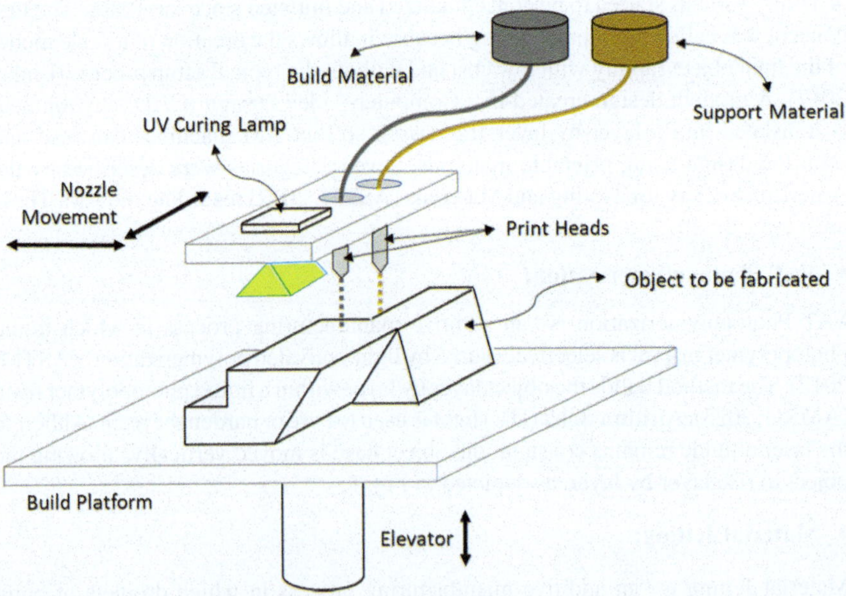

Fig. 2 Material jetting printing process

process relies on a powder-based material and a binder. The binder is usually liquid and acts as an adhesive between powder layers deposited through the nozzle (print head), building material in powder form (AMRG, 2023). The print head moves horizontally along the x and y-axis of the build surface, depositing both the build material and the binding material simultaneously. After the completion of each layer, the build platform is lowered by an elevator underneath the build platform. Since both build and binder material are deposited simultaneously during the printing process, unbound powder remains in position surrounding the object after printing. Thus, the remaining unbound powder surrounding the printed object is removed at the end of the printing process (Alexandrea, 2019). The spread of powder causes a dusty environment, requiring the use of the binder jetting method in an enclosed environment (Fig. 3).

- **Material Extrusion:**

Material Extrusion is "an additive manufacturing process in which material is selectively dispensed through a nozzle or orifice" (ASTM, 2012). Material is drawn by a nozzle where it is heated and then deposited layer by layer in a semi-liquid format so that after dispensing, laid material is solidified immediately to allow the build of the object. The nozzle can move horizontally on the x and y axis, while the build

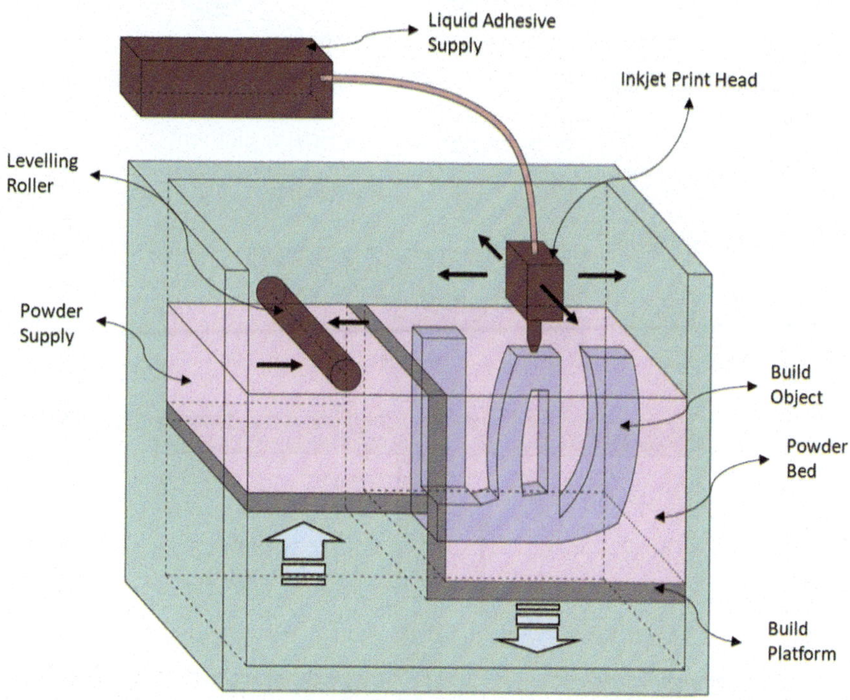

Fig. 3 Binder jetting printing process

platform can move vertically (AMRG, 2023). When a layer is deposited, the build platform moves down to allow the deposition of upper layers. The material extrusion method is the cheapest and, thus, commonly used technique for domestic and hobby 3-D printers (Fig. 4).

- **Powder Bed Fusion:**

Powder Bed Fusion (PDF) is "an additive manufacturing process in which thermal energy selectively fuses regions of a powder bed" (ASTM, 2012). PDF is a demonym to describe the use of any of the following techniques (AMRG, 2023): Direct metal laser sintering (DMLS), Electron beam melting (EBM), Selective heat sintering (SHS), Selective laser melting (SLM) and Selective laser sintering (SLS). A laser or electron beam is directed to melt and fuse material powder. Like the binder jetting technique, the PDF spreads the powder material over previous layers. On the other hand, different than binder jetting, PDF uses thermal energy to melt or fuse the material to form a shape (Fig. 5).

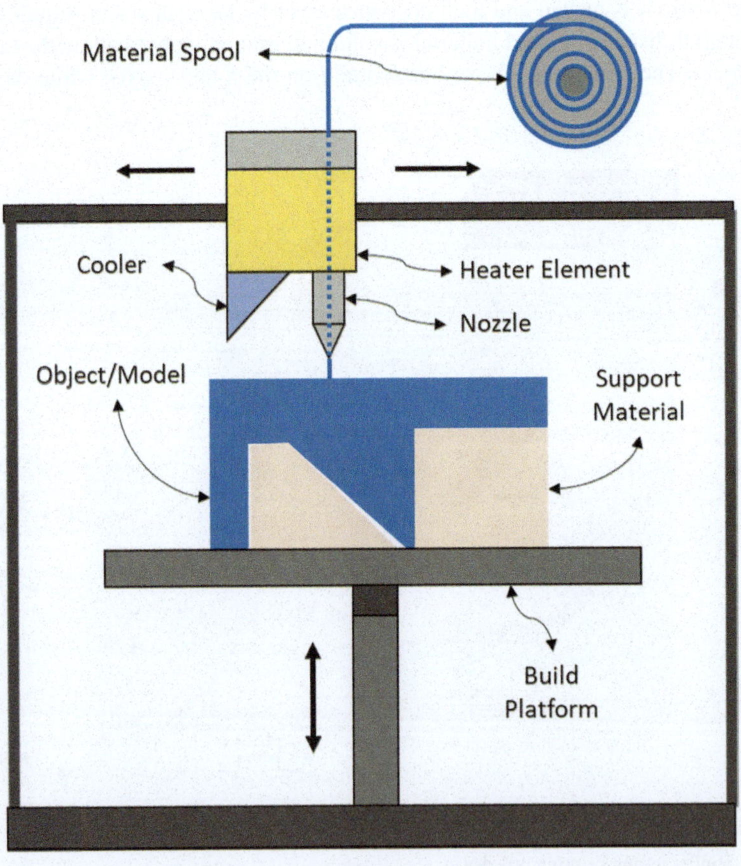

Fig. 4 Material extrusion technique printing process

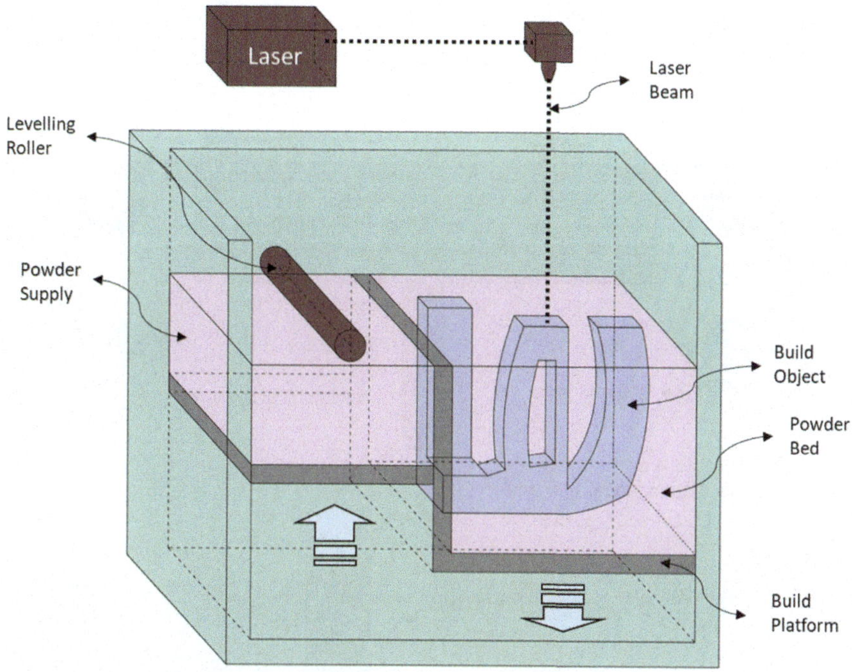

Fig. 5 Powder bed diffusion printing process

- **Sheet Lamination:**

Sheet Lamination is "an additive manufacturing process in which sheets of material are bonded to form an object." (ASTM, 2012). The sheet lamination method incorporates ultrasonic additive manufacturing (UAM) and laminated object manufacturing (LOM). The UAM utilizes ultrasonic welding to bind sheets or ribbons of metal (AMRG, 2023). The LOM, on the other hand, facilitates paper as sheet and adhesive as welding. The UAM objects could be used for structural purposes, while the LOM objects could be used for visual and aesthetic models (Fig. 6).

- **Directed Energy Deposition:**

Directed Energy Deposition (DED) is "an additive manufacturing process in which focused thermal energy is used to fuse materials by melting as they are being deposited" (ASTM, 2012). DEP includes a couple of terminologies: laser-engineered net shaping, directed light fabrication, direct metal deposition, and 3D laser cladding. A typical DED method utilizes a nozzle that can move in multiple directions and deposit melted material onto a specific surface (AMRG, 2023). The deposited material in powder or wire is melted by laser or electron beam and then solidified on the specified surface (Fig. 7).

The usable materials, advantages, and disadvantages of each AM method were represented and listed in Table 1. The AM methods presented in this section are

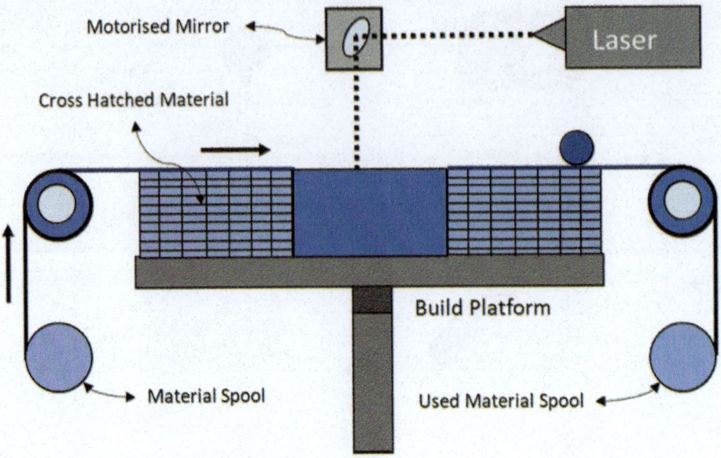

Fig. 6 The sheet lamination printing process

Fig. 7 Directed energy deposition printing process

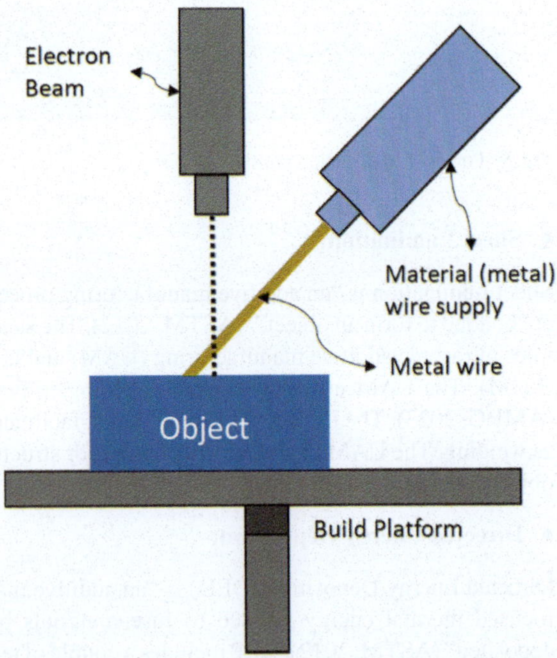

specific to the manufacturing industry since the first use of AM was rapid prototyping. Regarding AM's benefits and prevailing advantages, the construction industry looks for AM's adoption capabilities and opportunities for building construction.

Although the manufacturing industry supplies the construction industry with particular components and assemblies, till the adoption of AM, the manufacturing

Table 1 Comparison of additive manufacturing (AM) methods

AM methods	Materials	Advantages	Disadvantages
Vat photo-polymerization	Plastics and polymers: • Polymers: UV-curable Photopolymer resin • Resins: Visijet range (3D systems)	• High level of accuracy and good finish • Relatively quick process • Typically, large build areas: object 1000: 1000 × 800 × 500 and max model weight of 200 kg	• Relatively expensive • Lengthy post-processing time and removal from resin • Limited material use of photo-resins • Often requires support structures and post-curing for parts to be strong enough for structural use
Material jetting	Plastics and Polymers: • Polymers: Polypropylene, HDPE, PS, PMMA, PC, ABS, HIPS, EDP	• The process benefits from a high accuracy of deposition of droplets and, therefore, low waste • The process allows for multiple material parts and colors under one process	• Support material is often required • A high accuracy can be achieved, but materials are limited, and only polymers and waxes can be used
Binder jetting	• Metals: Stainless steel • Polymers: ABS, PA, PC • Ceramics: Glass	• Parts can be made with a range of different colors • Uses a range of materials: metal, polymers, and ceramics • The process is generally faster than others • The two-material method allows for many different binder-powder combinations and various mechanical properties	• Not always suitable for structural parts due to the use of binder material • Additional post-processing can add significant time to the overall process
Material extrusion	Plastics and Polymers: • Polymers: ABS, Nylon, PC, PC, AB	• Widespread and inexpensive process • ABS plastic can be used, which has good structural properties and is easily accessible	• The nozzle radius limits and reduces the final quality • Accuracy and speed are low when compared to other processes, and the accuracy of the final model is limited to material nozzle thickness • The constant pressure of material is required in order to increase the quality of finish

(continued)

Table 1 (continued)

AM methods	Materials	Advantages	Disadvantages
Powder bed fusion	Especially powder-based materials are used. Some metals and polymers are: • SHS: Nylon DMLS, SLS, SLM: Stainless Steel, Titanium, Aluminium, Cobalt Chrome, Steel • EBM: titanium, Cobalt Chrome, ss, al, and copper	• Relatively inexpensive • Suitable for visual models and prototypes • (SHS) Ability to integrate technology into small-scale, office-sized machine • Powder acts as an integrated support structure • Large range of material options	• Relatively slow speed (SHS) • Lack of structural properties in materials • Size limitations • High power usage • The finish is dependent on the powder grain size
Sheet lamination	Paper, plastic, and some sheet metals, such as A4 paper, can be rolled	• Benefits include speed, low cost, ease of material handling, but the strength and integrity of models is reliant on the adhesive used • Cutting can be fast because the cutting route is the shape outline, not the entire cross-sectional area	• Finishes can vary depending on paper or plastic material but may require post-processing to achieve the desired effect • Limited material use • Fusion processes require more research to advance the process into a more mainstream positioning
Directed energy deposition	Metals and Ceramics: • Cobalt Chrome, Titanium	• Advantages of DED • Ability to control the grain structure to a high degree, which lends the process to repair work of high quality, functional parts • A balance is needed between surface quality and speed, although with repair applications, speed can often be sacrificed for high accuracy and a pre-determined microstructure	• Finishes can vary depending on paper or plastic material but may require post-processing to achieve the desired effect • Limited material use • Fusion processes require more research to advance the process into a more mainstream positioning

and construction industries were managed by separate approaches. For example, the manufacturing process requires a controlled environment for maintaining the standard production of items. At the same time, the construction site consists of multiple participants requiring high-level management skills, making the construction process more chaotic and, thus, hard to create a controlled environment.

On the other hand, production excellence depends on standardized services and products. Therefore, there is an effort to change the construction site's uncontrollable and chaotic nature by adopting particular systems and methods. Prefabrication was an important effort to achieve such purposes where construction components were manufactured at the atelier and then assembled in the construction site to increase the control over the production while decreasing the workmanship in the construction site.

Rapid prototyping experiences in AM raise particular attention to the adoption of AM in the construction industry. In particular, due to the autonomous construction process, AM presents significant opportunities for colonizing foreign planets. Many research and development efforts were devoted to achieving this purpose.

1.2 From Construction to Printing of Building Projects

Scholars and practitioners have argued that printing building projects is like object printing. Since building projects were designed and created in the virtual environment, the model data could be transferred to a 3-D printing device for layer-by-layer construction of building projects. On the other hand, printing a building is more complex and challenging than printing an object regarding the following issues:

- The scale of the building projects is significantly higher than manufacturing projects. AM devices print the object in a controlled and enclosed space; thus, AM devices have space limitations to print an object. At this condition, large AM devices must be produced to enable the print of the building projects.
- Manufacturing projects could be printed using one or a couple of materials, while nature and state of the art of building projects include many building components generated by many construction stakeholders. A new approach is required for building design, eliminating the amount of building components to allow 3-D printing building projects.
- Construction projects include many disciplines, such as architectural, structural, mechanical, and electrical design, whereas manufacturing industries consist of much simpler disciplines.
- The construction site is located in a physical world impacted by environmental conditions such as sun, rain, wind, and dust, preventing the creation of a controlled environment required for manufacturing.
- The printing process requires deposition, melting, or fusion of printing material to form a design in the build platform. Special materials and methods were discovered and used to print an object. However, the construction industry has

no material opportunities for printing building projects. Particular investment is needed to be devoted to discovering new materials and methods for enabling the printing of building projects.

Although aforementioned challenges exist for the adoption of additive manufacturing in the construction industry, there are the following opportunities:

- The construction industry seeks to enable a controlled environment and standardized production and services to maintain the quality of the process. Although much effort was spent throughout the years, the existing construction process limits these efforts to reach more excellence. A paradigm shift is required. AM can cause a paradigm shift in the regular pattern of the construction process.
- The regular project delivery approach follows that an architectural project is generated, constructed by constructors, and occupied by the user. Time and cost-consuming activity in the available delivery approach is the construction process due to its many building components, assemblies, and materials. On the other hand, AM can eliminate the multi-component, assembly, and material nature of the construction process. A building project created in a virtual environment could be directly printed using AM devices, eliminating the involvement of building components, assemblies, and materials compared to traditional construction processes.
- Many construction materials have problems with lack of recyclability and being carbon footprint depository. Especially, widely used materials such as concrete still lack recyclability, causing serious environmental problems during pouring, curing, demolishing, and reusing. Methods and materials proposed for 3-D printing of building premises are more sustainable solutions than existing building construction approaches.
- Local materials could be used for printing buildings by discovering particular adhesives. This option allows the construction of regional architecture in the world and other planets.
- Eliminating human involvement during construction enables automatization, transforming the construction process into production and, thus, colonization of other planets.

The challenges and opportunities promised by AM motivated scholars and professionals to discover and test new building materials, 3-D printer configurations, application solutions, and unique structural system designs (Besklubova et al., 2021). The literature is divided into four categories of research and development studies in 3-D printing in construction. These are (i) research about extraterrestrial construction (Bulger & Skonieczny, 2016; Carrato, 2021; Covey & Metzger, 2018; Roman et al., 2016; Troemner et al., 2022), (ii) new method and technic proposal (Gosselin et al., 2016; Na et al., 2022; Prasittisopin et al., 2021; Wu et al., 2018), (iii) new material development for 3-D printing (Brooks & Zhou, 2021; Buchanan & Gardner, 2019; Ting et al., 2021; Zareiyan & Khoshnevis, 2017), (iv) BIM and 3-D printing integration (Borrmann et al., 2018; Ding et al., 2014; Pessoa et al., 2021; Sakin & Kiroglu,

2017) and, (v) critical review and evaluation of existing studies (Bedarf et al., 2021; Paul et al., 2018; Tay et al., 2017; Wu et al., 2016).

1.3 3-D Printing in Construction

Scholars and professionals have presented solid proposals in the last decades to overcome the deficits and handicaps of 3-D printing adoption in the construction industry. A typical 3-D printing device in construction has three common characteristics, as stated in Fig. 8. These are (i) a printing material depository, a computer to create the virtual building model and slice the project form into 3-D printable layers, and (iii) a 3-D printing device. The virtual building project model is created in the computer and transferred to an intermediary software, collaborating the virtual model with a 3-D printing device. The intermediary software evaluates the building project form and then slices it into layers. The thicknesses of the layers are between 2 and 5 cm, according to the available printing technologies. Later, the software creates a printing path for the device to follow the path during the printing. The software simulates the designated printing layers and paths at the end of the progress. Once the 3-D printing operator confirms the simulation, the 3-D printing device starts printing the layers. The material is deposited from the material depository in the Material Deposition Method (MDM) to the device's nozzle, whereas the 3-D printing process of binder jetting differs from MDM. An adhesive material is spread upon the powder in binder jetting in each printing layer.

Although AM has seven methods, only two were adopted by the construction industry, as illustrated in Fig. 9. These are the material deposition method (MDM) and binder jetting. The binder jetting method is patented by an entity known as the D-shape method. MDM has further been developed, and various techniques were discovered. Contour crafting, stick dispensers, digital construction platforms, concrete printing, flow-based fabrication, mini-builders, and mesh mold are the current techniques actively tested and used in the 3-D printing of building projects.

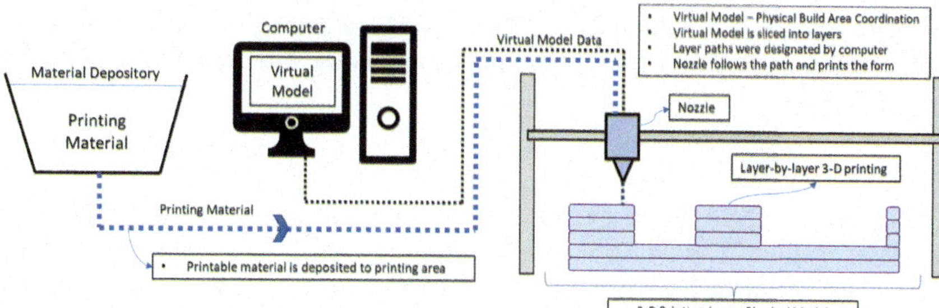

Fig. 8 Basic components of a typical 3-D printing device in construction. Adopted from: (Bos et al., 2016)

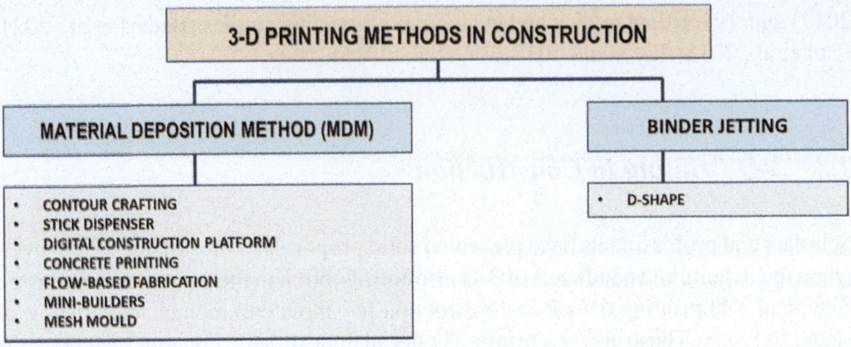

Fig. 9 3-D printing methods used in construction. Adopted from: (Tay et al., 2017)

The specific attributes, advantages, and handicaps of each technique were explained in the following sections (Table 2 and Fig. 10).

1.3.1 Material Deposition Method (MDM) in 3-D Printing of Building Projects

Seven techniques are available in current literature and practice as a sub-category of MDM. Each technique is explained in the following sub-sections.

- ***Contour Crafting:***

The contour crafting technique extrudes materials from the material depository layer by layer. The key feature of the technique is the use of trowels attached to the nozzle, as illustrated in Fig. 11. The trowel, which can be deflected at different angles, guides the printed material to create particularly smooth and accurate surfaces. The CC allows the creation of double-skin walls where utility conduits could be replaced within the void between the double-skin walls. This makes the automated construction of plumbing, electrical, and structural steel networks within the structure as much as possible. When proper adhesive is found, in-situ materials could be used. Thirteen mm slice thickness is used in current technologies, allowing smooth wall surfaces. Furthermore, a two hundred m2 area, single-story house project could be printed in a day (Fig. 12).

- ***Stick Dispenser:***

Yoshida et al. (2015) used chopstick material composites to develop the stick dispenser method. The tool is a specially designed hand-held printing device that allows consistent feed of chopsticks. The chopsticks coated with wood glue are dropped randomly, forming an aggregated porous structure (Yoshida et al., 2015). A real-time depth camera and a projector guide the stick dispenser. A sample structure printed using the stick dispenser method is presented in Fig. 13.

Table 2 Advantages, handicaps and future potentials of 3-D printing in construction

Advantages
• Less work, workforce, vehicle, device, technique, material, and energy are required • Almost no waste of material • Better integration of mechanical and electrical systems into the building
Current handicaps
• Unsuitability of the available automated fabrication technologies for large-scale products • Conventional design approaches are not suitable for automation • Economic unattractiveness of expensive automated equipment • Limitations in the materials that an automated system could employ • A significantly smaller ratio of the quantity of final products as compared with other industries • Managerial issues • The nozzle idle time cannot be long; otherwise, concrete may solidify and block the machine • The lower layer must be able to support the upper layer; therefore, the time interval between depositing subsequent layers cannot be shorter than the minimum curing time • Subsequent layers must be able to adhere; therefore, the interval between depositing subsequent layers should not exceed a critical limit • The printing nozzles cannot be allowed to collide with the previously deposited layer or other nozzles when traveling because if this happens, then when moving between the endpoints of wall segments, the nozzle may not be able to travel in a straight line in order to avoid obstacles • Initial equipment cost is high
Future potentials
Integration with BIM • Less waste • Direct construction of virtual model • Decreased fragmentation of design discipline **Labor** • Less labor skill is required • There is a growing skill shortage all over the World **Speed** • A 200 m2 house takes two days to finish

Source (Gomaa et al., 2021; Tay et al., 2017)

- ***Digital Construction Platform (DCP):***

The digital construction platform (DCP) is a research project developed by a research group at the Massachusetts Institute of Technology (MIT) for on-site sensing, analysis, and fabrication (Keating et al., 2014). The DCP consists of a compound robotic arm system comprised of a 5-axis hydraulic mobile boom arm for gross positioning attached to a 6-axis robotic arm for fine positioning and oscillation correction, respectively. The research study concluded that the DCP presents successful outcomes for the fabrication of nonstandard architectural forms, integration of real-time on-site sensing data, improvements in construction efficiency, enhanced resolution, lower error rates, and increased safety (Fig. 14).

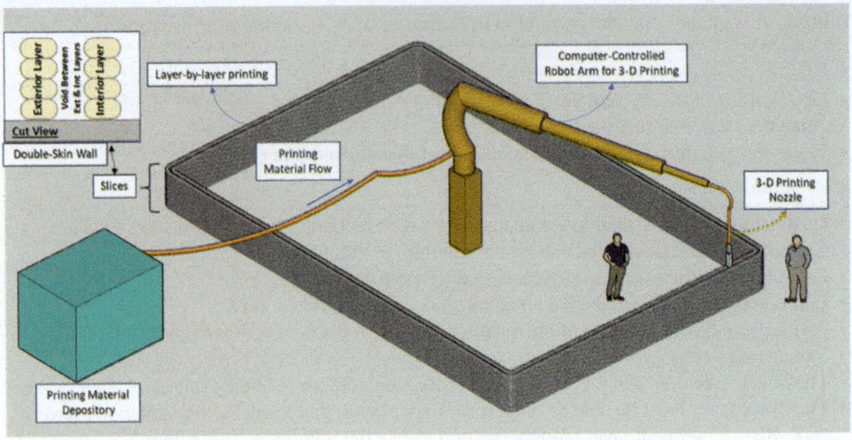

Fig. 10 3-Dimensional view of a typical 3-D printing method in construction

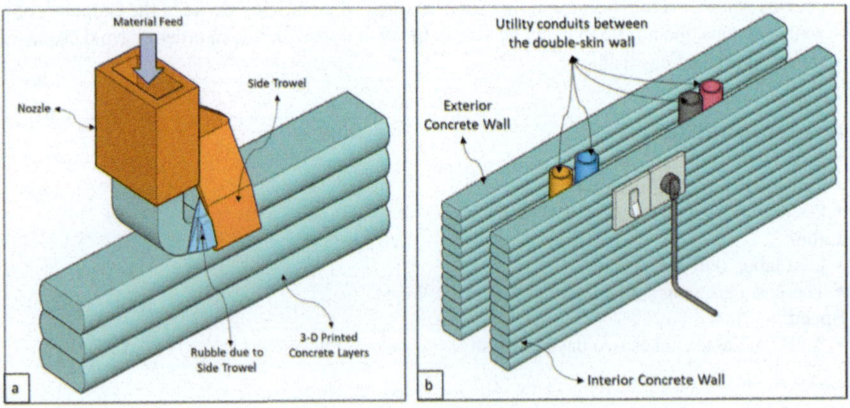

Fig. 11 **a** Side trowel mechanism and **b** utility conduits in the contour crafting

- ***Concrete Printing:***

Concrete printing shares similar printing characteristics with contour crafting. However, concrete printing does not utilize a trowel attached to the nozzle. Cement mortar is extruded from a nozzle in a layer-by-layer process. The printing process could be completed without using labor-intensive formwork and can incorporate functional voids into the structure where utility conduits could be placed. Since the method does not use a trowel, a smaller resolution of material deposition is required to achieve greater 3D freedom and smoothness in the wall surface. Nevertheless, a

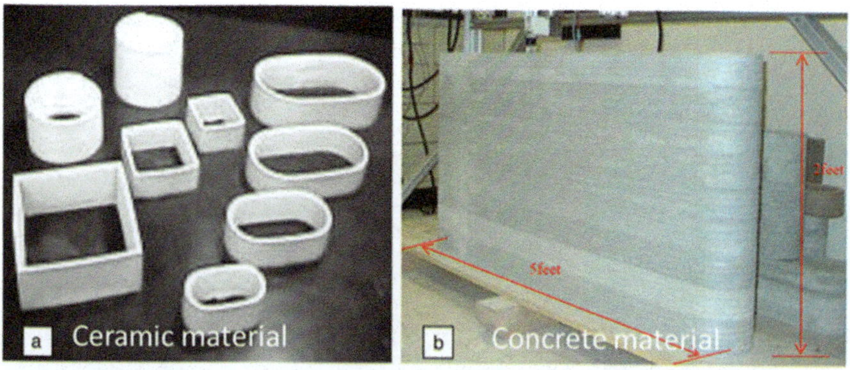

Fig. 12 Sample contour crafting studies created in lab environments. Retrieved from: (Tay et al., 2017)

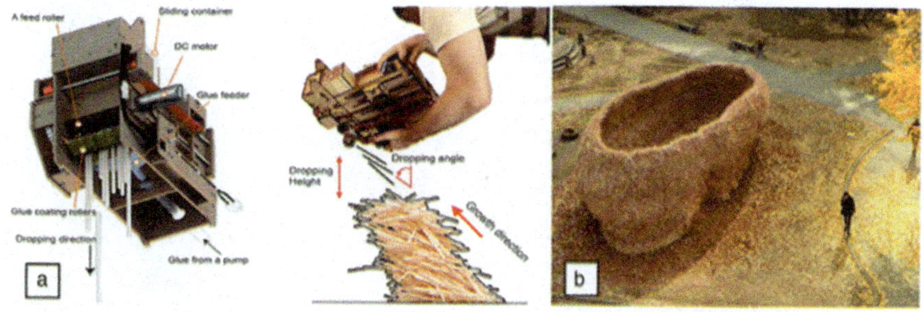

Fig. 13 The stick dispenser tool (**a**) and a sample structure (**b**). Retrieved from: (Yoshida et al., 2015)

ribbed finish is a characteristic of the wall surfaces due to the involvement of the trowel during the printing (Fig. 15).

- ***Flow-Based Fabrication:***

Water-based polysaccharide gels and natural composites with a single pneumatic extrusion system are attached to the end effector of a 6-axis robotic arm. The design and advanced manufacturing of heterogeneous materials and anisotropic structures result in high stiffness, lower weight, high wear, and resistance (Fig. 16).

- ***Mini-Builders:***

Mini-builders can be autonomous during the deposition, curing and setting the nozzle speed. A coordinated system of three individual robots, which are compact, lightweight, and have autonomous mobility, have different functions during the printing process. A two-component resin material was developed for the robots as the movement and speed of the robot determine the extrusion rate and movement speed.

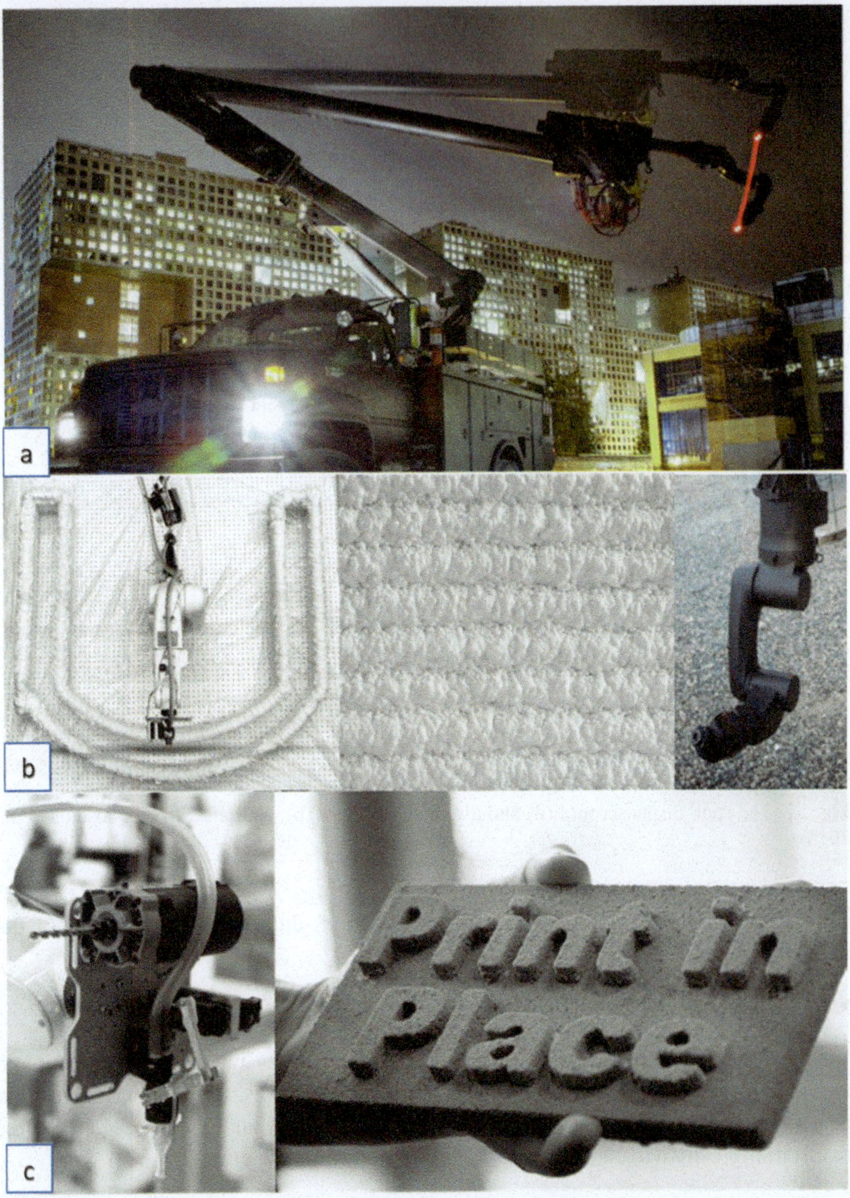

Fig. 14 The digital construction platform is a (**a**) mobile system having the capability to spray (**b**) and carve (**c**) to form a shape. Adopted from: (Keating et al., 2014)

Fig. 15 A sample house of concrete printing (**a**) and ribbed finish at wall surface (**b**). *Sources* a: (1), b: (2)

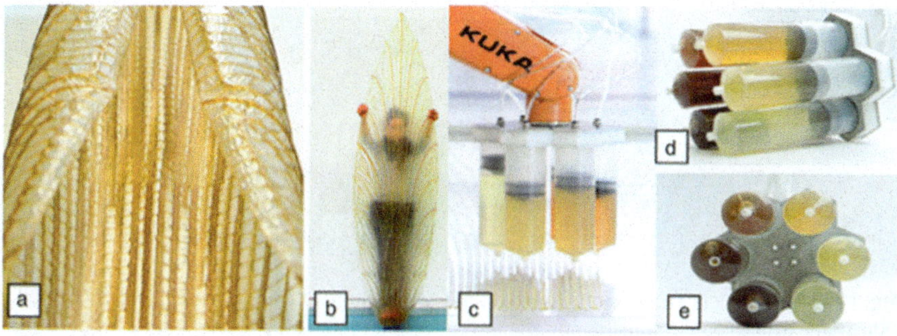

Fig. 16 Sample structures (**a** and **b**) printed by use of flow-based fabrication (**c**), using a special type of resin (**d** and **e**). Retrieved from: (Tay et al., 2017)

There is an option to add a heat source depending on weather conditions to expedite the chemical reaction and reduce the printed material's curing time (Fig. 17).

- ***Mesh Mould:***

A large 6-axis robot is used to extrude thermoplastics polymer to print in situ structures freely in 3D space. Pressured air from pinpoint cooling attached to the nozzle is directed to the printing surface, allowing for a high level of control, thus facilitating the weaving of wireframe structures freely in space shown in Fig. 18a. The structure is then used as reinforcement inside the concrete, pouring concrete over this formwork and later trowelling manually to smooth the surface (Fig. 18b). The proposed methodology reduces the time required to fabricate complex structures, making large-scale applications feasible. Mesh form could be printed in diverse densities required by the acting forces on the structure (Fig. 18c). Therefore, the presence of the mesh increases the tensile force of concrete, ultimately becoming a possible replacement for conventional steel reinforcement.

Fig. 17 Mini-builders (**a**) as grip, foundation, and vacuum robots. Mini-builders connected to industrial extruder on the construction site as foundation robot (**b**), close-up of grip robot (**c**), grip robot finishing structure (**d**). Retrieved from: (Nan, 2015)

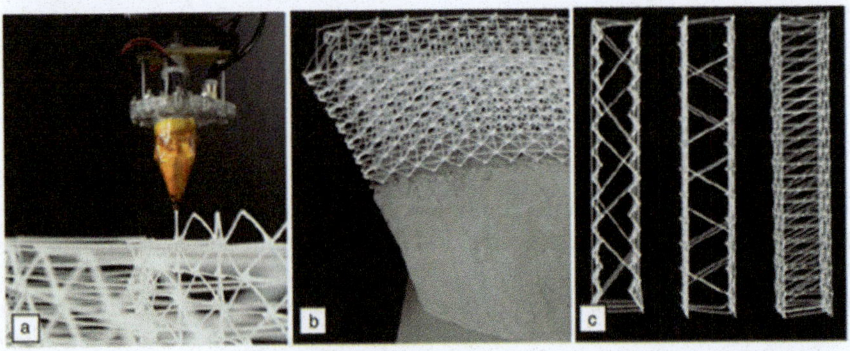

Fig. 18 Mesh-mould printing (**a**) and sample structures (**b** and **c**). Retrieved from: (Hack & Lauer, 2014)

1.3.2 Binder Jetting Method in 3-D Printing of Building Projects

- D-SHAPE

D-SHAPE is the patented trademark of the binder jetting method, where the object is created by depositing the binder layer by layer over a powder bed. The binder is ejected in droplet form onto a thin powder material spread on the build tray. This method incrementally glues 2D cross-sections of the intended component to each material powder layer. The unbound material can be removed from the print bed using a vacuum cleaner, recycled, and deployed for another printing task. The method promotes designs with voids and overhanging features in complex geometrical forms. The high resolution of the printing method results in a good surface finish and a more natural look for the viewer (Fig. 19).

Fig. 19 A binder jetting device (**a**), a printed object with unbounded materials (**b**), and unbounded materials are removed (**c**). Retrieved from: (Dini, 2009)

1.4 Sample 3-D Printed Building Structures

The 3-D printing constructs the building layer-by-layer, which is out of traditional component-by-component construction. In the component-by-component method, initially, the structure of the building is constructed. Following the construction of secondary building components, finishings of the construction are applied. On the other hand, different from the component-by-component method, the form of the building design is sliced into layers so that a 3-D printing device can manufacture the form layer-by-layer.

During the printing process, each layer is overlapped and bonded. Although the described method is unique and specialized to 3-D printing, the method results with particular limitations when the entire form or shell structure wants to be 3-D printed.

Two methods have emerged in practice to overcome the limitation, as illustrated in Fig. 20. The first method is printing the entire structure with 3-D printing methods by designing the horizontal parts of the structure as a vault or dome-shaped, where layer-by-layer construction would be enabled. The second method utilizes prefabrication of horizontal parts of the structure in the atelier and then montage in the site upon the 3-D printed shell structure. The first method presents undesired ceiling voids due to the vault and dome-shaped form; thus, it is not as common as with the hybrid 3-D printing method.

The existing samples in the current market of 3-D printing in construction utilize either the first or second option. Since the technology is at an ad-hoc level, it could be anticipated that scholars and professionals could discover new methods and techniques to fully overcome constructability limitations to use 3-D printing in construction.

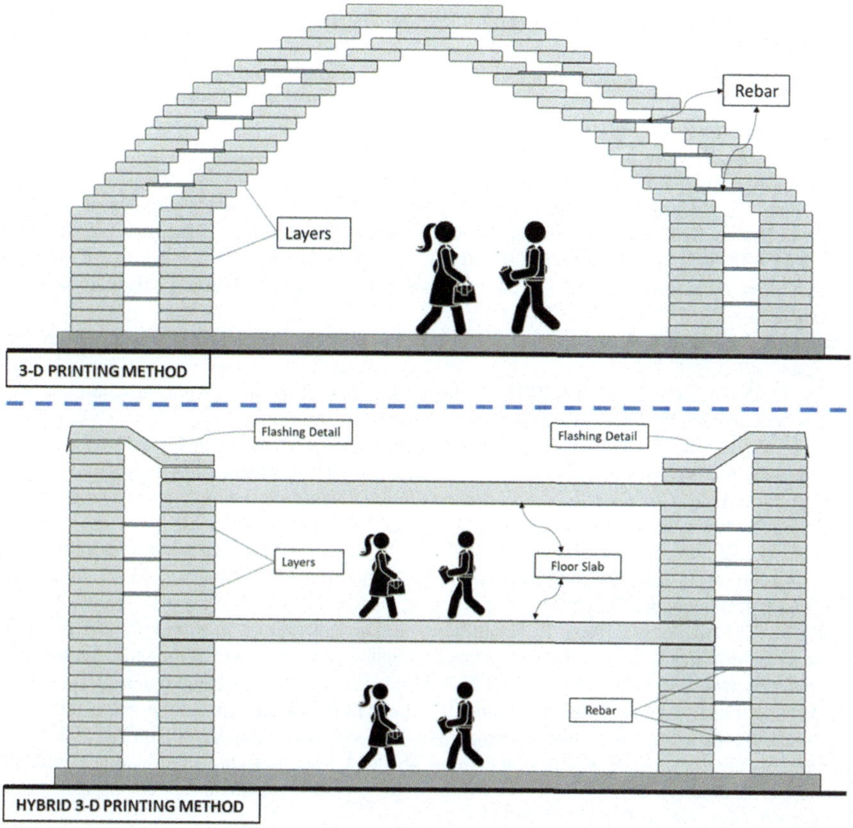

Fig. 20 3-D printing methods in building construction. The slices (layers) are exaggerated to increase the methods' understanding and perception

References

Alexandrea, P. (2019). *The complete guide to binder jetting in 3D printing*. 3D Natives. https://www.3dnatives.com/en/powder-binding100420174/#!

AMRG. (2023). *The 7 categories of additive manufacturing*. Additive Manufacturing Research Group. https://www.lboro.ac.uk/research/amrg/about/the7categoriesofadditivemanufacturing/

ASTM. (2012). *Standard terminology for additive manufacturing technologies*. https://www.astm.org/f2792-12.html

Bedarf, P., Dutto, A., Zanini, M., & Dillenburger, B. (2021). Foam 3D printing for construction: A review of applications, materials, and processes. *Automation in Construction, 130*(July), 103861. https://doi.org/10.1016/j.autcon.2021.103861

Besklubova, S., Skibniewski, M. J., & Zhang, X. (2021). Factors affecting 3D printing technology adaptation in construction. *Journal of Construction Engineering and Management, 147*(5), 04021026. https://doi.org/10.1061/(asce)co.1943-7862.0002034

Borrmann, A., König, M., Koch, C., & Beetz, J. (2018). BIM for 3D printing in construction. In A. Borrmann, M. König, C. Koch, & J. Beetz (Eds.), *Building information modeling: Technology foundations and industry practice* (pp. 1–25). Springer. https://doi.org/10.1007/978-3-319-92862-3

Bos, F., Wolfs, R., Ahmed, Z., & Salet, T. (2016). Additive manufacturing of concrete in construction: Potentials and challenges of 3D concrete printing. *Virtual and Physical Prototyping, 11*(3), 209–225. https://doi.org/10.1080/17452759.2016.1209867

Brooks, A., & Zhou, H. Y. (2021). Developing 3D printable lightweight functional cementitious composite. *Earth and Space, 2021*, 52–58.

Buchanan, C., & Gardner, L. (2019). Metal 3D printing in construction: A review of methods, research, applications, opportunities and challenges. *Engineering Structures, 180*(October 2018), 332–348. https://doi.org/10.1016/j.engstruct.2018.11.045

Bulger, S., & Skonieczny, K. (2016). Towards mobile 3D printing for planetary construction. *Earth and Space, 2016*, 324–332.

Carrato, P. J. (2021). Use of BIM and 3D printing in mars habitat design challenge. *Earth and Space, 2021*, 780–790.

Covey, S. D., & Metzger, P. T. (2018). Extruded clay-based regoliths for construction on mars, Phobos, and NEAs. *Earth and Space, 2018*, 570–579.

Ding, L., Wei, R., & Che, H. (2014). Development of a BIM-based automated construction system. *Procedia Engineering, 85*, 123–131. https://doi.org/10.1016/j.proeng.2014.10.536

Dini, E. (2009). *D-SHAPE - The 21st century revolution in building technology has a name* (pp. 1–16). http://www.cadblog.pl/podcasty/luty_2012/d_shape_presentation.pdf

Gomaa, M., Jabi, W., Veliz Reyes, A., & Soebarto, V. (2021). 3D printing system for earth-based construction: Case study of cob. *Automation in Construction, 124*(January), 103577. https://doi.org/10.1016/j.autcon.2021.103577

Gosselin, C., Duballet, R., Roux, P., Gaudillière, N., Dirrenberger, J., & Morel, P. (2016). Large-scale 3D printing of ultra-high performance concrete—a new processing route for architects and builders. *Materials and Design, 100*, 102–109. https://doi.org/10.1016/j.matdes.2016.03.097

Hack, N., & Lauer, W. V. (2014). Mesh-mould: Robotically fabricated spatial meshes as reinforced concrete formwork. *Architectural Design, 84*(3), 44–53. https://doi.org/10.1002/ad.1753

Keating, S., Spielberg, N. A., Klein, J., & Oxman, N. (2014, March). Robotic fabrication in architecture, art and design 2014. *Robotic fabrication in architecture, art and design 2014* (pp. 99–110). https://doi.org/10.1007/978-3-319-04663-1

Linke, R. (2017). *Additive manufacturing, explained*. MIT SLOAN SCHOOL OF MANAGEMENT. https://mitsloan.mit.edu/ideas-made-to-matter/additive-manufacturing-explained

Na, S., Kim, S., & Moon, S. (2022). Additive manufacturing (3D Printing)-applied construction: Smart node system for an irregular building façade. *Journal of Building Engineering, 56*(June), 104743. https://doi.org/10.1016/j.jobe.2022.104743

Nan, C. (2015). A new machinecraft: A critical evaluation of architectural robots. *Communications in Computer and Information Science, 527*, 422–438. https://doi.org/10.1007/978-3-662-47386-3_23

Paul, S. C., van Zijl, G. P. A. G., & Gibson, I. (2018). A review of 3D concrete printing systems and materials properties: Current status and future research prospects. *Rapid Prototyping Journal, 24*(4), 784–798. https://doi.org/10.1108/RPJ-09-2016-0154

Pessoa, S., Guimarães, A. S., Lucas, S. S., & Simões, N. (2021). 3D printing in the construction industry—A systematic review of the thermal performance in buildings. *Renewable and Sustainable Energy Reviews, 141*(September 2020). https://doi.org/10.1016/j.rser.2021.110794

Prasittisopin, L., Sakdanaraseth, T., & Horayangkura, V. (2021). Design and construction method of a 3D concrete printing self-supporting curvilinear pavilion. *Journal of Architectural Engineering, 27*(3), 1–9. https://doi.org/10.1061/(asce)ae.1943-5568.0000485

Roman, M. C., Eberly, E. A., Mueller, R. P., & Deutsch, S. (2016). NASA centennial challenge: three dimensional (3D) printed habitat. *Earth and Space, 2016*, 333–342.

Sakin, M., & Kiroglu, Y. C. (2017). 3D printing of buildings: construction of the sustainable houses of the future by BIM. *Energy Procedia, 134*, 702–711. https://doi.org/10.1016/j.egypro.2017.09.562

Tay, Y. W. D., Panda, B., Paul, S. C., Noor Mohamed, N. A., Tan, M. J., & Leong, K. F. (2017). 3D printing trends in building and construction industry: A review. *Virtual and Physical Prototyping, 12*(3), 261–276. https://doi.org/10.1080/17452759.2017.1326724

Ting, G. H. A., Tay, Y. W. D., & Tan, M. J. (2021). Experimental measurement on the effects of recycled glass cullets as aggregates for construction 3D printing. *Journal of Cleaner Production, 300*, 126919. https://doi.org/10.1016/j.jclepro.2021.126919

Troemner, M., Ramyar, E., Meehan, J., Johnson, B., Goudarzi, N., & Cusatis, G. (2022). A 3D-printing centered approach to mars habitat architecture and fabrication. *Journal of Aerospace Engineering, 35*(1), 1–13. https://doi.org/10.1061/(asce)as.1943-5525.0001359

Wu, P., Wang, J., & Wang, X. (2016). A critical review of the use of 3-D printing in the construction industry. *Automation in Construction, 68*, 21–31. https://doi.org/10.1016/j.autcon.2016.04.005

Wu, P., Zhao, X., Baller, J. H., & Wang, X. (2018). Developing a conceptual framework to improve the implementation of 3D printing technology in the construction industry. *Architectural Science Review, 61*(3), 133–142. https://doi.org/10.1080/00038628.2018.1450727

Yoshida, H., Igarashi, T., Obuchi, Y., Takami, Y., Sato, J., Araki, M., Miki, M., Nagata, K., Sakai, K., & Igarashi, S. (2015). Architecture-scale human-assisted additive manufacturing. *ACM Transactions on Graphics, 34*(4), 1–8. https://doi.org/10.1145/2766951

Zareiyan, B., & Khoshnevis, B. (2017). Effects of interlocking on interlayer adhesion and strength of structures in 3D printing of concrete. *Automation in Construction, 83*(November 2016), 212–221. https://doi.org/10.1016/j.autcon.2017.08.019

https://www.abc.net.au/news/2021-05-01/3d-printed-home-in-dutch-city/100109484

https://all3dp.com/1/3d-concrete-printing-guide/

Bibliography

Abdelrazaq, A. (2012). Validating the structural behavior and response of Burj Khalifa. *International Journal of High-Rise Buildings, 1*(1), 37–51. www.ctbuh.org

AIA California Council. (2007). Integrated project delivery: A guide. *American Institute of Architects*, 1–62. https://doi.org/10.1016/j.autcon.2010.09.002

Alexandrea, P. (2019). *The complete guide to binder jetting in 3D printing*. 3D Natives. https://www.3dnatives.com/en/powder-binding100420174/#!

AMRG. (2023). *The 7 categories of additive manufacturing*. Additive Manufacturing Research Group. https://www.lboro.ac.uk/research/amrg/about/the7categoriesofadditivemanufacturing/

ASTM. (2012). *Standard terminology for additive manufacturing technologies*. https://www.astm.org/f2792-12.html

Bedarf, P., Dutto, A., Zanini, M., & Dillenburger, B. (2021). Foam 3D printing for construction: A review of applications, materials, and processes. *Automation in Construction, 130*(July), 103861. https://doi.org/10.1016/j.autcon.2021.103861

Besklubova, S., Skibniewski, M. J., & Zhang, X. (2021). Factors affecting 3D printing technology adaptation in construction. *Journal of Construction Engineering and Management, 147*(5), 04021026. https://doi.org/10.1061/(asce)co.1943-7862.0002034

Borrmann, A., König, M., Koch, C., & Beetz, J. (2018). BIM for 3D printing in construction. In A. Borrmann, M. König, C. Koch, & J. Beetz (Eds.), *Building information modeling: Technology foundations and industry practice* (pp. 1–25). Springer. https://doi.org/10.1007/978-3-319-92862-3

Bos, F., Wolfs, R., Ahmed, Z., & Salet, T. (2016). Additive manufacturing of concrete in construction: Potentials and challenges of 3D concrete printing. *Virtual and Physical Prototyping, 11*(3), 209–225. https://doi.org/10.1080/17452759.2016.1209867

Bridges, A. H. (2010). Building construction handbook. In *The construction net*. https://doi.org/10.4324/9780203476826_chapter_9

Brooks, A., & Zhou, H. Y. (2021). Developing 3D printable lightweight functional cementitious composite. *Earth and Space, 2021*, 52–58.

Buchanan, C., & Gardner, L. (2019). Metal 3D printing in construction: A review of methods, research, applications, opportunities and challenges. *Engineering Structures, 180*(October 2018), 332–348. https://doi.org/10.1016/j.engstruct.2018.11.045

Bulger, S., & Skonieczny, K. (2016). Towards mobile 3D printing for planetary construction. *Earth and Space, 2016*, 324–332.

Carrato, P. J. (2021). Use of BIM and 3D printing in mars habitat design challenge. *Earth and Space, 2021*, 780–790.

R. Sarı and E. B. Çalışkan, *Building Construction Methods and Systems*, https://doi.org/10.1007/978-3-031-50043-5

CAT. (2011). *Tmmob mimarlar odasi mimarlik hizmetleri şartnamesi ve en az bedel tarifesi (28.12.2011).* http://www.mimarist.org/include/uploads/2015/11/mimarlik-hizmetleri-sartnamesi-en-az-bedel-tarifesi.pdf

Covey, S. D., & Metzger, P. T. (2018). Extruded clay-based regoliths for construction on mars, Phobos, and NEAs. *Earth and Space, 2018*, 570–579.

Davies, I., & Davies, I. (2020). The RIBA plan of work 2013. *Contract Administration*, 10–11. https://doi.org/10.4324/9780429347177-2

Ding, L., Wei, R., & Che, H. (2014). Development of a BIM-based automated construction system. *Procedia Engineering, 85*, 123–131. https://doi.org/10.1016/j.proeng.2014.10.536

Dini, E. (2009). *D-SHAPE - The 21st century revolution in building technology has a name* (pp. 1–16). http://www.cadblog.pl/podcasty/luty_2012/d_shape_presentation.pdf

Eren, Ö. (2014). *Büyük Açıklıklı Çelik Yapılar*. Arı Sanat.

Feller, R. M. (2017). Enhancing the stone. *Architectural Record, 205*(1), 150–151.

Gomaa, M., Jabi, W., Veliz Reyes, A., & Soebarto, V. (2021). 3D printing system for earth-based construction: Case study of cob. *Automation in Construction, 124*(January), 103577. https://doi.org/10.1016/j.autcon.2021.103577

Gosselin, C., Duballet, R., Roux, P., Gaudillière, N., Dirrenberger, J., & Morel, P. (2016). Large-scale 3D printing of ultra-high performance concrete—a new processing route for architects and builders. *Materials and Design, 100*, 102–109. https://doi.org/10.1016/j.matdes.2016.03.097

Hack, N., & Lauer, W. V. (2014). Mesh-mould: Robotically fabricated spatial meshes as reinforced concrete formwork. *Architectural Design, 84*(3), 44–53. https://doi.org/10.1002/ad.1753

Keating, S., Spielberg, N. A., Klein, J., & Oxman, N. (2014, March). Robotic fabrication in architecture, art and design 2014. *Robotic fabrication in architecture, art and design 2014* (pp. 99–110). https://doi.org/10.1007/978-3-319-04663-1

Linke, R. (2017). *Additive manufacturing, explained*. MIT SLOAN SCHOOL OF MANAGEMENT. https://mitsloan.mit.edu/ideas-made-to-matter/additive-manufacturing-explained

Mehta, M., Scarborough, W., & Armpriest, D. (2013). *Building construction; principles, materials, systems* (Second). Pearson.

Na, S., Kim, S., & Moon, S. (2022). Additive manufacturing (3D Printing)-applied construction: Smart node system for an irregular building façade. *Journal of Building Engineering, 56*(June), 104743. https://doi.org/10.1016/j.jobe.2022.104743

Nan, C. (2015). A new machinecraft: A critical evaluation of architectural robots. *Communications in Computer and Information Science, 527*, 422–438. https://doi.org/10.1007/978-3-662-47386-3_23

Paul, S. C., van Zijl, G. P. A. G., & Gibson, I. (2018). A review of 3D concrete printing systems and materials properties: Current status and future research prospects. *Rapid Prototyping Journal, 24*(4), 784–798. https://doi.org/10.1108/RPJ-09-2016-0154

Pessoa, S., Guimarães, A. S., Lucas, S. S., & Simões, N. (2021). 3D printing in the construction industry—A systematic review of the thermal performance in buildings. *Renewable and Sustainable Energy Reviews, 141*(September 2020). https://doi.org/10.1016/j.rser.2021.110794

Prasittisopin, L., Sakdanaraseth, T., & Horayangkura, V. (2021). Design and construction method of a 3D concrete printing self-supporting curvilinear pavilion. *Journal of Architectural Engineering, 27*(3), 1–9. https://doi.org/10.1061/(asce)ae.1943-5568.0000485

Roman, M. C., Eberly, E. A., Mueller, R. P., & Deutsch, S. (2016). NASA centennial challenge: three dimensional (3D) printed habitat. *Earth and Space, 2016*, 333–342.

Sakin, M., & Kiroglu, Y. C. (2017). 3D printing of buildings: construction of the sustainable houses of the future by BIM. *Energy Procedia, 134*, 702–711. https://doi.org/10.1016/j.egypro.2017.09.562

Tay, Y. W. D., Panda, B., Paul, S. C., Noor Mohamed, N. A., Tan, M. J., & Leong, K. F. (2017). 3D printing trends in building and construction industry: A review. *Virtual and Physical Prototyping, 12*(3), 261–276. https://doi.org/10.1080/17452759.2017.1326724

Ting, G. H. A., Tay, Y. W. D., & Tan, M. J. (2021). Experimental measurement on the effects of recycled glass cullets as aggregates for construction 3D printing. *Journal of Cleaner Production, 300*, 126919. https://doi.org/10.1016/j.jclepro.2021.126919

Troemner, M., Ramyar, E., Meehan, J., Johnson, B., Goudarzi, N., & Cusatis, G. (2022). A 3D-printing centered approach to mars habitat architecture and fabrication. *Journal of Aerospace Engineering, 35*(1), 1–13. https://doi.org/10.1061/(asce)as.1943-5525.0001359

Wu, P., Wang, J., & Wang, X. (2016). A critical review of the use of 3-D printing in the construction industry. *Automation in Construction, 68*, 21–31. https://doi.org/10.1016/j.autcon.2016.04.005

Wu, P., Zhao, X., Baller, J. H., & Wang, X. (2018). Developing a conceptual framework to improve the implementation of 3D printing technology in the construction industry. *Architectural Science Review, 61*(3), 133–142. https://doi.org/10.1080/00038628.2018.1450727

Yoshida, H., Igarashi, T., Obuchi, Y., Takami, Y., Sato, J., Araki, M., Miki, M., Nagata, K., Sakai, K., & Igarashi, S. (2015). Architecture-scale human-assisted additive manufacturing. *ACM Transactions on Graphics, 34*(4), 1–8. https://doi.org/10.1145/2766951

Zareiyan, B., & Khoshnevis, B. (2017). Effects of interlocking on interlayer adhesion and strength of structures in 3D printing of concrete. *Automation in Construction, 83*(November 2016), 212–221. https://doi.org/10.1016/j.autcon.2017.08.019

GPSR Compliance

The European Union's (EU) General Product Safety Regulation (GPSR) is a set of rules that requires consumer products to be safe and our obligations to ensure this.

If you have any concerns about our products, you can contact us on ProductSafety@springernature.com

In case Publisher is established outside the EU, the EU authorized representative is:

Springer Nature Customer Service Center GmbH
Europaplatz 3
69115 Heidelberg, Germany